An Introduction to
Economic Geography

Visit the *An Introduction to Economic Geography* Companion
Website at **www.pearsoned.co.uk/mackinnon** to find key annual
updates in Economic Geography

We work with leading authors to develop the strongest
educational materials in geography, bringing cutting-edge
thinking and best learning practice to a global market.

Under a range of well-known imprints, including
Prentice Hall, we craft high-quality print and electronic
publications that help readers to understand and apply
their content, whether studying or at work.

To find out more about the complete range of our
publishing, please visit us on the World Wide Web at:
www.pearsoned.co.uk

An Introduction to Economic Geography

Globalization, Uneven Development and Place

Danny Mackinnon and
Andrew Cumbers

PEARSON
Prentice
Hall

Harlow, England • London • New York • Boston • San Francisco • Toronto • Sydney • Singapore • Hong Kong
Tokyo • Seoul • Taipei • New Delhi • Cape Town • Madrid • Mexico City • Amsterdam • Munich • Paris • Milan

Pearson Education Limited
Edinburgh Gate
Harlow
Essex CM20 2JE
England

and Associated Companies throughout the world

Visit us on the World Wide Web at:
www.pearsoned.co.uk

First published 2007

© Pearson Education Limited 2007

ISBN 978-0-13-129316-8

British Library Cataloguing-in-Publication Data
A catalogue record for this book is available from the British Library

Library of Congress Cataloging-in-Publication Data
A catalog record for this book is available from the Library of Congress

10 9 8 7 6 5 4 3 2 1
10 09 08 07

Typeset in 9.75pt Minion by 3
Printed and bound in Great Britain by Ashford Colour Press, Hampshire

The publisher's policy is to use paper manufactured from sustainable forests.

Brief contents

'This is an excellent and comprehensive introduction to the diverse field of economic geography. Its clear style, engaging case studies and well-constructed summaries of the major debates make it an invaluable resource. It should be essential reading for students at all levels.'

Andrew Jones, Birkbeck College, University of London, and member of the Economic Geography Research Group

'A thoughtful, stimulating, accessible introduction to the range of approaches used by economic geographers to understand and explain the patterns and processes of contemporary globalization and uneven development. An excellent platform for more advanced exploration of some key themes in contemporary economic geography.'

Peter Daniels, University of Birmingham

'A stimulating and accessible introduction to a core area of the discipline. Ranging from traditional topics such as regional development, multinational firms and labour markets, to new concerns focused on knowledge-based economies, learning regions, spaces of consumption, and the Internet economy, *An Introduction to Economic Geography* conveys a clear sense of the diversity and vitality of contemporary economic geography.'

Neil Wrigley, University of Southampton, and Editor of Journal of Economic Geography

Contents

Part 1 INTRODUCTION

Contents

Supporting resources

Visit **www.pearsoned.co.uk/mackinnon** to find valuable online resources

Companion Website for students
- Key annual updates in Economic Geography

For more information please contact your local Pearson Education sales representative or visit **www.pearsoned.co.uk/mackinnon**

List of tables

List of figures

Preface

Economic geography has become a highly diverse and open field of research in recent years, incorporating a wide range of research topics, theories and methodologies. The influence of the so-called cultural and institutional 'turns' in particular has enlivened the subject, exposing it to new ideas and concerns (Thrift, 2000). New areas of interest such as consumption, corporate cultures and gender relations in the workplace have been embraced by economic geographers, alongside work on more 'traditional' topics such as regional development, large firms and labour markets, often informed by new perspectives. As a result, the notion of the economy as a self-evident and self-contained entity has been destabilized, leading to an extension of its boundaries and the forging of new linkages with a range of other subject areas. The profusion of approaches and methodologies means that there is no single approach or 'paradigm' that dominates the field. Such diversity and pluralism has generated considerable excitement and vitality among researchers and students. At the same time, however, it has raised concerns about the coherence, identity and purpose of economic geography (see *Antipode*, 2001). One of the major challenges is that of how to communicate the diversity of the subject to the relatively uninitiated, particularly economic geography students, many of whom may expect to be introduced to a single 'right' approach or a clear set of core concerns (Barnes, 2006).

Our purpose in writing this textbook is to convey some of the diversity and vitality of contemporary economic geography to students. It is intended to work as an introductory text for undergraduate geography students taking courses in economic geography at the equivalent of Levels 1 and 2 in England and Wales (Levels 2 and 3 in Scotland). Our decision to write this book was prompted by the apparent lack of an introductory general textbook for British students, certainly when compared to other areas of human geography, such as political geography, 'concepts/approaches' or 'methods', which seem to have witnessed a profusion of textbooks in recent years. Three main types of existing texts in economic geography can be identified, alongside chapters in introductory human geography collections such as Cloke *et al.* (2005) and Daniels *et al.* (2005) and more specialized books on particular topics (e.g. consumption or labour markets). First, there are those that focus on a core theme such as globalization, of which Dicken (2003a) and Knox *et al.* (2003) are the established favourites. Second, a number of edited collections in the form of readers or companions presenting overviews of key topics or selections of 'classic' papers have been published in recent years (Barnes and Sheppard, 2000; Barnes *et al.*, 2004; Bryson *et al.*, 1999). Third, more advanced research-level texts such as Hudson (2005) are also available. This book aims to provide a text that is more accessible and student-friendly than the second or third of these types while incorporating a broader range of topics than the first. It is designed to guide students through key debates and issues in an integrated fashion.

Three main theoretical approaches can be identified within contemporary economic geography: spatial analysis, political economy and cultural economy (Chapter 2). As Hudson (2005, p.15) argues, the latter two should be seen as potentially complementary rather than alternative approaches, providing different 'analytic windows' from which to view the economy. While the book is underpinned by our favoured political economy approach, we have sought to connect this to some of the cultural and institutional insights that have informed research in recent years. Rather than having 'a 1970s/1980s feel' about it – as one reviewer commented on the original book proposal – the 'new' or revised form of political economy that we have adopted has moved beyond the rather clunky and deterministic nature of earlier versions to become more flexible and open to the importance of context, difference and identity (Peck, 2005, p.166). To adopt Hudson's terms, our approach can be described as 'culturally sensitive political economy' rather than a 'politically sensitive cultural economy' (Hudson, 2005, p.15).

This book is underpinned by three main thematic concerns, highlighted in the subtitle: globalization, uneven development and place. Globalization is one of the key forces reshaping the geography of economic activity, driven by multinational corporations, financial institutions, international economic organizations and governments. It has sparked a wave of protests from 'anti-globalization' activists and groups since the late 1990s. In many ways, globalization provides a key contemporary vehicle for examining the longer-standing concerns of economic geography, many of which can be related to the overarching concepts of uneven development and place. We view uneven development as an inherent characteristic of the capitalist economy, reflecting the tendency for growth and prosperity to be geographically concentrated in particular locations (Smith, 1984). The theme of place, in turn, reflects geographers' traditional interest in distinctive localities. Crucially, however, such local distinctiveness must be seen as the product of interaction with wider economic processes, not isolation. In this sense, one of the main geographical effects of globalization is the forging of closer linkages between the economies of distant countries and regions.

In selecting the topics and issues covered in the book, we have tried to reflect contemporary concerns in economic geography. Relatively 'new' areas (to economic geographers), such as consumption, services and the cultural industries, are incorporated alongside more 'traditional' ones such as regional development, agglomeration and labour. The book is global in terms of its geographical range and scope, and we have tried to include research and case studies drawn from a range of countries and regions. Economic development in the 'global South' is the subject of a discrete chapter, and the connections between developed and developing countries are highlighted throughout the book. Such is the breadth and diversity of the subject, however, that significant omissions are inevitable. Perhaps the most obvious of these is the set of relations between the economy and the environment (see Hudson, 2005, pp.38–56), though individual readers will identify others (transition economies receive little direct attention, for instance). In general terms, the outlook and scope of the book are inevitably limited by our UK residence, reliance on English-language materials and immersion in the concerns of Anglo-American human geography. More specifically, our personal research interests in areas such as regional development, labour markets and state restructuring in developed economies have probably exerted some (indeterminate) influence over the selection of topics and approach.

Please note that bold text indicates an entry in the Glossary.

Acknowledgements

The idea for the book emerged from a discussion between Andrew Taylor of Pearson Education and one of the authors. Andrew has subsequently been the editor responsible for the book and we are grateful to him, Sarah Busby and the rest of the team at Pearson for their assistance and patience. Most of the figures in the book were drawn by the production team at Pearson while Alison Sandison and Jenny Johnston at Aberdeen provided more specialist cartographic support. We would also like to thank the various reviewers of both the original proposal and, especially, the draft chapters, particularly the three reviewers of the entire text – Professor Peter Daniels at Birmingham, Dr Michael Punch at University College Dublin and Dr Pete North at Liverpool. We have tried to incorporate their suggestions wherever possible and the book is undoubtedly much better as a result. Special thanks are due also to Keith Chapman, for the support and encouragement he has provided to both of us in our careers, and to close colleagues in the Geography Departments at Aberdeen and Glasgow, for general camaraderie. Andy would also like to acknowledge the help and advice of Ray Hudson from PhD days onwards, and lastly give a big thanks to Fran and Anna for their continuing support, tolerance and love.

Publisher's acknowledgements

We are grateful to the following for permission to reproduce copyright material:

Figures 1.1, 3.1 and 9.8, reprinted by permission of Sage Publications Ltd from Castree, N., Coe, N., Ward, K. and Samers, M., *Spaces of Work: Global Capitalism and Geographies of Labour*, copyright (© Noel Castree, Neil M. Coe, Kevin Ward and Michael Samers, 2004); Figures 1.2, 3.5, 5.2a, 5.2b, 5.4, 5.6, 5.7, 6.7, 7.2 and 7.4 and Table 7.6 reprinted by permission of Sage Publications Ltd from Dicken, P., *Global Shift: Reshaping the Global Economic Map in the Twenty-first Century*, 4th edn, copyright (© Peter Dicken 2003); Figure 1.4 from www.ideas-forum.org.uk, illustration drawn by Jan Nimmo; Figure 1.6 from *Environment and Planning*, 2002, A 34, p. 1583, PION Limited; Figure 1.9 drawn by Ken Byrne, in Gibson-Graham, J.K. 2006, *A Postcapitalist Politics*, Minneapolis and London: University of Minnesota Press, p. 70; Figure 1.11 from 'Industrial districts', in *A Companion to Economic Geography*, edited by Sheppard, E. and

Barnes, T.J., Blackwell Publishing (Amin, A. 2000); Figure 2.1 reproduced with permission from Lee, R., *Progress in Human Geography*, 26, copyright (© Sage Publications, 2002), by permission of Sage Publications Ltd; Figure 2.3 from *Industrial Location*, John Wiley & Sons, Inc. (Smith, D.M. 1981); Figure 2.5 from *Geography and Geographers: Anglo-American Human Geography since 1945*, 6th edn, Johnston, R.J. and Sidaway, J.D., Edward Arnold (Publishers) Ltd, 2004, reproduced by permission of Edward Arnold (Publishers) Ltd; Figure 3.3 from 'Financing entrepreneurship: venture capital and regional development', Mason, C.M. and Harrison, R.T., in *Money and the Space Economy*, Martin, R. (ed.), copyright 1999, copyright John Wiley & Sons Limited, reproduced with permission, taken from PricewaterhouseCoopers LLP/National Venture Capital Association MoneyTree report based on data from Thomson Financial; Table 3.3 from *Geographies of Economies*, Lee, R. and Wills, J., Edward Arnold (Publishers) Ltd, 1997, reproduced by permission of Edward Arnold (Publishers) Ltd; Figure 3.4 from *Wrecking a Region*, Hudson, R., 1989, p. 4, PION Limited; Figures 3.6, 4.5 and 4.8 from *Atlas of Industrializing Britain 1780–1914*, Lawton, P., copyright © 1986, Methuen, reproduced by permission of Taylor & Francis Books UK; Figure 4.2 'Map: Nature's Metropolis with American Railroads, 1861', from *Nature's Metropolis: Chicago and the Great West*, by William Cronon, copyright © 1991 by William Cronon, used by permission of W. W. Norton & Company, Inc.; Table 4.2 from Massey, D., *Spatial Divisions of Labour: Social Structures and the Geography of Production*, 1994, Macmillan, reproduced with permission of Palgrave Macmillan; Figure 4.3 from 'Annihilating space? The speed-up of communications', in *A Shrinking World? Global Unevenness and Inequality*, edited by Allen, J. and Hammett, C. (Leyshon, A. 1995), by permission of Oxford University Press; Figure 4.4 from *Industrial Location*, 2nd edn, Blackwell Publishing (Chapman, K. and Walker, D. 1991); Figure 4.6 from *Peaceful Conquest: The Industrialisation of Europe 1760–1870* (Pollard, S. 1981), by permission of Oxford University Press; Figures 4.7, 4.9 and 11.1 from *The Geography of the World Economy*, 4th edn, Knox, P., Agnew, J. and McCarthy, L., Edward Arnold (Publishers) Ltd, 2003, reproduced by permission of Edward Arnold

(Publishers) Ltd; Figure 4.7 from *Geography and the Urban Environment*, Vol. IV, Herbert, D.T. and Johnston, R.J. (eds) after Conzen, M., copyright 1981, copyright John Wiley & Sons Limited, reproduced with permission; Figure 4.11 adapted from the Rand McNally *World Atlas*, 1992; Figures 5.1, 11.2, 11.5 and 11.6 from *Geographies of Development*, 2nd edn, Potter, R.B., Binns, T., Elliott, J.A. and Smith, D., Pearson Education Limited; Table 5.1 from *Global Transformations: Politics, Economics and Culture*, Polity (Held, D., McGrew, A., Goldblatt, D. and Perraton, J. 1999); Table 5.2 from *The Geography of the World Economy*, Knox, P. and Agnew, J., Edward Arnold (Publishers) Ltd, 1989, reproduced by permission of Edward Arnold (Publishers) Ltd; Table 5.3 adapted from Table 3-1c, Shares of world GDP, 1000–1998, *The World Economy: A Millennial Perspective*, © OECD 2001; Table 5.4 adapted from Table 3-2b, Merchandise Exports as Per Cent of GDP in 1990 Prices, World and Major Regions, 1870–1998, *The World Economy: A Millennial Perspective*, © OECD 2001; Figure 5.5 from 'Brave new world', in *Geographical Magazine*, 67(1), Circle Publishing (Evans R. 1995); Figure 5.6 from *Industrial Development: Global Report, 1997*, United Nations Industrial Development Organization (1997); Table 5.7 from *Millennium Development Goals: A compact among nations to end human poverty*, Human Development Report, UNDP (2003); Table 5.8 from *International Cooperation at a Crossroads: Aid, Trade and Security in an Unequal World*, Human Development Report, UNDP (2005); Table 6.1 from 'The rise of the workfare state', in *Geographies of Global Change: Remapping the World*, 2nd edn, edited by Johnston, R.I., Taylor, P. and Watts, M., Blackwell Publishing (Painter, J. 2002); Figure 6.2 from 'UK regional policy: an evaluation', in *Regional Studies*, 31, Taylor & Francis Ltd (Taylor J. and Wren C. 1997), http://www.tandf.co.uk/journals; Figure 6.3 reprinted from *Progress in Planning*, 44, Tuppen, J.N. and Thompson, I.B., 'Industrial restructuring in contemporary France: spatial priorities and policies', p. 126, copyright (1994), with permission from Elsevier; Table 6.3 from *Capitalism since 1945*, Blackwell Publishing (Armstrong, P., Glyn, A. and Harrison, J. 1991); Figure 6.4 from *The Condition of Postmodernity*, Blackwell Publishing, (Harvey, D. 1989); Table 6.5 from

Environment and Planning A, 2004, 36, p. 2098, PION Limited; Figure 6.8 from HM Treasury (2005) *Public Expenditure Statistical Analyses 2005*, reproduced under the terms of the Click-Use Licence; Figure 6.10 from *A United Kingdom? Economic, Social and Political Geographies*, Mohan, J., Edward Arnold (Publishers) Ltd, 1999, reproduced by permission of Edward Arnold (Publishers) Ltd; Figure 6.11 from 'Devolution and economic governance in the UK: uneven geographies, uneven capacities?', in *Local Economy*, 17, Taylor & Francis Ltd (Goodwin, M., Jones, M., Jones, R., Pett, K. and Simpson, G. 2002), http://www.tandf.co.uk/journals; Table 7.2 from *Environment and Planning D: Society and Space*, 1985, 3, p. 37, PION Limited; Tables 7.3a and 7.3b from *World Investment Report 2005: Transnational Corporations and the Internationalization of R and D*, UNCTAD (2005); Table 7.4 from 'Changing relationships between multinational companies and their host regions?: A case study of Aberdeen and the international oil industry', in *Scottish Geographical Journal*, 117(31), Royal Scottish Geographical Society (Cumbers, A. and Martin, S. 2001); Figure 8.3 from *Measuring Trade in Services*, WTO (2006); Tables 8.3, 8.4 and 8.6 from *World Investment Report 2004: The Shift Towards Services*, UNCTAD (2004); Figures 8.4, 8.5 and 8.6 and Table 8.2 from *International Trade Statistics 2005*, WTO (2005); Figure 8.7 from Figure B.5.4. Share of the service sector in the total inward FDI positions of OECD countries, OECD, International Direct Investment Statistics, June 2005, reproduced in *Measuring Globalization OECD Economic Globalization Indicators*, © OECD 2005; Figure 8.8 and Table 8.5 from International Financial Services London (IFSL); Figure 8.9 from *Environment and Planning A*, 1996, 28, pp. 1209–32, PION Limited; Figure 8.10 reprinted from *Cities*, 16(6), Beaverstock, J.V., Smith, R.G. and Taylor, P.J., 'A roster of world cities', copyright (1999), with permission from Elsevier; Figure 8.13 from *Environment and Planning A*, 2001, 33, p. 69, PION Limited; Figure 8.14 reprinted by permission of Sage Publications Ltd from Dicken, P., *Global Shift: Transforming the World Economy*, copyright (© Peter Dicken 1998), taken from 'Telecommunications and the changing geographies of knowledge transmission in the late twentieth century', in *Urban Studies*, 32, Taylor & Francis Ltd (Warf, B. 1995), http://www.tandf.co.uk/journals, and from *World Investment Report 1996: Investment, Trade and International Policy Agreements*, UNCTAD (1996); Figure 9.4 and Table 9.6 from 'A cross-country study of union membership', in *Institute of Labour Discussion Paper 2016*, IZA (Blanchflower, D. 2006); Table 9.4 from 'Places of work', in *A Companion to Economic Geography*, edited by Sheppard, E. and Barnes, T., Blackwell Publishing (Peck, J. 2000); Table 9.5 from *World Employment Report 2004–05*, p. 191, copyright © International Labour Organization; Figure 9.6 reprinted by permission of Sage Publications Ltd from Allen, J. and Massey, D. (eds), *The Economy in Question*, copyright (© 1988), taken from *Flexibility, Uncertainty and Manpower Management*, Institute of Manpower Studies Report 89, Institute of Employment Studies, (Atkinson, J. 1984); Figure 9.7 from Labour Force Survey, TUC; Table 9.7 from *International Framework Agreements: Increasing the Effectiveness of Core Labour Standards*, Global Labour Institute, www.global-labour.org. (Gibb, E. 2005); Figure 9.8 from Castles, S. and Miller, M., *The Age of Migration*, 2nd edn, 1998, Macmillan, reproduced with permission of Palgrave Macmillan; Figure 10.1 'Investment in knowledge', *OECD Factbook 2006: Economic, Environmental and Social Statistics*, © OECD 2006; Figure 10.6 from *The Knowledge-creating Company: How Japanese Companies Create the Dynamics of Innovation*, (Nonaka, I. and Taekuchi, H. 1995), by permission of Oxford University Press, Inc.; Figure 10.7 from Trends Business Research *et al.*, 2001; Figure 10.8 from 'Locations, clusters and company strategy', in *The Oxford Handbook of Economic Geography* edited by Clark, G.L, Feldman, M. and Gertler, M. (Porter, M.E. 2000), by permission of Oxford University Press; Figure 10.9 from 'A new map of Hollywood: the production and distribution of American motion pictures', in *Regional Studies*, 36, Taylor & Francis Ltd (Scott, A.J. 2002), http://www.tandf.co.uk/journals; Figures 10.10 and 10.11 reprinted from *Geoforum*, 31, Henry, N. and Pinch, S., 'Spatialising knowledge: placing the knowledge community of Motor Sport Valley', copyright (2000), with permission from Elsevier; Figure 10.12 reprinted by permission of Sage Publications Ltd from Bathelt, H., Malmberg, A. and Maskell, P., *Progress in Human Geography*, copyright (© 2004 Sage

Publications); Figure 11.3 from *International Cooperation at a Crossroads: Aid, Trade and Security in an Unequal World*, Human Development Report, UNDP (2005); Figure 11.6 adapted from R.B. Potter, *Urbanisation in the Third World* (OUP, 1992), by permission of Oxford University Press; Figure 11.8 from *Introducing Human Geographies*, 2nd edn, Cloke, P., Crang, P. and Goodwin, M. (eds), Edward Arnold (Publishers) Ltd, 2005, © 2005 Hodder Arnold, reproduced by permission of Edward Arnold (Publishers) Ltd; Figure 11.10 from *Making New Technologies Work for Human Development*, Human Development Report, UNDP (2001); Figure 11.13 from 'Net ODA in 2004 as a percentage of GNI, aid rising sharply, according to the latest OECD figures (13 December 2005)', © OECD 2005; Figure 12.1 © *Imagining Scotland: Tradition, Representation and Promotion in Scottish Tourism since 1750*, Gold, J.R. and Gold, M., 1995, Ashgate; Table 12.1 from *Tourism Highlights, 2005*, World Tourism Organization; Table 12.2 from *Consuming Places*, Urry, J., Routledge, Thomson Publishing Services, North Way, Hanover HANTS SP10 5BE; Figure 12.3 reprinted from *Annals of Tourism Research*, 27, Waitt, G., 'Consuming heritage: perceived historical authenticity', copyright (2000), with permission from Elsevier; Figure 12.6 ONS (2005) *Travel Trends 2004*, reproduced under the terms of the Click-Use Licence; Figure 12.7 © *Contemporary Portugal: Dimensions of Economic and Political Change*, Syrett, S. (ed.), 2002, Ashgate; Figure 12.8 reprinted by permission of Sage Publications Ltd from Shaw, G. and Williams, A.M., *Tourism and Tourism Spaces*, copyright (© Gareth Shaw and Allan M. Williams 2004); Figure 12.13 reprinted by permission of Sage Publications Ltd from Rodriguez, A., Martinez, A. and Guenga, G. 'Uneven development: new urban policies and socio-spatial fragmentation in metropolitan Bilbao', in *European Urban and Regional Studies*, 8, copyright © 2001 by SAGE Publications.

Photographs

Figure 1.7 John Lawrence, Getty Images; Figure 1.8 © Hulton-Deutsch Collection/Corbis; Figure 2.6 Robin MacDougall, Getty Images; Figure 2.7 Franco Zecchin, Getty Images; Figure 2.8 © David Cooper/Toronto Star/Corbis; Figure 3.7 Source: Mary Evans Picture Library;

Figure 9.3 Corbis; Figure 9.5 © Sherwin Crasto/Reuters/Corbis; Figure 9.9 © Danny Lehman/Corbis; Figure 10.2 © Jim Craigmyle/Corbis; Figure 10.5 © Dana Hoff/Beateworks/Corbis; Figure 11.11 Luciney Martins, http://www.mst.org.br; Figure 12.11 © National Museums Scotland, Licensor www.scran.ac.uk; Figure 12.10 from *From Mahsuri to Mahathir*, Insan, (B. Bird 1989)

We are grateful to the Financial Times Limited for permission to reprint the following material:

Figure 5.3 redrawn from 'Global capital, trade and foreign currency transactions' (Lee, R., 2002) © *Financial Times*, 20 March 2002, p. 29; Figure 8.12 from 'Rock 'n' roll suicide' (Sanghera S., 2001), © *Financial Times*, 15 November 2001, p. 19.

In some instances we have been unable to trace the owners of copyright material, and we would appreciate any information that would enable us to do so.

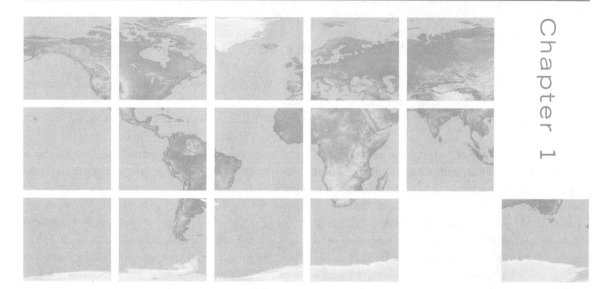

Chapter 1

Introducing economic geography

Chapter map

In the introduction to this chapter, we highlight some of the key questions addressed in this book and relate these to contemporary debates on globalization and economic development. This is followed by a discussion of each of the three main themes of the book: globalization, uneven development and place. In section 1.3, we provide a brief definition of the economy and a basic introduction to economic geography as a distinct subject area. This is followed by an outline of the political economy perspective that informs our approach in this book. Finally, section 1.5 describes the contents of the remainder of the book.

1.1 Introduction

Our aim in this book is to examine the changing geography of economic activity within the contemporary world economy. As economic geographers, we are particularly concerned with the location of different types of activity, the economies of particular regions and the economic relationships between different places. A number of key questions are addressed in the book.

How is economic activity organized across different geographical scales? In what ways has the economy become more globally integrated in recent decades? How is the growth of information and communications technology affecting the location of economic activity? In what ways is economic development geographically uneven and what are the reasons for this? What kinds of economic activity are found in different types of places? To what extent do locally specific factors and influences shape processes of economic development?

One of the key forces shaping the geography of economic activity is the much-discussed phenomenon of **globalization**. This term refers to the growing connections and linkages between people and firms located in different places, manifested in increased flows of goods, services, money, information and people across national and continental borders. It has become a key buzzword of the new millennium, promoted and supported by many business people, politicians, journalists and academics, and actively contested and resisted by 'anti-globalization' groups and activists. Several prominent writers and academics have argued that globalization is bringing about the 'end of geography' with advanced information technology and communications making location and distance irrelevant since businesses can locate production anywhere and still maintain close contact with suppliers and customers (O'Brien, 1992). In a more recent version of this argument, the *New York Times* columnist Thomas Friedmann (2005) argues that new technologies are creating a flat world in which work can be relocated to a wide variety of locations, as manifest in the shift of service employment to developing countries such as India.

While these provocative arguments appear to capture important changes in the organization of the economy, they are rather simplistic and overstated, relying upon an impoverished view of geography. New information technologies are certainly enabling aspects of service employment such as call centres to be relocated to developing countries where costs are lower, but this trend is focused on particular countries like India with others being left behind. Rather than having the same effects everywhere, globalization has different outcomes in different places. Instead of a 'flat' world in which advanced technologies are bringing everyone together on equal terms, as Friedmann claims, the world is characterized by huge inequalities in wealth, as indicated by the contrasting economic fortunes of China and Africa in recent years (see Box 1.2). While the speed and density of global economic connections has certainly increased, distinct forms of production are still associated with particular regions. Indeed, the evidence from dynamic growth regions such as Silicon Valley in California, the City of London and Baden-Württemberg in southern Germany suggests that competitive advantages in the global economy can often be tied to local cultures of production and specialist forms of skills and knowledge that are difficult to transfer elsewhere (Storper, 1997). In other words, regions or places remain important in a global system and we need to examine the interaction between particular places and globalizing processes.

In assessing this process of interaction, we need to pay equal attention to global processes and locally specific factors (Johnston, 1984). The decision to locate a major new investment, for example a car plant or a call centre, in a particular region is not only a product of global economic imperatives; it is also shaped by local conditions and factors (market position, wage rates, skills, site availability and financial incentives from government agencies). Similarly, while the actions of firms, governments and workers based in particular places are structured by the basic pressures of capitalist production, requiring firms to make profits, workers to earn a living and governments to encourage growth, they also tend to be informed by locally specific values and beliefs. The conservative culture of the southern states of the United States, for example, is associated with anti-trade union attitudes and practices among business and state-level officials, helping to attract business into the region in recent decades, in contrast to the strong union traditions of the manufacturing heartland in the North-east and Midwest, which has lost much of its traditional industrial base since the 1970s (pp. 454–6).

1.2 Key themes: globalization, uneven development and place

In this section, we focus on the three main themes of the book: globalization, uneven development and place. Our selection of these themes is informed by the basic geographical concepts of **scale**, **space** and **place**. **Scale** refers to the different geographical levels of human activity, from the local to the regional, national and global (Figure 1.1). **Space** is simply an area of the earth's surface such as, for example, that contained within the boundaries of a particular region or country. **Place** refers to a particular area (space), usually occupied, to which a group of people have become attached, endowing it with meaning and significance. The geographer Tim Cresswell (2004, p.8) illustrates the distinction between space and place by referring to an advertisement in a local furniture shop entitled 'turning space into place', reflecting how people use furniture and interior decor to make their houses meaningful, turning them from empty

locations into personalized and comfortable homes. This domestic transformation of space into place is something with which we are all familiar, perhaps from decorating rooms in university halls of residence or shared flats.

1.2.1 Globalization and connections across space

The first underlying theme that runs through this book is that economic activities are connected across space through flows of goods, money, information and people. These connections are not new: trading relations between distant people and places involving the exchange of goods have existed throughout much of human history. The notion of globalization, however, emphasizes that the volume and scope of global flows has increased significantly in recent decades. Increased trade and economic interaction between distant places is dependent on technology in terms of the ease of movement and communication across space. In this context, space is understood in terms of the distance between two points and the time it takes to move between them.

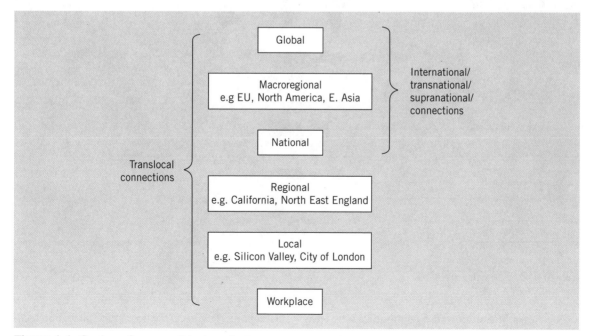

Figure 1.1 Scales of geographical analysis.
Source: Castree *et al.*, 2005, p.xvix.

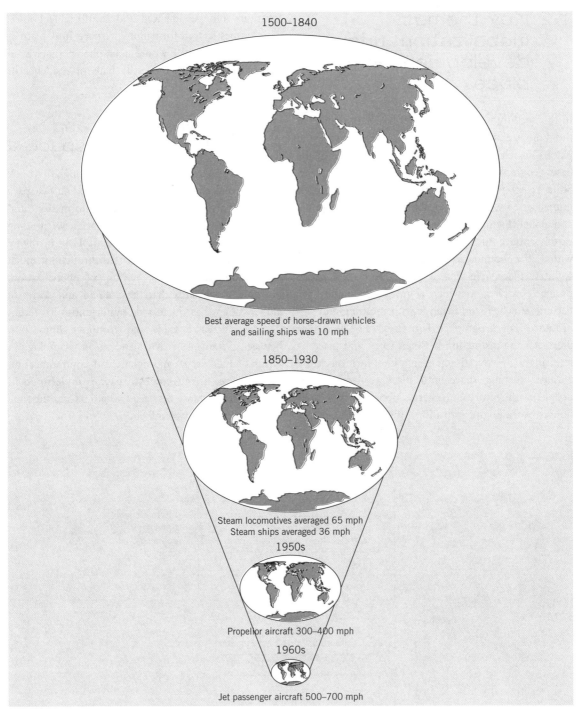

Figure 1.2 'A shrinking world'.
Source: Dicken, 2003a, p.92.

A new set of transport and communications technologies has emerged since the 1960s, leading to a large expansion in the volume of spatial movement and interaction. The effects of these 'space-shrinking technologies' have brought the world closer together, effectively reducing the distance between places in terms of the time and costs of movement and communication (Figure 1.2). The growth of jet aircraft since the 1960s has facilitated the growth of business travel, making it easier for executives to oversee and coordinate economic activities in different countries and continents. On top of this, the growth of budget carriers such as easyJet and Ryanair in Europe or Southwest Airlines in the United States over the last decade or so has underpinned the continuing expansion of international tourism. Another key trend has been the development of containerized shipping, which now accounts for approximately 90 per cent of total world trade, greatly reducing the cost and time of transporting goods over long distances (Dicken, 2003a, p.91).

At the same time, new information and communications technologies (ICTs) such as the Internet, email and mobile telephones have spread readily, to the extent that many people and businesses are now dependent on them, relying on a crucial but often unseen mass of hardware consisting of terminals, fibre optic cables and networks. The last decade or so has witnessed the rapid growth of the Internet, which had an estimated 140 million users world wide in mid-1998, increasing to an estimated 600 million in 2002 (UNDP, 1999, p.5; 2004, p.183). This new ICT infrastructure has made it possible for large volumes of information to be exchanged at a fraction of the previous cost (Figure 1.3).

Information and communications technologies have effectively shrunk the distance between places, resulting in 'time–space compression' as it becomes much easier and cheaper to move money and information across space. The term was introduced by Harvey (1989a) who argued that the process of 'time–space compression' has been driven by the development of the economy, requiring geographical expansion in search of new markets and supplies of labour and raw materials. By overcoming the constraints of geography (distance and space) through investments in transport and communications infrastructure, corporations have reduced the effects of distance as it becomes easier and cheaper to transmit information, money and goods between places. As such, time and space are effectively being compressed through the development of new technologies. This is not an entirely novel process; a previous 'round' of time–space compression occurred towards the end of the nineteenth century through inventions such as railways, steamships, the telegraph and the telephone which allowed goods, information and money to be moved far more rapidly (section 4.2).

Another crucial set of connections across space is provided by flows of commodities, defined as products

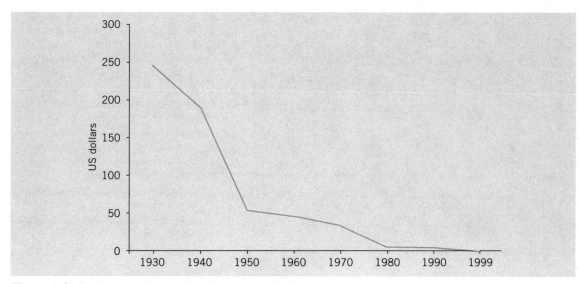

Figure 1.3 Falling cost of London–New York telephone calls.
Source: UNDP, 1999, pp.28, 30.

Box 1.1

The 'Secret Life of a Banana' (Vidal, 1999)

The humble banana is the world's most traded fruit, with 5 billion sold every year in the UK (I. Cook et al., 2002). Recently, for example, one of the authors bought a bunch of four bananas for 66p in his local branch of Morrison's, a British supermarket chain. In making this purchase, he was, in common with many other consumers, primarily concerned with the price and physical appearance of the banana. Hidden beneath these aspects, however, lies a complex geography of production and distribution which links different people and places together (Watts, 2005, pp.537–8). While individual consumers may remain unaware of such linkages, they create real social relationships between people in different places, for example consumers in countries such as the UK and banana farmers in tropical countries in Africa, the Caribbean and Latin America.

Bananas are grown by either individual small farmers, sometimes working under contract, or on large industrial plantations run by MNCs. Either way, the fruit must be grown, picked and packed and transported to the nearest port from where it is shipped in special temperature-controlled compartments to another port in the destination country (I. Cook et al., 2002, p.4). The fruit is then ripened in special ripening centres before being sent to the supermarket by truck. Bananas exported from the tiny West Indian island of St Vincent to the UK, for instance, are transported to Southampton by the Geest line shipping company, taking roughly two weeks (Vidal, 1999).

The complex chain of linkages involved in the production, distribution and exchange of any particular commodity creates real conflicts of interest between groups of people over who captures the most added value from the product (Watts, 2005, pp.534–6). Figure 1.4 shows how the price of a 30p banana is distributed between the various actors in the production chain. The banana has been the subject of several trade disputes, most recently between Europe and the US over the European Union's (EU) system of preferential trade with former colonies in Africa and the Caribbean. This meant that imports from Latin America, where production is controlled by large American agri-business multinationals like Chiquita, faced a range of taxes and restrictions. The US government successfully appealed to the World Trade Organization, threatening to decimate the economies of some Caribbean islands where small producers cannot compete with large 10,000-acre industrialized plantations in Central America (Vidal, 1999).

Even without such competition, small farmers were facing a real struggle to survive due to reduced prices, longer hours and stricter quality specifications from buyers (I. Cook et al., 2002, p.1). By contrast, the supermarkets, multinational distributors and assorted 'middle men' appear to be making substantial profits. For the small Caribbean farmers, then, and the communities that are economically dependent on the export of bananas, the future looks bleak, unless supermarkets and consumers can be persuaded to alter their buying habits, paying higher prices for fair trade rather than free trade fruit (Vidal, 1999).

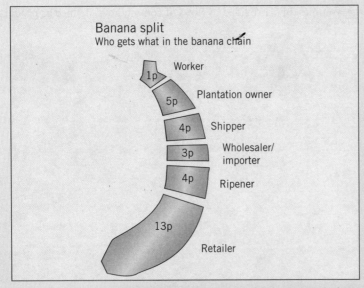

Banana split
Who gets what in the banana chain

- 1p — Worker
- 5p — Plantation owner
- 4p — Shipper
- 3p — Wholesaler/importer
- 4p — Ripener
- 13p — Retailer

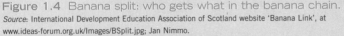

Figure 1.4 Banana split: who gets what in the banana chain.
Source: International Development Education Association of Scotland website 'Banana Link', at www.ideas-forum.org.uk/Images/BSplit.jpg; Jan Nimmo.

or services that are sold commercially. The modern economy involves the production and consumption of a vast array of commodities, spanning everything from iPods to package holidays. The commodity is so basic to the workings of the economy that Karl Marx – who began his famous work *Capital* with an examination of its properties – described it as the 'economic cell form' of capitalism. 'It is as if he is saying that in the same way that the DNA sequence holds the secrets to life, so the commodity is the economic DNA, and hence the secret of modern capitalism' (Watts, 2005, p.532).

Commodity chains link together the production and supply of raw materials, the processing of these materials, the production of components, the assembly of finished products, and the distribution, sales and consumption of these products. They involve a range of different organizations and actors, for example farmers, mining or plantation companies, component suppliers, manufacturers, subcontractors, transport operators, distributors, retailers and consumers. Commodity chains have a distinct geography, linking together different stages of production carried out in different places (Watts, 2005, pp.537–8). Some parts of the production process add more value or profit, creating tensions between the different participants over who

captures the value-added. The role of powerful **multinational corporations** (MNCs) in controlling the production and distribution processes across national boundaries has been the subject of particular scrutiny in recent years (Gereffi, 1994). Exploring the production and consumption of particular commodities, then, helps us to trace and uncover economic connections between places, linking different localities to global trading networks (see Box 1.1).

1.2.2 Uneven Development

A basic feature of the process of economic development that we emphasize in this book is its geographical unevenness. **Uneven development** is an inherent feature of the capitalist economy, reflecting the tendency for growth and investment to become concentrated in particular locations. These areas may be favoured by a particular set of advantages such as geographical position, resource base, availability of capital or the skills and capabilities of the workforce. Once growth begins to accelerate in a particular area, it tends to 'suck in' investment, labour and resources from surrounding regions. Capital is attracted by the opportunities for profit while workers are drawn by

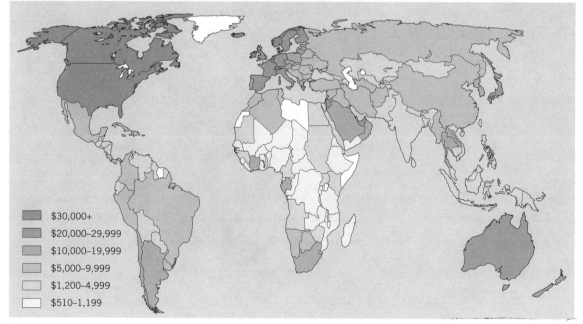

Figure 1.5 Gross Domestic Product (GDP) per capita (PPP), 2003.
Source: UNDP, 2005, pp.219–22.

7

abundant job opportunities and high wages. Surrounding regions are often left behind, relegated to a subordinate role supplying resources and labour to the growth area.

One key aspect of the process of uneven economic development is that it occurs at different geographical scales (Figure 1.1). This can be illustrated with reference to three key scales of activity: global, regional and local.

➤ At the global level, there is a marked divergence between the 'core' in North America, Japan and Western Europe and the 'periphery' in the 'global South' of Asia, Latin America and Africa (see Figure 1.5). This pattern reflects the legacy of colonialism, whereby the core countries in Europe and North America produced high-value manufactured goods and the colonies produced low-value raw materials and agricultural products. While a number of East Asian countries, including China, have been able to overcome this legacy, experiencing rapid growth and rising prosperity over the past 25 years, others, particularly in sub-Saharan Africa, have been left behind, experiencing conditions of extreme deprivation and poverty (Box 1.2).

➤ Within individual countries, too, economic disparities between regions are evident. The rapid economic development of China, for instance, since the late 1970s has opened up a growing divide between the booming coastal provinces in the South and East and a poor, underdeveloped interior (Figure 1.6). Developed countries are also characterized by regional disparities, such as the persistent North–South divide that has characterized the economic geography of the United Kingdom since the 1930s (Amin *et al.*, 2003, p.13).

➤ Even on a local level within cities, uneven development is present in the form of social polarization between rich middle-class neighbourhoods and poorer inner-city areas and public housing schemes. Thus a city like London contains some of the highest property prices and salaries on earth only a few hundred metres away from inner-city districts where over a quarter of the workforce is out of work and living on state benefits.

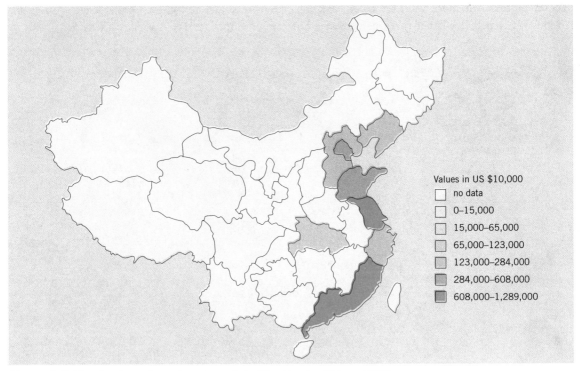

Values in US $10,000
- no data
- 0–15,000
- 15,000–65,000
- 65,000–123,000
- 123,000–284,000
- 284,000–608,000
- 608,000–1,289,000

Figure 1.6 Foreign investment and capital utilization in China, 1999.
Source: Pannell, 2002, p.1583.

Box 1.2

The contrasting economic fortunes of China and Africa

Figure 1.7 Skyscrapers in Shanghai.
Source: © John Lawrence, Getty Images.

Since the Communist regime opened up its economy to attract foreign investment in 1978, China has experienced rapid economic development, with the economy growing at an average rate of 9.5 per cent per year between 1980 and 2003 (Wolf, 2005). Such sustained high growth is almost unparalleled in human history (Harvey, 2005, p.1). This 'second industrial revolution', as it has been termed, has transformed the cities and coastal areas of southern and eastern China, demonstrated by the construction of gleaming skyscrapers and shopping centres dominated by Western brands (Figure 1.7). Such rapid economic growth has attracted more and more people from China's vast rural interior, generating what is predicted to be the largest mass migration in human history. At the same time, however, modernization has brought huge social and environmental costs with the gap between rich and poor widening markedly while China has been found to

Figure 1.8 Poverty in Africa.
Source: © Hulton-Deutsch Collection/Corbis.

Box 1.2 (continued)

contain 16 of the world's 20 most polluted cities (World Bank, 1998). By contrast, Africa has failed to develop since independence in the 1950s and 1960s. Since 1980, its position relative to other world regions has actually deteriorated, with the continent appearing to have almost 'fallen out of the world economy' (Agnew and Grant, 1997). By the year 2000, half of the world's poor were in Africa compared with 10 per cent in 1970 (BBC News, undated). The human costs of this sorry tale of poverty and underdevelopment are evident in terms of disease, famine and war with HIV and AIDS having decimated the adult populations of many countries in sub-Saharan Africa in recent years (Figure 1.8). The plight of Africa has become a political issue in the West in recent years with the British Prime Minister, Tony Blair, describing the continent as a 'scar on the conscience of the world' in a much-publicized speech in October 2001 (quoted in Seldon, 2001, p.500). A wide range of development and 'anti-globalization' groups and activists have subsequently helped to focus attention on Africa, more recently coming together in the 'Make Poverty History' campaign during the summer of 2005, prompted by the G8 economic summit in Gleneagles, Scotland.

The process of economic development is highly dynamic in nature as new technologies are developed, new forms of customer demand emerge and work practices change. Over time, patterns of uneven development are periodically restructured as capital moves between different locations, investing in those that offer the highest rate of return (profit). As a result, new growth regions emerge while established ones can experience stagnation and decline. As broader market conditions and technologies change, the specialized economic base of formerly prosperous regions can be undermined by reduced demand, rising costs, competition and the invention of new products and methods of production. On a global scale, the most dramatic change in patterns of uneven development is the emergence of East Asia as a dynamic growth region over the last 30 years or so. Within developed countries, established industrial regions dependent on nineteenth- and early twentieth-century industries such as coal, steel and shipbuilding have experienced decline while new growth centres have emerged in regions such as the south and east of the US (the so-called 'sunbelt'), southern Germany and north-eastern Italy.

1.2.3 The importance of place

The role of **place** in shaping economic activity is a third key theme of this book. As we suggested in the previous section, processes of uneven geographical development have created distinctive forms of production in particular places. During the nineteenth and early twentieth century, highly **specialized industrial regions** emerged in Europe and North America. As a result,

> distinct places are associated with sectoral and functional divisions of labour. In the United States, for example, 'Pittsburgh meant steel, Lowell meant textiles, and Detroit meant automobiles' (Clark *et al.*, 1986, p.23), while in the United Kingdom, 'one finds metal workers in the Midlands, office professionals in London, miners in South Wales, and academics in Oxford' (Storper and Walker, 1989, p.156).
>
> (Peck, 1996, p.14.)

Although some of these specific associations have been weakened by deindustrialization, the general point about distinctive forms of production being associated with particular places remains important. The City of London, for instance, continues to be associated with finance and business services, Silicon Valley in California with semiconductors, Los Angeles with movies and Milan with clothing design and fashion. Such variety is continually reproduced through the interaction between wider processes of uneven development and local political, social, economic and cultural conditions. These conditions reflect the economic history of a place in terms of the particular industries found there and the institutions and practices associated with them. In any one place, the interaction between established industries and institutions and contemporary processes of change shapes and moulds the economic landscape (Massey, 1984).

It has become increasingly clear in recent years that globalization is a differentiated and uneven process, generating different outcomes in different places. In particular, globalization seems to be associated with a resurgence of certain regions as economic units. The success of dynamic growth regions such as the City of London (financial and business services), Silicon Valley (advanced electronics), southern Germany (vehicles and electronics) and north-eastern Italy (machine tools, textiles), for example, is rooted in the specialized production systems that have flourished there. Geographical proximity seems to encourage close linkages and communication between firms, enabling them to share information and resources. The existence of a large pool of skilled labour is a crucial feature of such regions, allowing firms to recruit easily and workers to move jobs without leaving the local area. These aspects of the local production system encourage innovation and entrepreneurship, enhancing the competitiveness of such regions within a global economy.

While globalization is not leading to the erasure of place as a significant dimension of economic life, it does undermine traditional notions of places as homogenous and clearly bounded local areas. As such, there is a need to rethink place in terms of connections and relations across space. It is in this sense that the British geographer Doreen Massey's work on the development of a 'global sense of place' is of particular interest. Massey develops a new conception of place as a meeting place, a kind of node or point where wider social relations and connections come together:

> what gives place its specificity is not some long internalised history but the fact that it is constructed out of a particular constellation of social relations, meeting and weaving together at a particular locus. ... Instead ... of thinking of places as areas with boundaries around them, they can be imagined as articulated moments in networks of social relations and understandings ... and this in

turn allows a sense of place which is extroverted, which includes a consciousness of its links with the wider world, which integrates in a positive way the global and the local (Massey, 1994, pp.154–5).

From this perspective, place can itself be regarded as a process rather than seen as some static and unchanging essence. Places are connected and linked through wider processes of uneven development operating through flows of capital, goods, services, information and people. Movement of particular commodities like bananas, for example, link different parts of the UK to the economies of certain Caribbean islands (see Box 1.1).

Reflect

> Do you agree that specific places (regions) remain important within a global economy? Justify your answer.

1.3 The economy and economic geography

1.3.1 The capitalist economy

'The economy' refers to the interrelated processes of production, circulation, exchange and consumption through which wealth is generated (Hudson, 2005, p.1). It is through such processes that people strive to meet their material needs, earning a living in the form of wages, profits or rent. Production involves combining land (including resources), capital, labour and knowledge – commonly known as the **factors of production** – to make or provide particular commodities. It relies on a supply of resources from nature, meaning that economic activities have a direct impact on the environment. The commodities produced are then either directly consumed by the producers or, much more commonly, sold to individual consumers and households through the market.

Human societies have tended to organize and structure their economic activities through overarching **modes of production**. These can be defined as economic and social systems that determine how resources

11

are deployed, how work is organized and how wealth is distributed. Economic historians have identified a number of modes of production, principally subsistence, slavery, feudalism, capitalism and socialism. Each of these creates distinctive relationships between the main factors of production. **Capitalism** is clearly the dominant mode of production in the world today, operating at an increasingly global scale. It is defined by individual ownership of the means of production – factories, equipment and money capital – and the associated need for most people to sell their labour power to employers or capitalists in order to earn a wage. This allows them to purchase commodities produced by other firms, creating the market demand that underpins the capitalist system. Compared to earlier modes of production, production and consumption are often geographically separate under capitalism, creating a need for extensive transport and distribution networks.

A key underlying point here concerns the fact that the principal features of the modern capitalist economy

– such as the role of the market, profits and competition – are not natural and eternal forces that determine human behaviour, as mainstream economists and business commentators tend to assume. Instead, capitalism is a historically specific mode of production that has emerged from its roots in early modern Europe in the sixteenth and seventeenth centuries to encompass virtually the entire globe today. It has been superimposed on a complex mosaic of pre-capitalist societies and cultures, resulting in great regional variation as pre-existing local characteristics interact with broader global processes (Johnston, 1984).

While capitalism is clearly the dominant mode of production in the world today, it does not follow that all economic activity is capitalist in nature. In reality, the formal, capitalist economy based on striving to maximize profits or earnings coexists with a range of other economic activities and motivations such as domestic work, volunteering, the exchange of gifts and cooperatives. Gibson-Graham (2006) represents this in terms of an 'iceberg' model with the capitalist economy masking a wide range of other forms of economic activity (Figure 1.9). The two categories are not separate in practice, however, with non-capitalist activities interacting with capitalism in a variety of ways. Think of the relationship between domestic work and paid employment, for example, or the role of gift-buying *within* capitalism. Gibson-Graham (1996) emphasizes the existence of **diverse economies**, criticizing the preoccupation with the formal, capitalist economy among economists and economic geographers. This critique has informed a number of studies of 'diverse' or 'alternative' economies such as informal work, local currencies and cooperatives (Leyshon *et al.*, 2003).

1.3.2 An economic geography perspective

Economic geography is concerned with concrete questions about the location and distribution of economic activity, the role of uneven geographical development and processes of local and regional economic development. It asks the key questions of 'what?' (the type of economic activity), 'where?' (location), 'why?' (requiring explanation) and 'so what?' (referring to the

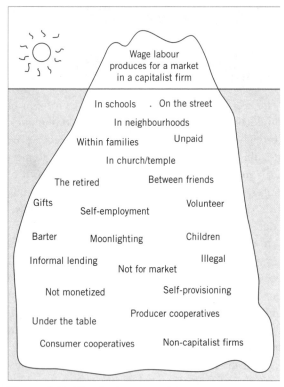

Figure 1.9 The iceberg model.
Source: Drawn by Ken Byrne. Gibson-Graham, J.K., 2006, *A Postcapitalist Politics*, Minneapolis and London: University of Minnesota Press, p.70.

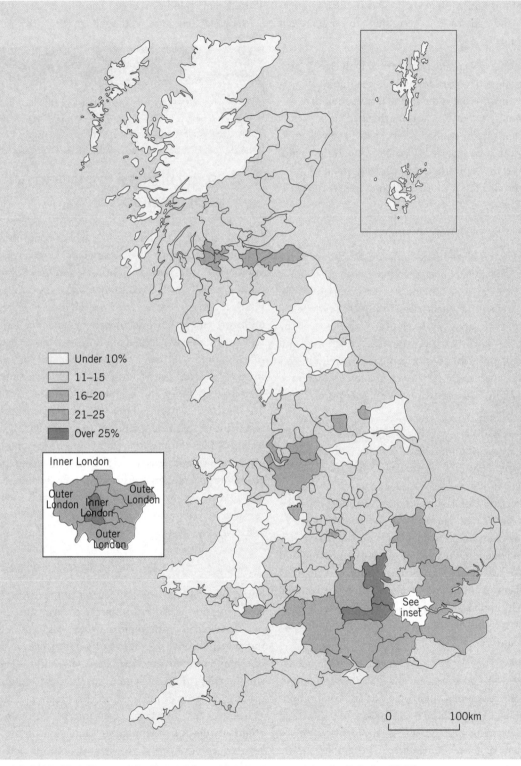

Figure 1.10 Employment in financial and business services in Britain, 2005.
Source: ONS, 2006.

implications and consequences of particular arrangements and processes). According to one recent definition:

> Economic geography is concerned with the economics of geography and the geography of economics. What is the spatial distribution of economic activity? How is it explained? Is it efficient and/or equitable? How has it evolved, and how can it be expected to evolve in the future? And what is the appropriate role of government in influencing this evolution? (Arnott and Wrigley, 2001, p.1).

Three key themes emerge from this:

a) The geographical distribution and location of economic phenomena. Figure 1.10 provides an example of such a distribution, showing variations in employment in financial and business services – a key growth sector in developed countries over the past 25 years – across the UK. Describing and mapping distributions of economic activity in this way can perhaps be seen as the first task of economic geography, addressing the basic questions of 'what' and 'where'.

b) The next stage is to explain and understand these spatial distributions and patterns of economic activity. This is a more demanding task, which brings in theory and requires us to have some appreciation of history. It involves addressing the more advanced questions of 'why' and 'how'. In trying to explain why financial and business services employment is clustered in south-east England and one or two other areas, then, we would need to draw on theories of the spatial concentration of economic activity and to have some understanding of the basic economic history of the UK since the industrial revolution.

c) A third issue is that of engagement with policymakers in government and the private sector, making recommendations and offering advice about particular geographical issues and problems. As well as describing and explaining the distribution of certain economic phenomena, geographers have also sought to outline how the economic geography of particular countries and regions *should* be organ-

ized. This role has sparked periodic debates about the social 'relevance' of the subject (Peck, 1999).

1.4 A political economy approach

In this section, we build on the above definition of economic geography by introducing our favoured approach to the subject: **political economy**. Political economy analyses the economy within its social and political context, rather then seeing it as a separate entity driven by its own set of rules based on individual self-interest. It is concerned not only with the exchange of commodities through the market, but also with production and the distribution of wealth between the various sections of the population (Barnes, 2000b). This focus on questions of production and distribution distinguishes political economy from modern economics. While the classical political economy of the nineteenth century was not concerned with geography, economic geographers have sought to apply a political economy framework to geographical questions such as regional development and urban growth (Harvey, 1982; Massey, 1984).

1.4.1 Social relations

Social relations provide the general link between the economy and society. In contrast to mainstream economics, which is underpinned by the assumption of individual rationality, political economists believe that economic activity is grounded in social relations. These simply refer to the relationships between different groups of people involved in the economy such as employers, workers, consumers, government regulators, etc. A key defining social relationship is that between employers and workers, structuring activity within the workplace. The two groups are commonly regarded as different classes within society, a position set out with particular force by Marx, although for some commentators the growth of a large middle class of 'white collar' workers has blurred the distinction considerably. Other important social relations are those between producers and consumers, different firms (e.g. manufacturers and suppliers), different groups of workers (e.g. supervisors and ordinary employees) and government agencies and firms. As our case study of the banana indicated (Box 1.1), the production of a simple commodity creates complex social relations between people based in different places. Even if some of the parties – for example, consumers at large supermarkets in the UK or US – are not aware of such relationships, they still exist.

Once we have accepted that the economy is structured by social relations, it is important to recognize that these relations will change over time as society evolves. We have already pointed to a transformation occurring in Western societies from around the fifteenth century onwards where a system of market capitalism gradually replaced feudalism as the underlying mode of production. In other words, feudal ties and values of the peasantry and the nobility have given way to market competition, the pursuit of profit and the wage relationship between capital (employers) and labour (workers) as the key social relationship shaping economic and indeed human development. Increasingly intensive competition between firms is also a critical relationship, driven by the relentless pursuit of profit and the subsequent drive to eliminate rivals to gain a greater share of the market.

1.4.2 Power

Having accepted that the economy is structured by social relationships, it is also important to recognize the role of **power** in underpinning these relationships. Ultimately all human social relations are underpinned by power, in the sense of the ability or capacity to take decisions that involve or affect other people (see Allen, 2003). Economic relations in this sense are no different. Power percolates through economic relationships at all geographic levels: from that of the household in terms of who makes decisions regarding the domestic budget, and who 'goes out to work' and who 'stays at home'; at the level of the firm in terms of the share of the wealth

generated in production that accrues to employers rather than employees; in the relationships between firms within particular industries with large retailers and manufacturers often able to dictate prices and terms to their suppliers (e.g. the cost-cutting strategies of the large supermarket chains such as Tesco in the UK have reduced prices for farmers); and at the international level, in the way that some institutions and governments – the World Trade Organization (WTO) or the US – have greater power to set the rules of trade than others.

The 'secret life of a banana' (Box 1.2) indicates how the social relationships between different groups of people located in different places are structured by power. At a very basic level, it is clear that some actors in the chain, particularly the supermarkets and the multinational firms that coordinate the production and distribution processes, are in a more powerful position that the small Caribbean farmers or the labourers in the large Central American plantations. Unequal power relations in this sense are central in understanding the concept of uneven development.

1.4.3 Institutions and the construction of markets

Another crucial element in our political economy perspective is a view of markets as socially constructed entities that require social and political regulation rather than representing naturally occurring phenomena capable of self-regulation. Unlike mainstream economics, which holds that markets, if left to their own devices, will return to an equilibrium position where supply equals demand and waste is eliminated, we believe that unregulated ('free') markets are destabilizing and socially destructive. While the notion of the *free* market retains its ideological and political power, in reality virtually all economies are mixed, containing substantial public sectors. Following Karl Polanyi (Box 1.3), we emphasize the **institutional foundations of markets**, recognizing that the economy is shaped by a wide range of institutional forms and practices. These include cultural rules, habits and norms that structure the social relations between individuals, helping to generate the trust that underpins legal and contractual relationships, and the direct

15

Box 1.3

Karl Polanyi: The economy as an instituted process

The work of Karl Polanyi (1886–1964) on the institutional foundations of economic processes is well known within the social sciences, making a crucial contribution to the development of wider institutional and sociological perspectives on the economy. Polanyi's broad conception of the economy and his insights into the underlying nature of market society have come to inform debates about contemporary processes of globalization and the social 'embeddedness' of the economy (Block, 2003).

In *The Great Transformation*, published in 1944, Polanyi explores the origins and development of market society during the nineteenth century. He identified land, labour and money as 'fictitious commodities' because they are not actually produced for sale on the market like true commodities, a crucial point neglected by orthodox economic analyses, which assume that the price mechanism will balance supply and demand in the

normal fashion. The identification of these fictitious commodities is important to Polanyi's argument, indicating that a pure self-regulating market economy is impossible since state intervention is required to match the supply and demand for land, labour and money (Polanyi, 1944). The book demonstrates how the construction of competitive markets depended upon state action through the upholding of property rights and contracts, the introduction of labour legislation, the establishment of measures to ensure a stable food supply and the regulation of the banking system.

In the 1950s, Polanyi turned away from modern market economies to the analysis of primitive and archaic economies. This substantive focus was closely informed by his enduring interest in the scope of the economy and the role of institutions in shaping economic processes, as demonstrated by his famous analysis of 'the economy as an instituted process'. As

this phrase suggests, Polanyi views institutions as constitutive of the economy:

> The instituting of the economic process vests that process with unity and stability; it produces a structure with a definite function in society; it shifts the place of the process in society, thus adding significance to its history; it centres interest on values, motives and policy.
>
> (Polanyi [1959] 1982, pp.243–70)

Institutions are both economic and non-economic with the inclusion of the latter regarded as vital since religion or government, for example, may be as important in underpinning the operation of the economy as the monetary system or the development of new labour-saving technologies. From this perspective, then, the research agenda is one of examining the manner in which 'the economic process is instituted at different times and places' (ibid., p.250).

intervention of the state in managing the economy and in running systems of social welfare.

Key forms of **institution** at the national level include firms, markets, the monetary system, business organizations, the state and a wide range of state agencies and trade unions. These are not merely organizational structures; they also tend to incorporate and embody specific practices, strategies and values that evolve over time. Economic geographers are particularly interested in how local and regional economies are shaped by distinctive institutional arrangements (Martin, 2000). Important forms of institutions in this respect include local authorities

and development agencies, employers' organizations, business associations and chambers of commerce, local political groupings, trade union branches and voluntary agencies. According to Amin and Thrift (1994) institutional 'thickness' or density is an important factor shaping local economic success, referring to the capacity of different organizations and interests to work together in the pursuit of a common agenda. While far from typical of the structure of contemporary local economies, the **industrial districts** of central and north-eastern Italy offer a striking illustration of the role of institutions in shaping local economic development (Box 1.4).

Box 1.4

Institutions and local economic development in the 'Third Italy'

The notion of industrial districts is derived from the writings of the English economist Alfred Marshall in the early twentieth century. Marshall drew attention to the existence of highly specialized industrial regions based on networks of small firms in northern England such as the Sheffield cutlery district or the various woollen textile areas in west Yorkshire. He emphasized the inseparability of economic life from local society and culture (Amin 2000, p.153). Over the course of the twentieth century, such areas experienced decline, eclipsed by the growth of mass production and large, integrated corporations.

In the 1980s, however, the renaissance of industrial districts was highlighted by writers like Piore and Sabel (1984) who spoke of a 'second industrial divide' where mass production was giving way to new forms of flexible specialization, based on increased demand for design-intensive and customized products. Flexible specialization was defined in terms of the flexible use of machinery and labour within a highly decentralized production system where individual firms concentrated on specific tasks. Industrial districts within advanced economies like Italy, France, Japan, Denmark and Spain were experiencing rapid growth in the 1980s and 1990s.

The industrial districts of central and north-eastern Italy are scattered across the regions of Tuscany, Emilia-Romagna and Veneto. They became known as the 'Third Italy' to distinguish this new growth region from the rich, industrial North and the poor, agrarian South. Particular districts include the internationally famous centres of Prato (textiles), Modena (machine tools), Santa Croce (leather tanning), Capri (knitwear) and Sassuolo (ceramics) (Figure 1.11) (Amin, 2000, p.154). In general, the growth of these districts has been

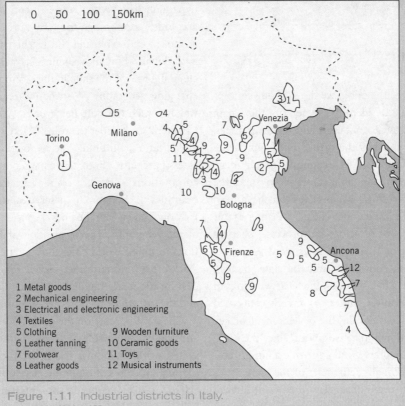

1 Metal goods
2 Mechanical engineering
3 Electrical and electronic engineering
4 Textiles
5 Clothing
6 Leather tanning
7 Footwear
8 Leather goods
9 Wooden furniture
10 Ceramic goods
11 Toys
12 Musical instruments

Figure 1.11 Industrial districts in Italy.
Source: Amin, 2000, p.155.

Box 1.4 (continued)

based on a revival of local craft traditions and skills, harnessing these to
modern production methods to meet
growing market demand.

As a number of commentators have
noted, the success of the Italian districts is deeply rooted in local culture.
The strongly communitarian political
cultures of central and north-eastern
Italy (socialist in Tuscany and Emilia-
Romagna, Catholic in Veneto)
provided a basis for inter-firm collaboration with the communist and

socialist parties in Emilia-Romagna,
for example, having a strong influence over the trade unions, craft
associations and cooperatives of
small entrepreneurs. Local authorities
offered business premises and a
range of services to small firms from
the 1970s, while a dense network of
labour unions, industry associations
and chambers of commerce developed a sophisticated reservoir of
knowledge, skills and resources for
the use of members. Close ties

between individual firms within a
highly specialized production system,
underpinned by a common culture,
has generated high levels of trust.
The sharing of knowledge and ideas
between firms has facilitated incremental forms of continuous
innovation, although questions
remain about the Italian districts'
capacity to adjust to the prospect of
radical change in market demand or
technological development (Asheim,
1996).

Reflect

➤ In what ways do institutions shape the process of
economic development?

1.5 Outline of the book

In this first part of the book, we develop our conceptual approach and identify some of the main
underlying features of capitalism and its geographies.
In Chapter 2 we consider the main sets of approaches
that have been adopted by economic geographers,
focusing particularly on political economy and the
institutional and cultural perspectives that have been
embraced over the last decade. The next chapter examines the main groups of actors and processes shaping
the development of the capitalist economy, focusing
on capital, labour, consumers and the state in turn.
Whereas Chapter 3 focuses upon the basic workings of
the capitalist economy, Chapter 4 is directly concerned
with its geography, outlining the development of key
spaces of production and consumption since the nineteenth century. Chapters 3 and 4 provide an important
historical context for the remainder of the book.

Guided by our three substantive themes of uneven
development, place and the flows connecting places,
and informed by the more abstract categories and defi-

nitions provided in Chapters 1–4, we explore various
key contemporary topics in economic geography in
Chapters 5–12. Chapter 5 focuses on globalization,
contrasting the approach of neoliberal economists and
hyperglobalists with more sceptical and critical
accounts, examining the structure of the global
economy and exploring the recent emergence of the
anti-globalization movement. Chapter 6 explores the
changing role of the state in the economy, assessing the
changing geography of economic regulation and the
way in which the state has evolved. We then go on in
Chapter 7 to consider the role of MNCs as key agents of
globalization, outlining their growth over time and
assessing claims about the emergence of truly global
corporations in the last couple of decades. Chapter 8
examines the geography of services, focusing particularly on financial and business services. The twin trends
of the geographical concentration of high-level functions in major world cities and the global dispersal of
'back office' and customer-service functions to lower-
cost locations are highlighted, along with digitization
and the growth of the Internet economy.

Chapter 9 focuses on the geography of employment
and the changing role of labour within a more integrated world economy. The active role of workers in
shaping processes of economic development is stressed
(Herod, 1997). Chapter 10 concentrates on processes of
innovation and learning at the regional level, reviewing
contemporary debates on clusters and learning regions.

Chapter 11 proceeds to examine development in the Global South, identifying the main approaches to development, examining broad patterns of inequality and highlighting contemporary policy debates on trade, aid and debt. Chapter 12, the last substantive chapter, then assesses the development of the tourist industry, emphasizing its importance as a vehicle for the consumption and promotion of place and assessing its use as a means of economic development in both developed and developing countries.

Finally, a brief conclusion in Chapter 13 pulls the main themes together and considers some key public policy issues.

Exercises

Think of a commodity that you have recently consumed. This could be something you ate for lunch or breakfast or an item of clothing that you have recently bought. A jar of coffee or a pair of training shoes could be an example.

1. When you purchased this commodity, were you primarily concerned with the price and physical qualities of the good?

2. Was there anything, a label, to indicate the geographical origin of this good?

3. Why would it have been produced in that particular region or country?

4. Under what conditions do you think it would have been produced (e.g. in a factory, by craft workers, on a farm)?

5. What main actors would have been involved in its production (e.g. MNCs, small firms, farmers)?

6. How might the profits be distributed among these main actors?

Key reading

Barnes, T.J. and Sheppard, E. (2000) 'The art of economic geography', in Sheppard, E. and Barnes, T.J. (eds) *A Companion to Economic Geography*, Oxford: Blackwell, pp.1–8.
A brief introduction to economic geography as an academic subject area. Highlights the key questions that economic geographers address and the changing perspectives that inform their work.

Castree, N., Coe, N., Ward, K. and Samers, M. (2004) *Spaces of Work: Global Capitalism and Geographies of Labour*, London: Sage, pp.1–23.
The introduction to a recent textbook on the geography of employment and labour. Begins with brief examples of 'working lives', stressing how the experiences of people in different locations are linked together under global capitalism.

Leyshon, A. (1995) 'Annihilating space?: The speed-up of communications', in Allen, J. and Hamnett, C. (eds) *A Shrinking World?: Global Unevenness and Inequality*, Oxford: Oxford University Press, pp.11–54.
Introduces the key concept of time–space compression, emphasizing the speed-up of global communications since the early 1980s. Explains this in terms of the geographical expansion of capitalism over time, facilitated by successive advances in transport and communications technologies.

Taylor, P.J., Watts, M. and Johnston, R.J. (2002) 'Geography/globalization', in Johnston, R.J., Taylor, P. and Watts, M. (eds) *Geographies of Global Change: Remapping the World*, 2nd edn, Oxford: Blackwell, pp.1–17, 21–28.
A stimulating introduction to the relationships between geography as a discipline and the issue of globalization. Outlines the rise of globalization as a key topic of interest, discusses the uneven geography of globalization and highlights recent political debates about the nature of globalization.

Watts, M. (2005) 'Commodities', in Cloke, P., Crang, P. and Goodwin, M. (eds) (2005) *Introducing Human Geographies*, 2nd edn, London: Arnold, pp.527–46.
A review of the commodity as a key topic of interest to geographers. Highlights the economic importance of the commodity within capitalism and the role of commodities in linking production in different places by means of commodity chains and networks.

Useful websites

http://www.theglobalsite.ac.uk/global-library/index.html
Provides a very accessible and comprehensive introduction to key aspects of globalization. The companion website for the *Global Transformations* textbook (Held *et al.*, 1999).

http://www.exchange-values.org
The website of a social sculptures project by Shelley Sacks in collaboration with the banana growers of the Windward Islands and a range of representative organizations. Contains a number of useful articles under the 'texts, debates, discussion' link and images. The articles by the geographers Ian Cook and Luke Desforges are particularly recommended along with 'banana wars', 'banana lives' and 'unfair trade'.

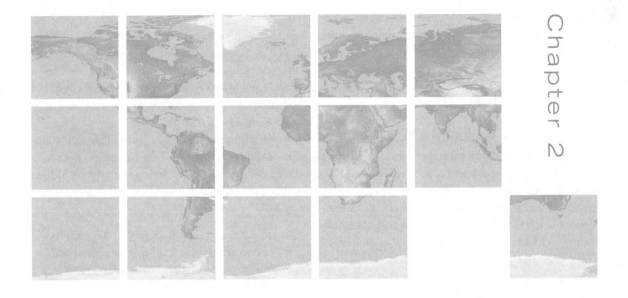

Chapter 2

From commercial geography to the 'cultural turn'? Approaches to economic geography

Topics covered in this chapter

➤ The development of economic geography as an academic discipline.

➤ The main approaches adopted by economic geographers, covering:

• traditional approaches, specifically commercial geography and regional geography;

• spatial analysis, emphasizing scientific methods and quantitative modelling;

• Marxist political economy;

• cultural and institutional perspectives.

➤ Our favoured approach, which can be characterized as a modified political economy, informed by the insights generated by cultural and institutional approaches.

Chapter map

Having defined economic geography in section 1.3, in this chapter we set out the main approaches that economic geographers have adopted over time. This serves to frame and position the political economy perspective that informs this book. We begin by considering the relationship between economic geography and the neighbouring discipline of economics. In the remainder of the chapter we examine the different approaches that have been adopted in economic geography, gaining favour at different times. Four main approaches are identified: traditional economic geography, which was dominant from the late nineteenth century to the 1950s; spatial analysis, embraced in the 1960s and 1970s; political economy, which became popular in the 1970s and 1980s; and a set of cultural and institutional frameworks towards which economic geographers have turned since the early 1990s. The latter two approaches are considered in greater detail because of their importance in shaping debates over the direction of the discipline in recent years.

2.1 Introduction

A key starting point for this chapter is to recognize that no academic subject has a natural existence. Instead, as Barnes argues, subjects must be 'invented' in the sense of being created by people at particular times: the first economic geography course was taught at Cornell University in 1893, the first English-language textbook, George G. Chisholm's *Handbook of Commercial Geography*, was published in 1889 and the journal *Economic Geography* was established in 1925 (Barnes, 2000a, pp.14–15). The neighbouring discipline of economics was also established in the late nineteenth century, along with a number of other social sciences. From the start, however, the two disciplines assumed different characteristics.

Economics views the economy as governed by market forces that basically operate in the same fashion everywhere, irrespective of time and space. The market is composed of a multitude of buyers and sellers – the forces of demand and supply – who dictate how scarce resources are allocated through their decisions about what to produce and consume. Mainstream neoclassical economics is underpinned by the idea of 'economic man', assuming that people act in a rational

and self-interested manner, continually weighing up alternatives on the basis of cost and benefits, almost like calculating machines. The market is viewed as an essentially self-regulating mechanism, tending towards a state of equilibrium or balance through the role of the price mechanism in mediating between the forces of demand and supply (Figure 2.1).

While economics developed as a theoretical discipline adopting the methods of natural sciences such as physics and chemistry, economic geography established itself as a strongly factual and practical enterprise (see section 2.2):

> As a discipline it [economic geography] grew less out of concerns by economists to generalise and theorise, than the concerns of geographers to describe and explain the individual economics of different places, and their connections one to another.
>
> (Barnes and Sheppard, 2000, pp.2–3)

In general, a clear contrast can be drawn between the formal and theoretical approach of economics and geography's more open-ended ethos and more substantive concerns. While geography can be seen as synthetic in nature, focusing on the relationships between, rather than the separation of, processes and things (Lee, 2002, p.333), economics is analytic, seeking to separate the economy from its social and cultural context. Key features of each of the four approaches to economic geography examined in this chapter are set out in Table 2.1, providing an important backdrop to the ensuing discussion.

2.2 Traditional economic geography

What we can term the **traditional approach to economic geography** held sway from the late nineteenth century to the 1950s. It was factual and descriptive in nature, focusing on the compilation of information about economic conditions and resources in particular regions. Economic geography or **commercial geography** was highly prominent from the 1880s to the 1930s, a period in which geography assumed a key role

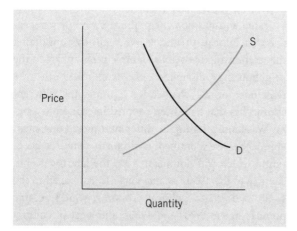

Figure 2.1 Demand and supply curves.
Source: Lee, 2002, p.337.

Table 2.1 Approaches to economic geography

	Traditional	Spatial Analysis	Political Economy	Institutional / Cultural
Supporting philosophy	Empiricism.[a]	Positivism (Box 2.1).	Dialectical materialism.[b]	Postmodernism (Box 2.6) and institutionalism.
Main source of ideas	Classical German geography, anthropology, biology.	Neoclassical economics.	Marxist economics, sociology and history.	Cultural studies, institutional economics, economic sociology.
Conception of the economy	Closely integrated with the natural resources and culture of an area.	Driven by rational choices of individual actors.	Structured by social relations of production. Driven by search for profit and competition.	Importance of social context. Informal conventions and norms shape economic action.
Geographical orientation (place or wider processes)	a) Commercial geography stressed global trading system; b) regional geography highlighting unique places (regions).	Wider forms of spatial organization.	Wider processes of capitalist development. Places passive 'victims' of these wider processes.	Emphasis on individual places in context of globalization.
Geographical focus	a) Colonial territories; b) distinctive regions, mainly in Europe and North America, often rural and geographically marginal.	Urban regions in North America, Britain and Germany.	Major cities in industrial regions in Europe and North America. Cities and regions in developing countries, especially Latin America.	Growth regions in developed countries. Global financial centres. Key sites of consumption.
Key research topics	Effects of the natural environment on production and trade; identifying distinctive regional economies.	Industrial location; urban settlement systems; spatial diffusion of technologies; land use patterns.	Urbanization processes; industrial restructuring in developed countries; global inequalities and underdevelopment.	Social and institutional foundations of economic development; consumption; work identities; financial services; corporate cultures.
Research methods	Direct observation and fieldwork.	Quantitative analysis based on survey results and secondary data.	Reinterpretation of secondary data according to Marxist categories. Interviews.	Interviews, focus groups, textual analysis, ethnography, participant observation.

[a] Emphasis on gaining knowledge directly from the senses, particularly through observation, and 'the facts'.
[b] Dialectical means that social change is seen in terms of a struggle between opposing forces (Box 2.2). Materialism stresses the real social and economic conditions of existence (production, labour, class relations, technology, resources) over ideas and culture, effectively privileging matter over mind.

as the 'handmaiden' of empire, providing useful knowledge for colonial administrators and traders (Livingstone, 1992, pp.216–59). Commercial geography was particularly useful in providing knowledge about the colonial territories of the European powers in Africa, Asia and Latin America, detailing the key resources and identifying the types of crop that could be grown there.

Commercial geography was based on the 'great geographical fact' that different parts of the world yield

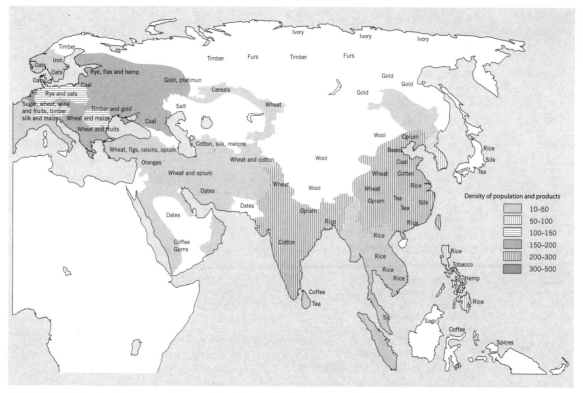

Figure 2.2 Regional economic specialization in Asia under colonialism.
Source: Chisholm, 1889, pp.302–3.

different products, underpinning the system of global commerce (Barnes, 2000a, p.15). The *Handbook of Commercial Geography* emphasized factual detail over abstract theory, with George G. Chisholm, credited as having wished the 'love of pure theory to the devil', displaying, instead, a 'meticulous mastery of detail' (MacLean, 1988, p.25). Chisholm provided detailed statistics on trade, compiling information about the production and exchange of a wide range of commodities and resources (Figure 2.2). His numerous maps, figures and tables played an important role in offering a concrete expression of imperialism as an economic system, rendering the complex flow of goods and resources more visible and tangible (Barnes, 2000a, p.15). The development of commercial geography was crucial in establishing economic geography as a distinct sub-discipline, helping to define many of its enduring characteristics such as an avoidance of theory, an emphasis on factual detail, a celebration of numbers and a reliance on geographical categories made visible by the map (ibid., p.16).

By the 1930s, however, the empires of the major powers were in decline, with policy debates becoming increasingly focused upon internal national issues such as unemployment and poverty. Partly in response, the focus of economic geography 'shifted from the general commercial relations of a global system to the geography of narrowly bounded, unique regions, especially those close to home' (ibid., p.18). This was the era of **regional geography**, defined by Hartshorne (1939, p.21) as a project of 'areal differentiation', which describes and interprets the variable character of the earth's surface, expressed through the identification of distinct regions. These were classified and described through works of synthesis that presented the key characteristics of an area in a logical order, beginning with the physical landscape and proceeding through to human settlement and culture (Livingstone, 1992, pp.260–303). Economic geography was largely subsumed within this broader regional enterprise, emphasizing regional uniqueness, the compilation of regional typologies and the need for first-hand fieldwork (Barnes, 2000a, pp.20–1).

2.3 Spatial analysis in economic geography

By the mid-1950s, considerable dissatisfaction was being directed towards the traditional approach. A new generation of researchers came increasingly to reject the idea that regional synthesis was the proper goal of geography, seeking to develop a more scientific approach. The attack on the traditional establishment found particularly cogent expression in a 1953 paper by Fred K. Schaefer, who called for geographers to employ **scientific methods** in a search for general theories and laws of location and spatial organization (Scott, 2000, p.485). This argument drew directly on the established **positivist** philosophy of science (Box 2.1)

Schaefer's philosophical arguments fitted with a new style of practical research being developed at the Universities of Iowa and Washington, Seattle, where younger geographers were using statistical and mathematical methods to analyse problems of industrial location, distance and movement (Barnes, 2000a, pp.21–2). A vibrant spatial analysis research programme developed at Seattle, for example, focusing on issues of industrial location and land use patterns, urbanization and central place theory, transport networks and the geographical dynamics of trade and social interaction. The new scientific geography spread to other departments in the US while, in the 1960s, the Universities of Cambridge and Bristol became key centres in the UK. Progress was such that Burton (1963) could proclaim the 'quantitative revolution' complete, defined as 'a radical transformation of the spirit and purpose of geography'. Economic geography was at the forefront of this movement, viewed as an area that was particularly suited to the application of quantitative methods.

Real world conditions were increasingly favourable to this new approach in the late 1950s and 1960s as policymakers focused on economic and urban problems in developed countries, providing funds for research and a demand for academic analysis and advice. A period of sustained economic growth and an underlying faith in science and technology created an

Box 2.1

Positivism

This is a philosophy of science originally associated with the French philosopher and sociologist Auguste Comte (1798–1857) and developed further by the Vienna Circle of thinkers in the 1920s and 1930s (Gregory, 2000, p.60). It seems to have gained broad acceptance as an account of the goals of natural science in the post-war period, although specific aspects of it sparked considerable argument. Since the 1970s particularly, some social scientists (including many human geographers) have rejected positivism.

Positivism holds that a real world exists independent of our knowledge of it. This real world has an underlying order and regularity that science seeks to discover and explain. Facts can be directly observed and analysed in a neutral manner. The separation of fact and value is a central tenet of positivism; personal beliefs and positions should not influence scientific research. The aim of science is to generate explanatory laws that explain and predict events and patterns in the real world. In the classic deductive method (moving from theory to practical research), scientists formulate hypotheses – formal statements of how a force or relationship is thought to operate in the real world – that are then tested against data collected by the scientist through experiment or measurement. Hypotheses that are supported by initial testing must then be verified or proved correct through objective and replicable procedures. If verification is successful, they gain the status of scientific laws.

10317967X

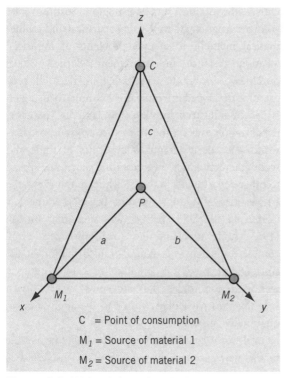

Figure 2.3 Weber's locational triangle.
Source: Knox and Agnew, 1994, p.77.

optimistic 'can-do' attitude, with urban and regional planning embraced as the means of addressing problems of location, land use management and transportation (Barnes, 2000a, pp.23–4). In this context, regional geography appeared increasingly backward and anachronistic, with its focus on rural backwaters and its concern with description and classification offering little of practical value to the planner or developer.

Neoclassical economic theory provided a ready source of concepts for quantitative economic geography in the 1960s. Geographers sought to apply the same style of deductive theorizing and analysis, beginning from simplifying assumptions to the development and testing of hypotheses and models against numerical data representing real-world conditions. As one of the pioneers of quantitative economic geography, Harold McCarty of the University of Iowa, put it, 'economic geography derives its concepts largely from the field of economics and its method largely from the field of geography' (McCarty, 1940, quoted in Barnes, 2000a, p.22).

The tradition of **German location theory** provided a

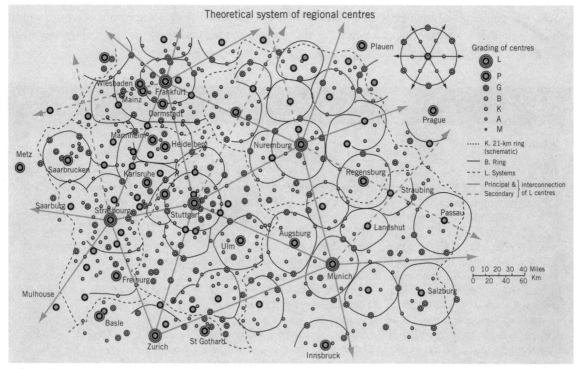

Figure 2.4 Central places in southern Germany.
Source: Knox and Agnew, 1994, p.72.

body of economic theory applied to geography that the new economic geography of the 1950s and 1960s could draw upon. The work of theorists such as Von Thunen, Weber, Christaller and Lösch was applied to the circumstances of North America and the UK in the 1960s, being used to explain and predict land use patterns, the location of industry and the organization of settlements and market areas. Weber's theory of industrial location emphasized the importance of transport costs in determining where a factory or plant would be located in relation to the sources of raw materials and the market area, represented in terms of a locational triangle (Figure 2.3). Point P is where the costs of transporting the material to the factory and the finished goods to market are minimized. If the raw materials

lost weight during manufacturing, the factory would be drawn towards the material sources. If, on the other hand, distribution costs are higher than the costs of transporting materials, the industry would be drawn towards its market. Weber formulated this model in 1929, a time in which heavy industries based in coalfield regions dominated the economic landscape.

Perhaps the best known of the German locational models is Christaller's **central place theory** (Johnston, 2000). Based on the assumption of economic rationality and the existence of certain geographical conditions such as uniform population distribution across an area, central place theory offers an account of the size and distribution of settlements within an urban system. The need for shop owners to select central locations produces

Box 2.2

The new geographical economics

The starting point for the 'new geographical economics' is the basic geographical fact that economic activity is unevenly distributed. It addresses questions such as why, and under what conditions, do industries concentrate, and how do centripetal (centralizing) forces favouring the geographical concentration of industries in a particular location interact with opposing forces favouring geographical dispersal to a range of locations.

In addressing these issues, the NGE applies the methods of mainstream economics, devising models based on a number of simplifying assumptions, described as 'silly but convenient' (Krugman, 2000, p.51) It retains much of the basic architecture of neoclassical economics, requiring explanations that are based in the rational decisions of individual actors. Some of the specific assumptions are different though, incorporating notions of imperfect (monopolistic) competition and

economies of scale where additional investment in production capacity brings firms additional profits (neoclassical economics assumes that there are no economies of scale).

Centripetal forces favouring geographical concentration are identified as the effects of market size, a large and specialized labour market and access to information from other firms located there. The main centrifugal forces encouraging dispersal, conversely, are immobile factors of production such as land and, to a considerable extent, labour, and the costs of concentration such as congestion (Krugman, 1998, p.8). Much actual research has been devoted to assessing the interaction between these forces under different conditions. Small changes in, for example, transport costs or technology can 'tip' the economy from a pattern of dispersal to one of concentration. The typical outcome of concentration processes will be a simple core–periphery pattern while

dispersal will create a number of specialized and relatively evenly sized centres of industry.

The new geographical economics, therefore, focuses on the same basic questions that have long interested economic geographers, but adopts a distinctive approach based on the methods of economics. For many economic geographers, it is simply a more sophisticated version of the spatial analysis popular in the 1960s, sharing its underlying limitations and generating 'a dull sense of déjà vu' (Martin, 1999a, p.70). A particular weakness of the new geographical economics is the characteristic tendency to 'focus on what is easier to model rather than on what is probably most important in practice' (Krugman, 2000, p.59). At the same time, while this approach can never hope to capture the complexity and richness of the real economic landscape, its analytical clarity and sense of purpose does perhaps carry some lessons for economic geographers (Martin, 1999b).

a hexagonal network of central places, organized into a distinct hierarchy of lower- and higher-order centres (Figure 2.4). More recently, a new version of spatial analysis has been developed by mainstream economists such as Paul Krugman. This **new geographical economics** – termed 'new economic geography' by Krugman himself – applies mathematical modelling techniques to analyse issues of industrial location (Box 2.2).

The so-called quantitative revolution transformed the nature of economic geography 'from a field-based, craft form of inquiry to a desk-bound technical one in which places were often analysed from afar' (Barnes, 2001, p.553). Instead of directly observing and mapping regions in the field, economic geographers now tended to use secondary information and statistical methods to analyse patterns of spatial organization from their desks. While not all practioners of economic geography adopted the new methods, spatial analysis came to occupy centre-stage with those who refused to follow its approach increasingly relegated to the sidelines (ibid., p.553). By the late 1960s, however, the mood was changing again with a growing number of geographers beginning to question 'the spirit and purpose' of this new quantitative geography.

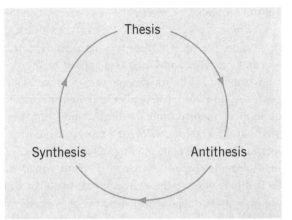

Figure 2.5 Dialectics.
Source: Johnston and Sidaway, 2004, p.231.

Reflect

➤ Do you think that economic geographers should adopt similar methods and perspectives as economists or that they should seek to differentiate themselves?

2.4 The political economy approach in economic geography

2.4.1 The origins of Marxist geography

In the late 1960s and early 1970s, quantitative geography was subject to criticism for its lack of social relevance and concern. To a new generation of young geographers, it seemed as if the discipline was narrowly focused on technical issues of urban and regional planning to the neglect of deeper questions about how society was organized. The key issues here were racial divisions in US cities, the Vietnam War (symbolizing the imperialism of US foreign policy), gender inequalities and the rediscovery of poverty in inner-city ghettos. Geography remained largely silent about such questions, leading Harvey to call for a revolution in geographic thought:

> The quantitative revolution has run its course, and diminishing marginal returns are apparently setting in . . . There is an ecological problem, an urban problem, an international trade problem, and yet we seem incapable of saying anything of depth or profundity about any of them.
>
> (Harvey, 1973, pp.128–9)

In response to these pressing social issues, a group of geographers in the US particularly sought to fashion a new **radical geography**. This movement began at Clark University in Massachusetts, where a group of postgraduate students, led by Richard Peet, launched *Antipode: A Radical Journal of Geography* in 1969.

This radical new geography turned to political economy for its intellectual foundations. The work of Marx in particular provided a framework for the critical analysis of advanced capitalism. Once again, economic geography was at the forefront of these developments. In general, **Marxism** emphasizes processes and relationships rather than fixed things, adopting a **dialectical perspective** that sees change as driven by the tensions between opposing forces, usually in the form of thesis–antithesis–synthesis (Figure 2.5). Capitalist society in particular is characterized by continual change and flux, driven by the search for profits in the face of competition:

> The bourgeoisie [capitalists] cannot exist without constantly revolutionising the instruments of production, and thereby the relations of production, and with them the whole relations of society ... Constant revolutionising of production, uninterrupted disturbance of all social conditions, everlasting uncertainty and agitation distinguish the bourgeois epoch from all earlier ones. All fixed, fast-frozen relations ... are swept away, all new formed ones become antiquated before they can ossify. All that is solid melts into air, all that is holy is profaned.
>
> (Marx and Engels [1848], 1967, p.83)

Box 2.3

The political economy tradition and the origins of a Marxist approach

It is important to set Marx's contribution to political economy within its historical context. Born in 1817, Marx's early academic career was as a student of German philosophy, influenced by Hegel's dialectical approach, in which human relations are conceived in terms of the struggle between opposing forces (Figure 2.5). During the 1840s, however, he became increasingly concerned with economic issues, fascinated by the rise of industrial and urban society. His big theoretical project, culminating in the three volumes, of *Capital*, became one of revising the classical political economy of Smith and Ricardo to understand the dramatic changes that were sweeping society in the mid-nineteenth century (Dowd, 2000, pp.86–90).

Smith, whose great work *The Wealth of Nations* was published in 1776, and Ricardo, whose *Principles of Political Economy and Taxation* appeared in 1817, were great advocates of the market and free trade. Smith's guiding philosophy was that greater wealth would accrue to the community of nations if governments desisted from intervening in the economy. Ricardo argued that a system of international trade without tariffs (import taxes) would encourage much greater efficiency since each country would specialize and produce goods that they were relatively more efficient at, therefore benefiting the broader commonwealth of nations (Box 4.1).

Both, however, were writing during the very early stages of industrial capitalism when feudalism allied to mercantilism – national protectionism – were viewed as impediments to social progress. For Smith, the development of market relations and competition would produce a more progressive and equal society than feudalism, liberating the individual from traditional social bonds. In this sense, he was very much in tune with the eighteenth-century enlightenment, which sought to overthrow the *ancien régime* and which influenced the French Revolution and American War of Independence. While Marx, writing over half a century after Smith, also recognized that capitalism was more progressive than feudal society, unlike Smith he was able to witness the negative consequences of industrial capitalism. The growth of the factory system, mass urbanization and the development of a large industrial working class, many living in conditions of appalling poverty and squalor, had transformed Smith's capitalist utopia into William Blake's 'dark satanic mills'.

According to Marx, class struggle is the basic driving force of history, from the relations of master and slave, to feudal lord and serf to capitalist and labourer. Under the capitalist mode of production, class relations are structured by the private ownership of capital (money, factories and equipment), creating a class of capitalists and a class of labourers who must sell their labour in exchange for a wage. The difference between the value of what the labourer actually produces and what he or she is paid in wages is retained and accumulated by the capitalist as surplus value or profit, representing the basis for class exploitation. The development of factories under capitalism brings large numbers of workers together in industrial cities, providing them with the means of organizing against the system that exploits them. When coupled with the inevitable overproduction that would result from the continual expansion of production, driven by competition, the growing power of the working class would eventually bring down the system and usher in a new socialist era. In this way, capitalism acts as its own gravedigger (Dowd, 2000, p.87)

Box 2.4

Neil Smith and the theory of uneven development

One of the most notable contributions to Marxist economic geography is **Neil Smith's theory of uneven development**. Processes of uneven development, according to Smith, are the result of a dialectic of **spatial differentiation and equalization** that is central to the logic of capitalism, transforming the complex mosaic of landscapes inherited from pre-capitalist systems. Capital moves to areas that offer high profits for investors, resulting in the economic development of these areas. The geographical concentration of production in such locations results in differentiation as they experience rapid development while other regions are left behind (see section 4.3). As a result, living standards and wage rates vary markedly between regions and, especially, countries. At the same time, the tendency towards equalization reflects the importance of expanding the market for commodities, implying

a need to develop newly incorporated colonies and territories so as to generate the income to underpin consumption.

The process of economic development in a particular region tends to undermine its own foundations, leading to higher wages, rising land prices, lower unemployment and the development of trade unions, reducing profit rates. In other regions, underdevelopment leads to low wages, high unemployment and the absence of trade unions, creating a basis for profit that attracts capital investment. Over time, capital will 'see-saw' from developed to underdeveloped areas, jumping between locations in its efforts to maintain profit levels. It is this movement of capital that creates patterns of uneven development. In this sense, 'capital is like a plague of locusts. It settles on one place, devours it, then moves on to

plague another place' (Smith, 1984, p.152).

According to Smith, the production of space under capitalism leads to the emergence of three primary geographical scales of economic and political organization: the urban, the national and the global. The dynamic nature of the uneven development process is most pronounced at the urban scale at which capital is most mobile, resulting in, for example, the rapid gentrification (upgrading through the attraction of investment and new middle-class residents) of previously declining inner-city areas like London's Docklands (Figure 2.6). Conversely, patterns of uneven development exhibit most stability at the global scale where the divide between developed and developing countries remains as wide as ever, although East Asia has risen to the core of the world economy through sustained economic growth since the 1960s.

Figure 2.6 Gentrification in London's Docklands.
Source: Robin MacDougall, Getty Images.

From this perspective, particular geographical objects, for example cities or a transport system, exist as expressions of wider relationships and are subject to transformation through the movement of broader forces of change (Swyngedouw, 2000, pp.45–6).

2.4.2 The development of Marxist theory

While the writings of Marx provided only a few scattered comments and insights into the geography of capitalism, Marxist geographers such as David Harvey have sought to build on these by developing a distinctively Marxist analysis of geographical change (Box 2.4). From this perspective, the economic landscape is shaped by the conflict-laden relationship between capital and labour, mediated by the state, providing a stark contrast to the harmonious equilibrium state of regional balance posited by neoclassical economic theory (Sheppard *et al.*, 2004, p.4).

The first phase of Marxist geography, beginning in the mid-1970s, concentrated on establishing how capitalism produces specific geographical landscapes (A. Smith [1776], 2001, p.9). In *The Limits to Capital*, Harvey (1982) identified a central **contradiction between the geographical fixity and motion of capital.** There is a need, on the one hand, for fixity of capital in one place for a sustained period, creating a built environment of factories, offices, houses, transport infrastructures and communication networks to enable production to take place. Such fixity is countered, on the other hand, by the need for capital to remain mobile, enabling firms to respond to changing economic conditions by seeking out more profitable locations (Box 2.4). This may require them to withdraw from existing centres of production in which they have invested heavily. Capital is never completely mobile, but must put down roots in particular places to be effectively deployed. Nonetheless, its relative mobility lends capital an important spatial advantage over labour, which is more place-bound (see section 3.3.1).

Harvey (1982) argues that capital overcomes the friction of space or distance through the production of space in the form of a built environment that enables production and consumption to occur. Indeed, such investment in the built environment can act as a 'spatial fix' to capitalism's inherent tendency towards overproduction by absorbing excess capital, performing an important displacement function. As economic conditions change, however, these infrastructures can themselves become a barrier to further expansion, appearing increasingly obsolete and redundant in the face of more attractive investment opportunities elsewhere. In these circumstances, capital is likely to abandon existing centres of production and establish a new 'spatial fix' involving investment in different regions (Box 2.4). The deindustrialization of many established centres of production in the 'rustbelts' of North America and Western Europe since the late 1970s and the growth of new industry in 'sunbelt' regions and the newly industrializing countries of East Asia can be understood in this light.

The second phase of Marxist geography began in the early 1980s, concentrating on developing Marxist analyses of specific situations and circumstances (A. Smith [1776], 2001, pp.10–11). One key research question was how particular places were affected by wider processes of economic restructuring. In a landmark text, Massey (1984) investigated the changing location of industry in Britain, developing the concept of 'spatial divisions of labour' (see section 4.3.2). In contrast to conventional location theory, which emphasizes the influence of resources, markets and transport costs, labour – its availability, cost and skills – is the key location factor from a Marxist perspective.

2.4.3 The regulation approach

Another strand of Marxist-informed research that was very influential in economic geography during the 1990s is the regulation approach. Derived from the work of a group of French economists in the 1970s, the regulation approach stresses the important role that wider processes of social regulation play in stabilizing and sustaining capitalist development. These wider processes of regulation find expression in specific institutional arrangements that mediate and manage the underlying contradictions of the capitalist system (see Chapter 3), expressed in the form of periodic crises,

Table 2.2 Fordist and post-Fordist modes of regulation

	Fordism	post-Fordism
The wage relation	Rising wages in exchange for productivity gains. Full recognition of trade union rights. System of national collective bargaining.	Flexible labour markets based on individual's position in market. Limited recognition of trade unions.
Forms of competition and business organization	Dominance of large corporations. Nationalization of key sectors, e.g. utilities.	Key role of small and medium-sized enterprises (SMEs) alongside large corporations. Privatization of state enterprises and general liberalization of the economy.
The monetary system	Monetary policy focused on demand management and maintaining full employment. Use of interest rates to facilitate economic expansion and contraction.	Focus on reducing inflation, relying on high interest rates if necessary. International monetary integration and coordination, e.g. the European single currency.
The state	Highly interventionist. Adoption of Keynesian policies of demand management. Provision of social services through welfare state. Goals of full employment and modest income redistribution.	Reduced state intervention in the economy. Abandonment of Keynesian policies for neoliberalism. Efforts to reduce welfare expenditure and privatize services.
The international regime	Cold war division into capitalist and communist blocs. Bretton Woods system of fixed exchange rates, anchored to the US dollar. Promotion of free trade, monetary stabilization and 'Third World' development.	Increased global economic integration. System of floating exchange rates. Renewed emphasis on free trade and openness to foreign investment. Imposition of neoliberal adjustment policies on developing countries.

enabling renewed growth to occur. This occurs through the coming together and consolidation of specific **modes of regulation**, referring to the institutions and conventions that shape the process of capitalist development. Regulation is focused on five key aspects of capitalism in particular: labour and the wage relation, forms of competition and business organization, the monetary system, the state and the international regime (Boyer, 1990). When these act in concert, a period of stable growth, known as a **regime of accumulation**, ensues (Jones *et al.*, 2004, p.59).

Two regimes of accumulation can be identified since the 1940s (Table 2.2). After the global economic depression of the 1930s, a new regime of **Fordism** emerged, named after the American car manufacturer Henry Ford, who pioneered the introduction of mass-production techniques. Fordism was based on a crucial link between mass production and mass consumption,

provided by rising wages for workers and increased productivity in the workplace. The state adopted a far more interventionist approach, informed by the theories of the British economist John Maynard Keynes, which stressed the important role of government in managing the overall level of demand in the economy to secure full employment. At the same time, governments provided a wide range of social services through the welfare state.

Fordism experienced mounting economic problems from the early 1970s, however, and elements of a new 'post-Fordist' regime of accumulation based on flexible production became prominent in the 1980s. Theories of **post-Fordism** emphasize the role of small firms, advanced information and communication technologies and more segmented and individualized forms of consumption (section 4.3.3). Whether these features amount to a distinctive and coherent regime of accumulation remains questionable, however, with mass production remaining important in some sectors. One key trend has been the abandonment of Keynesian policies since the 1970s in favour of a **neoliberal approach** that seeks to reduce state intervention in the

economy and embrace the free market, stressing the virtues of enterprise, competition and individual self-reliance (Harvey, 2005, pp.2–3).

2.4.4 The relevance and value of Marxist political economy

By the late 1980s, Marxist geography was becoming subject to increasing criticism, informed by the emergence of postmodern thought (Box 2.6). Marxism had

Reflect

> Do you agree that Marxism is still relevant to the analysis of global capitalism?

Box 2.5

Is Marxist political economy still relevant?

Since the early 1980s, there has been a wholesale abandonment of Marxist and socialist thinking in the face of the upsurge of neoliberal thinking in the West and the collapse of communism in the Soviet Union and Eastern Europe. The fall of the Berlin Wall in 1989 was heralded as bringing about the 'end of history' by the conservative American writer Francis Fukuyama in the sense that liberal democracy and the market had won the battle of ideas over socialism. The spirit of market triumphalism associated with globalization in the 1990s has, however, been punctured in recent years as the limitations of global capitalism have become increasingly apparent (Gray, 1999), highlighted by the anti-globalization movement in particular.

As part of the more sober climate of the late 1990s and early 2000s, Marx has been rediscovered. Triggered by the 150th anniversary of the publication of *The Communist Manifesto* in 1998 and the financial crises then engulfing East Asia and Russia, 'impeccably bourgeois magazines' such as the *Financial Times* and *New Yorker* published articles heralding Marx's thought (N. Smith, 2001, p.5). Writing in the *New*

Yorker, John Cassidy praised Marx as the 'next big thinker', citing his relevance to the workings of the global financial system and stating that his analyses will be worth reading 'as long as capitalism endures' (quoted in Rees, 1998). Although Wall Street's discovery of Marx was predictably short-lived, listeners voted Marx the greatest philosopher of all time in a BBC Radio 4 poll in 2005. The British historian, Eric Hobsbawm, explained this in terms of the 'stunning prediction of the nature and effects of globalization' found in *The Communist Manifesto* (quoted in Seddon, 2005).

We believe that Marxist political economy is still relevant because of its value as a framework for understanding the evolution of the global capitalist system. Marx's primary contribution to knowledge was as an analyst of capitalism, not as an architect of communism. While he offered only a few scattered comments about geography, this has been rectified by Marxist geographers like Harvey and Smith who have developed theories of uneven development (see Box 2.4). In the absence of other approaches that can match its historical–geographical reach and analytical

purchase, Marxism provides the most suitable framework for analysing the 'big questions' concerning the economic geography of global capitalism (Swyngedouw, 2000, pp.54–5). Marxism also retains a strong sense of social and political commitment, emphasizing issues of inequality and injustice and the need to change the world as well as interpret it.

At the same time, we recognize that Marxism does not hold all the answers, containing several limitations and weaknesses. This points to the need for a modified Marxist political economy that can incorporate insights from other perspectives, particularly the cultural and institutional approaches that economic geographers have turned towards in recent years (section 2.5). In particular, these approaches can provide a stronger sense of agency, sensitivity towards the cultural construction of the economy and a concern with the role of institutions and evolution (Hudson, 2006, pp.381–2). In short, we favour a kind of 'open' Marxism that does not claim to have a monopoly on truth, is receptive to insights from other perspectives and evolves in line with capitalism as its object of analysis.

also become rather out of touch with the 'new times' of the 1980s, marked by the dominance of neoliberal ideas, particularly in the UK and US. In the realm of left-wing politics too, the focus was shifting from the traditional 'politics of distribution' concerned with work, wages and welfare, to a 'politics of identity', concerned with asserting the rights of various groups, such as women, ethnic minorities and gay people, to recognition and justice (P. Crang, 1997, p.3). Such claims were channelled through broader social movements rather than the traditional labour movement.

Three main **criticisms of Marxism in geography** can be identified:

➤ Its apparent neglect of human agency in terms of an impoverished view of individuals and a failure to recognize human autonomy and creativity. Instead, Marxists tend to privilege wider social forces such as class and see people as bearers of class powers and identities, reading off their behaviour from this, not as unique individuals.

➤ Its emphasis on economic forces and relations. While the Marxist concept of production is, as we have seen, much broader than conventional notions of the economy, Marxists have been criticized for stressing the determining role of economic forces. Culture and ideas are often viewed as products of this economic base.

➤ Its overwhelming emphasis on class and neglect of other social categories like gender and race. Leading Marxists such as Harvey were attacked by feminist geographers in the early 1990s for their neglect of gender issues, accusing them of subsuming these within a class-based Marxist analysis.

These are important criticisms, although it is questionable whether Marxist geography really was as economically determinist as its critics allege (Hudson, 2006, pp.386–7). We do not think that they mean that Marxist political economy is no longer relevant or useful and should be abandoned (Box 2.5). As recent research on geographical scale and labour has shown, basic Marxist categories such as capital and labour remain highly relevant to an understanding of contemporary capitalism. In particular, such categories help us to address the 'big questions' of uneven devel-

opment, social justice and environmental degradation on an increasingly global scale (Swyngedouw, 2000, pp.54–5).

Partly in response to the above criticisms and reflecting its encounters with other philosophical perspectives like postmodernism, the political economy approach has become increasingly complex and diffuse in the 1990s and beyond. At the same time, the culturally and institutionally informed work of the past 15 or so years has raised important new questions about knowledge, identity and consumption. What seems to be needed is a framework for bringing together ideas from the two sets of perspectives, directing us towards an **'open' political economy approach** that is receptive to the insights offered by other perspectives (Box 2.5).

2.5 Cultural and institutional approaches in economic geography

2.5.1 The 'cultural turn'

A new set of approaches has emerged since the early 1990s, emphasizing the institutional and cultural foundations of economic processes. The key context here is the **'cultural turn'** that has taken place in human geography and the social sciences since the late 1980s, creating a 'new cultural geography'. Rather than the traditional view of culture as a possession of upper- and middle-class groups in society, the 'new cultural geography' has developed a broad and dynamic concept of culture. It is seen as a process through which individuals and social groups make sense of the world, often defining their identity against 'other' groups regarded as different according to categories such as nationality, race, gender and sexuality (Jackson, 1989). Meaning is generated through language that, instead of simply reflecting an underlying reality, actively creates that reality through **discourses** – networks of concepts, statements and practices that produce distinct bodies of knowledge (Barnes, 2000a, p.13). The cultural turn has been closely tied to the rise of **postmodernism** (Box 2.6).

Box 2.6

Postmodernism

Postmodernist ideas have attracted widespread interest since the 1980s, coming to exert considerable influence in architecture, the humanities and the social sciences. Ley defines postmodernism as

> a movement in philosophy, the arts and social sciences characterized by scepticism towards the grand claims and grand theory of the modern era, and their privileged vantage point, stressing in its place openness to a range of voices in social enquiry, artistic experimentation and political empowerment.
>
> (Ley, 1994, p.466)

As this quote indicates, pluralism is a key characteristic of postmodern thought in terms of embracing the knowledge claims of different social groups. Grand theories or 'meta-narratives' (big stories) claiming to uncover the changing organization of society are rejected as a product of the privileged position and authority of the observer rather than being accepted as objective representations of the realities that they purport to explain. Instead of functioning as a set of universal truths, then, knowledge should be regarded as partial and situated in particular places and times. Postmodernists reject conventional notions of scientific rationality and progress, favouring an open interplay of multiple local knowledges.

Rather than assuming that social life has an underlying order and coherence, postmodernists celebrate difference and variety. Difference and variety are held to be basic charac-teristics of the world, applied to a range of different phenomena including human groups and cultures, buildings, urban neighbourhoods, texts and artistic products. This basic attention to difference quickly attracted the interest of geographers, reflecting the fact that 'the discipline has always ... displayed a sensitivity to the specific kinds of differences to be found between different (and "unique") places, districts, regions and countries' (Cloke *et al.*, 1991, p.171). For Gregory (1996, p.231), postmodernism provides an opportunity for geographers to return to the notion of a real differentiation, emphasized by traditional regional geography, but armed with a new 'theoretical sensitivity' derived from work in cultural studies in particular.

2.5.2 Economy–culture links

While economic geography was, along with urban geography, regarded as the leading area of human geography during the 1960s, 1970s and 1980s when spatial analysis and Marxist political economy held sway, the 'new' cultural geography has come to be seen as the most exciting, 'cutting edge' area of the discipline since the late 1980s (Thrift, 2000, pp.692–3). In response, some economic geographers have sought to adapt their interests, approaches and methods, incorporating notions of difference, identity and language into their research (Wills and Lee, 1997, p.xv). The **links between economy and culture** are of central importance here with many observers agreeing that the economy has become increasingly cultural in terms of the growing importance of sectors such as entertainment, retail and tourism while culture has become increasingly economic, viewed as a set of commodities that can be bought and sold in the market (Box 2.7).

This has led to talk of a **'new' economic geography** that examines the links between economic action and social and cultural practices in different places (Thrift and Olds, 1996). As Wills and Lee (1997, p.xvii) put it, 'the point is to contextualise rather than undermine the economic, by locating it within the cultural, social and political relations through which it takes on meaning and direction'. Four main strands of culturally informed research in economic geography can be identified (Barnes, 2003):

➤ Consumption with studies focusing, for instance, on the creation and experience of particular landscapes of consumption such as shopping malls, supermarkets and heritage parks (see sections 3.4, 4.4).

➤ Gender, performance and identity in the workplace and labour market. This work focuses on how employers perform particular roles and tasks at

Box 2.7

Linking the economy and culture: the example of Christmas

The annual festival of Christmas offers an instructive example of the close links between the economy and culture, representing 'a cultural event of vast economic significance – or an economic event of vast cultural significance' (Thrift and Olds, 1996, p.331–2). For most people in developed countries particularly, Christmas is a time of leisure and consumption, associated with a holiday from work, usually spent with friends and family (Figure 2.7). As a cultural festival, Christmas draws together traditions from different countries: the Christmas tree from Germany, the practice of filling stockings from the Netherlands, the idea of Santa Claus or Father Christmas from the US and the Christmas card from Britain (Miller, 1993, cited in Thrift and Olds, 1996, p.332). As part of the global spread of a Western consumer culture, these practices are spreading to emerging economies like China.

The social customs associated with Christmas support the practice of gift-buying. This provides the economy with a crucial stimulus in the final quarter of the retail year with the Christmas shopping season typically accounting for 26–40 per cent of a store's annual sales (*Moneyweek*, 2004). In 2003–4 the British population was estimated to have spent £24 billion on gifts – an average of £330 each. Much of this is paid for by savings and debt with around 15 million people getting further into debt to fuel their spending (ibid).

Such expenditure benefits not only retailers, but also manufacturers, often located in areas distant from the main centres of global consumption. The Chinese city of Shenzhen, for example, where about 75 per cent of the world's toys are made, benefits hugely from Christmas with 1,000 containers a day leaving for Southampton, Rotterdam and New York (ibid.).

Figure 2.7 Christmas consumption.
Source: Franco Zecchin, Getty Images.

work, often informed by cultural and gendered norms. One of the most notable studies on this is Linda McDowell's research on work cultures in merchant banks in the City of London (see Box 2.8).

➤ Research on the importance of personal contact and interpretative skills in financial and business services. This has focused on the cultural practices that underpin communication and interaction, the sites in which they take place and the consequences for our understanding of financial markets. Research on financial centres such as the City of London, for example, has shown the importance of social networks and trust, encouraging geographical concentration through the need for regular face-to-face contact (Thrift, 1994).

➤ Corporate cultures and identities. Research has focused on how managers and workers create distinctive corporate cultures through particular discourses and day-to-day practices. Of particular interest here is Schoenberger's (1997) work on large American corporations such as Xerox, DEC and

Lockheed, showing the limitations of such cultures in dealing with a turbulent and unpredictable economic environment.

2.5.3 Institutions, embeddedness and path dependency

Economic geographers have drawn upon concepts from institutional economics that emphasize the social context of economic life and the evolutionary nature of economic development. Institutions are important because they link 'the economic' and 'the social' through a set of habits, practices and routines (Hodgson, 1993). Beyond the specific types of organization identified in section 1.4.3, institutions are defined broadly as informal conventions and norms that shape and influence the behaviour of economic actors. For example, modern capitalist societies are characterized by a set of norms that emphasize the importance of consumption in terms of personal fulfilment and

Box 2.8

Capital culture: gender at work in the City

Linda McDowell's research on the construction and performance of gender relations in the City of London represents one notable example of 'new economic geography' research, linking economic issues to wider social and cultural relations and focusing on questions of discourse, identity and power. McDowell aims to understand how an international financial centre like the City of London actually operates, 'viewing it through the lens of the lives and careers of individual men and women working in the City's merchant banks at the end of this period of radical change [the late 1980s and early 1990s] both in the global economy and in the city' (McDowell, 1997, p.4). In focusing on gender segre-

gation and identity at work, the research assesses men's experiences in addition to women's. At the same time, the extent to which the City has become more open to women and the career prospects it offers them remain key issues.

Capital Culture illustrates another key feature of the 'new economic geography' by employing a range of research methods (see above) and going beyond the conventional reliance on questionnaire and statistics to employ qualitative methods involving direct fieldwork. McDowell focused on the merchant bank sector, utilizing postal questionnaires sent out to all such banks in the City and detailed face-to-face interviews with employees of three of them.

This multi-method approach meant that information on changing employment trends and relations within companies could be combined with first-hand accounts of how individual employees have managed and negotiated processes of employment change within the workplace. Through the case study interviews in particular, the research examined people's everyday work experience, assessing how they assumed and performed particular gender roles. The importance of image, bodies and the presentation of self are major themes. A key conclusion is that changing work relations and power structures in the City still favour men as 'the cultural construction of the banking world remains elitist and masculinist' (ibid., p.207).

Figure 2.8 The sports utility vehicle (SUV).
Source: © David Cooper/Toronto Star/Corbis.

identity (section 4.3). At the same time, of course, increased consumption is vital in enabling retailers and manufacturers to generate profits, fuelling the process of economic growth (Box 2.6). Amin and Thrift (2004) cite the example of the sports utility vehicle (SUV), which has become particularly prominent in North America, addressing middle-class families' desire for 'safety', 'security' and 'status' (Figure 2.8).

A related set of ideas from economic sociology has also been influential in stressing that economic processes are grounded or 'embedded' in social relations (Granovetter, 1985). In contrast to the orthodox economic conception of the economy as a separate domain driven by the rational decisions of individual actors, economic sociologists, following Polanyi, argue that the economy is socially constructed with social norms and institutions playing a key role in informing and shaping economic action (see Box 1.3). One key concept that has been incorporated into economic geography in recent years is **'embeddedness'**, referring to the notion that economic action is always grounded in social relations. Geographers have 'spatialized' the concept of 'embeddedness', emphasizing how particular forms of economic activity are rooted in particular places.

In a study of advanced manufacturing technologies in southern Ontario, Gertler (1995) shows the importance of 'being there' in facilitating adoption of the technologies by ensuring close links between producers (organizations that actually develop, distribute and sell aspects of the technology, e.g. robots) and users (manufacturing firms). A shared 'embeddedness' in a distinctive industrial culture enabled users and producers to develop appropriate training regimes and industrial practices, involving the sharing of information and knowledge. This does not mean, however, that industrial cultures are always defined geographically or that learning and interaction cannot occur over longer distance, utilizing information and communication technologies. An awareness of the importance of this issue has fostered an interest in the development and organization of knowledge within large firms, raising questions about how such global and local knowledge are combined (see section 10.6).

Institutionalist ideas have encouraged the rise of a **'new regionalism'** in economic geography that examines the effects of social and cultural conditions within regions in helping to promote or hinder economic growth (Storper, 1997). In particular, inherited institutional frameworks and routines are held to be of considerable importance in influencing how particular regions respond to the challenges of globalization (Amin, 1999). Individual places have, as such, attracted renewed attention, as economic geographers have attempted to identify the social and cultural foundations of economic growth and prosperity in successful regions such as 'Silicon Valley' in California and Cambridge (UK), and 'Motor Sport Valley' in south-east England. In contrast to the political economy approaches of the 1980s, then, institutionalist perspectives emphasize the importance of internal conditions within regions, as opposed to external processes, in shaping their experience of economic development, and treat localities and regions as active participants in economic development rather than as passive arenas exploited by capital (e.g. large MNCs).

The idea that the process of economic development is '**path dependent**' is another key institutionalist idea adopted by economic geographers. This means that the ways in which economic actors respond to wider processes of economic change are shaped and informed by past decisions and experiences:

> One of the most exciting ideas in contemporary economic geography is that industrial history is literally embodied in the present. That is, choices made in the past – technologies embodied in machinery and product design, firm assets gained as patents or specific competencies, or labour skills required through learning – influence subsequent choices of methods, design and practices.
>
> (Walker, 2000, p.126)

The past is expressed and sustained through technology, machinery and equipment and organizations as well as a broader set of attitudes and habits that inform current practice. Regional culture is a product of the past in this sense.

The process of 'path dependence' can be illustrated with reference to 'old' industrial regions such as the Ruhr Valley in Germany and north-east England in which cultural factors have been closely associated with decline. The industrial cultures of these areas seemed to have become rigid and fossilized, meaning that firms and institutions were tied to obsolete production systems and methods, militating against more positive

responses to economic change in terms of generating new products and methods. These declining areas can be seen as:

> the hard-luck cases: once-successful places where local cultures fostered ties so strong, structures so rigid and attitudes so unbending that newcomers and new ways of doing things encountered insurmountable barriers to entry.
>
> (Gertler, 2003, p.134)

In this sense, declining regions tend to become 'locked in' to outmoded practices and habits, preventing them from acquiring new knowledge in an economically effective manner. In some cases, 'old' industrial regions have attempted to address these deficiencies by trying to transform their cultures through initiatives that encourage learning and collaboration between firms. Parts of the American Midwest, South Wales and the Basque Country are examples of such 'reclamation projects' (Gertler, 2003, p.134).

2.5.4 Assessing the cultural and institutional 'turns'

Opening up the 'Pandora's box' of culture has undoubtedly enriched and enlivened economic geography (Thrift, 2000), exposing it to new ideas and concerns. As a result, the notion of the economy as a self-evident and self-contained entity has been destabilized, generating a potentially bewildering array of research topics and questions. Research on 'traditional' issues such as regional development, large corporations and labour markets continues, often informed by these new perspectives, alongside a number of new areas of interest such as consumption, corporate cultures and gender relations. While much of this is to be welcomed, one limitation of recent cultural and institutional work is the neglect of some higher-level social relations and structures such as capitalism, class and the state (Peck, 2005, p.151). Wider processes of uneven geographical development have tended to be sidelined by a wealth of studies assessing the social and institutional conditions supporting growth in dynamic regional economies. The neglect of such issues emphasizes the need to 'rediscover a sense of political economy' in economic geography (Martin and Sunley, 2001, pp.155–7).

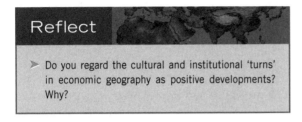

Reflect

➤ Do you regard the cultural and institutional 'turns' in economic geography as positive developments? Why?

2.6 Summary

Prevailing approaches to economic geography have changed considerably over time, as we have demonstrated, mirroring the development of geography more broadly. The traditional framework based on regional classification and description gave way to a more quantitative approach in the late 1950s and early 1960s, based on the development of concepts and techniques of spatial analysis. This was, in turn, replaced by Marxist political economy in the 1970s and 1980s before the cultural 'turn' led to the introduction of new cultural and institutional perspectives in the 1990s. These periodic shifts in intellectual orientation should be seen as changes in broad focus and direction rather than as wholesale transformations, with research in the spatial analysis and political economy traditions remaining important.

Our favoured approach in this book can be described as 'new' political economy, signalling that it has moved beyond the rather clunky and deterministic theories of the 1970s and early 1980s to become more flexible and open to the importance of context, difference and identity, partly as a result of encountering cultural and institutional approaches. As such, our position brings together concepts from the two traditions, but from a standpoint that is anchored in political economy. It represents, as such, a 'culturally sensitive political economy' rather than a 'politically sensitive cultural economy' (Hudson, 2005, p.15). Our approach can be described as broadly regulationist in nature, stressing the important role that wider processes of social regulation play in stabilizing and sustaining capitalism.

From a base in political economy, we examine the uneven development of capitalism, 'the individual economics of different places, and their connections one to another' (Barnes and Sheppard, 2000, pp.2-3) in this book. In adopting a broad, synthetic conception of the economy, we recognize the importance of the link between the economy and culture, and assess some connections between the two in specific geographical settings. Our approach also takes difference and variety seriously, emphasizing the distinctive character of individual places, but insisting that we need to consider how geographical difference is produced and reproduced through wider processes of economic development and state intervention.

Exercises

A major supermarket chain is proposing to open a new out-of-town superstore in the town or city where you live.

1. How would each of the three latter perspectives covered in this chapter (spatial analysis, political economy and the cultural/institutional approach) examine and understand this issue?

2. On what aspects of the development would each approach focus attention (e.g. people's experience of shopping, broader policies and practices of the corporation in question, finding the optimum location, links between consumption and identity, impact on local shops, analysing the characteristics of the local market, assessing why customers shop in major superstores, competition between supermarket chains, relationships with suppliers)?

3. On this basis, assess the strengths and weaknesses of each approach. Which, if any, offers the 'best' understanding of the issue? Why?

4. Is it appropriate to try to bring elements of the different approaches together? If so, which ones?

Key reading

Barnes, T.J. (2000a) 'Inventing Anglo-American economic geography', in Sheppard, E. and Barnes, T.J. (eds) *A Companion to Economic Geography*, Oxford: Blackwell, pp.11–26.
An engaging account of the formation and early development of economic geography. Stresses how academic subjects are

invented as specific projects at particular times and in particular places by groups of people. Covers the traditional approach and spatial analysis.

Lee, R. (2000) 'Economic geography', in Johnston, R.J., Gregory, D., Pratt, G. and Watts, M. (eds) (2000) *The Dictionary of Human Geography*, 4th edn, Oxford: Blackwell, pp.195–8.
Defines economic geography as 'the geography of people's struggles to make a living' (p.261). Outlines how the subject has evolved over time and highlights a range of examples of contemporary economic geography research.

Scott, A.J. (2000) 'Economic geography: the great half century', in Clark, G., Feldmann, M. and Gertler, M. (eds) *The Oxford Handbook of Economic Geography*, Oxford: Oxford University Press, pp.18–44.
An upbeat review of the development of economic geography in the post-war period. Focuses particularly on the spatial analysis and political economy approaches, highlighting specific research topics such as localities and the rediscovery of regions since the 1980s.

Swyngedouw, E. (2000) 'The Marxian alternative: historical–geographical materialism and the political economy of capitalism', in Sheppard, E. and Barnes, T.J. (eds) *A Companion to Economic Geography*, Oxford: Blackwell. pp.40–59.
A compelling account of Marxist political economy from one of its leading exponents. Identifies key aspects of the Marxist approach and its development within geography. Emphasizes its commitment to social and political change and relevance in addressing big questions associated with globalization, uneven development and environmental degradation.

Thrift, N. (2000) 'Pandora's box?: cultural geographies of economies', in Clark, G., Feldmann, M. and Gertler, M. (eds) *The Oxford Handbook of Economic Geography*, Oxford: Oxford University Press, pp.689–704.
A review and summary of the cultural 'turn' in economic geography. Emphasizes how it has broadened and enlivened the field, opening up an array of new questions. Identifies the main areas of research and speculates on future developments.

Thrift, N. and Olds, K. (1996) 'Refiguring the economic in economic geography', *Progress in Human Geography* 20: 311–37.
One of the key articles advocating a cultural 'turn' in economic geography research, helping to set the agenda for recent developments. Highlights the links between the economy and culture, using the example of Christmas (Box 2.7), and identifies four metaphors underpinning economic geography research.

Useful website

http://www.econgeog.org.uk/about.html
The website of the Economic Geography Research Group of the Royal Geographical Society (with the Institute of the British Geographers). This site is mainly used by academic researchers in the field so much of the material is likely to be difficult. It is worth exploring the site, however, to get a feel for the types of issues and topics on which economic geographers conduct research.

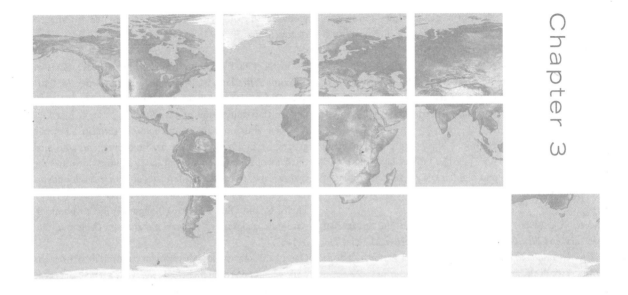

Chapter 3

Shaping the capitalist economy: key actors and processes

Chapter map

➤ The development of the capitalist economy is shaped by the actions of groups of people such as capitalists, workers, consumers and state officials. In this chapter, we consider the role of these key sets of actors and assess the main historical processes that have structured their actions. Following a brief introduction, section 3.2 examines the role of capital in the economy, focusing on the circuit of capital, the nature of the firm, the organization of finance and investment and processes of innovation and technological change. We then turn to labour, considering the position of labour within capitalism and assessing the division of labour within production. Section 3.4 is concerned with consumers and the process of consumption, focusing attention on the convergence between economy and culture (section 2.5). The role of the state in the economy is briefly examined in the penultimate part of the chapter, highlighting its importance in regulating and coordinating economic activities.

3.1 Introduction

Our approach in this book emphasizes the social relations structuring the economy, rejecting the individualism of mainstream economics. While we deal with each of the main groups of economic actors – capital, labour, consumers and the state – in turn in this chapter, it is the (social) relations between these groups that are crucial. Foremost among these is the relationship between capital and labour within the production process. The decisions of consumers about which goods and services to purchase also underpin the economic process. Increased consumption is crucial in allowing the economy to grow and expand. The state plays an important role in regulating the economy, mediating between the interests of capital and labour in an effort to create stable conditions that facilitate economic growth. We believe that the position of each of these groups within capitalism gives them a particular set of material interests that shapes their actions. At a general level, we can say that capital seeks to generate profits; labour strives to maximise wages and living conditions; consumers seek to further their interests and aspirations by purchasing goods and services; and the state aims to promote economic growth within its territory.

While people are not the wholly rational and self-interested actors that underpin conventional economic theory, they are knowledgeable individuals with their own interests and aspirations (Hudson, 2005, p.3). Indeed, one of the key characteristics of the 'new' economic geography is its emphasis on the specific knowledge and identity of economic actors, without assuming that their decisions are always rational (section 2.5). The fact that capitalism operates through the actions of individuals whose knowledge of alternatives and outcomes is always imperfect introduces an important element of uncertainty and openness into the economic process. At the same time, individual actors operate within the broader structures of the capitalist economy, meaning that they are confronted by pressures such as the need to earn a wage or to generate profits. The pressing need for people to earn an income to support themselves and their families has led one commentator to describe economic geography as 'the study of people's struggle to make a living' (Lee, 2000, p.195). The specific ways in which people

respond to wider economic imperatives will vary according to a wide range of factors, including their economic status, gender, family situation, age and cultural values. These responses can be expected to vary spatially, given the variety of local and regional environments that exist within the contemporary world economy, reflecting the interaction between capitalism and a range of pre-existing societies and cultures (Johnston, 1984).

3.2 Capital

Capital is a complex term, used in different ways by different people. For our purposes, however, it can simply be defined as money that is invested in production or financial markets. Capitalists are those people who have acquired capital, allowing them to own the means of production: land, materials, factories, offices and machines. The pressures of competition mean that the capitalist has to seek to expand the stock of capital by reinvesting it in production to generate higher profits. This reinvestment in order to generate more profits – which are in turn reinvested – is often referred to as the process of **capital accumulation** (Barnes, 1997, p.222). It lies at the heart of the capitalist system with profit-seeking representing the basic driving force for economic growth and expansion.

3.2.1 The circuit of capital

The basic economic process under capitalism can be understood in terms of the **circuit of capital** through which profits are generated (Harvey, 1982). The first stage (see Figure 3.1) is that capital in its money form (M) is transformed into commodity form (C) by purchasing the means of production (MP) – factories, machines, materials, etc. – and labour power (LP). The means of production (MP) and labour power (LP) are then combined in the production process, under the supervision of the owners of capital or their managers and representatives, to produce a commodity for sale (C*) – for example a car, a house or even a haircut. This commodity is sold for the initial money outlay plus a profit (Δ), representing what some economists call surplus value.

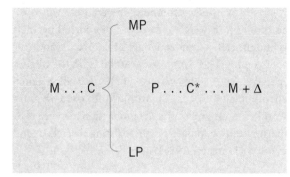

$$M \ldots C \left\{ \begin{array}{c} MP \\ \\ P \ldots C^* \ldots M + \Delta \\ \\ LP \end{array} \right.$$

Figure 3.1 The process of production under capitalism.
Source: Castree *et al.*, 2004, p.28.

Part of the money (M) realized from the sale of the commodity is reinvested back into the production process, which recreates the circuit anew. Following each circuit of capital, there is an expansion in the total amount of capital, forming the basis of capital accumulation. Clearly, the distribution of income at the end of each circuit becomes an important political question and, through taxation, governments play a regulatory role in diverting capital into welfare spending and other state functions (e.g. investment in transport infrastructure, housing).

3.2.2 The firm as grounded capital

The main organizational form that capital takes is **the firm**. Firms are legal entities owned by individual capitalists or, more commonly now, a range of shareholders. Along with the commodity, the firm could be described as a basic 'economic cell form' of capitalism (see section 1.2.1). Our approach here is informed by the **competence- or resource-based theory** of the firm. Derived from the work of the economist Edith Penrose (1959), the competence perspective views firms as bundles of assets and competencies that have been built up over time. A firm becomes distinctive from its competitors through the specific way in which it combines the resources of land, labour and capital in the production of certain commodities. Competencies can be seen as particular sets of skills, practices and forms of knowledge. The need for firms to focus on identifying and developing their 'core competencies' is a key theme of recent management theories. For example, the core competencies that have enabled Dell to acquire such a strong position in the personal computer market include the online 'bespoking' of each computer according to customer requirements and the high quality of its manufacturing and distribution, offering reliable products at competitive prices (tutor 2u net, undated).

The knowledge and skills developed by a firm become embedded within its organizational rules and routines (Taylor and Asheim, 2001, p.323). Firms acquire and develop knowledge and skills through processes of learning associated with the repetition of particular tasks. As such, the competence-based approach draws attention to the dynamic processes of learning and knowledge creation that occur within firms, providing much of the basis for understanding firm strategy (Nelson and Winter 1982). It is well suited to economic geography, indicating that firms can derive some of their competitiveness from factors present in the broader regional environment (Maskell, 2001, p.339), combining these with internal resources to create distinctive competencies.

In addition to this abstract theory of what firms are and how they operate, it is important to appreciate the diversity of firms in terms of size, ownership and structure. In practice, the nature of **business organization** is very dynamic, taking new forms over time. The heroic small firms and entrepreneurs of the nineteenth century have given rise to large transnational corporations or what some Marxist economists have termed monopoly capital (Baran and Sweezy, 1964). While mainstream economics had assumed the existence of a world of small firms with no power to influence prices in the marketplace, the emergence of these large multiplant corporations required more sophisticated analyses (Galbraith, 1967). This work looked at the internal organization of business, emphasizing how the growth of the large joint-stock company had resulted in a separation of ownership (shareholders) from control (managers) and the emergence of vast internal labour markets through which employment is organized (Doeringer and Piore, 1971). In the large public limited firm or joint-stock company, shares are typically distributed between a diverse range of interests that represent different sections of society (see Table 3.1). At the same time, small firms (of less than 100 employees)

Table 3.1 Ownership structure of share capital in the UK's privatized utilities, 1993

Investor	Share in privatized utilities %
Central government	9.2
Pension funds	27.6
Insurance companies	12.4
Unit trusts	4.5
Individuals	24.5
Overseas	8.3
Others	12.1

Source: Sawyer and O'Donnell, 1999, p.62

continue to exist and employ more people than large corporations, even in advanced national economies, as suppliers and competitors to larger firms.

3.2.3 Finance and investment

Two key functions of money within the capitalist economy can be identified. First, money acts as a measure and store of value, with the value of particular commodities expressed in term of their price. Similarly, the wealth of individuals is expressed in terms of the monetary value of their income and assets while share prices measure the fluctuating fortunes of large firms. Second, money is also the medium of exchange through which commodities are bought and sold, enabling the seller to buy other commodities with the money received. Money is advanced by banks and other organizations to firms and individuals as credit – effectively loans that must be paid back later – allowing them to invest in production or to buy goods that they would not otherwise have been able to do at this particular time (effectively bringing investment and consumption forward in time).

In order to grow and expand, firms need capital for investment. Some of this capital can be generated internally through the profits made from selling commodities in the marketplace, but many firms require access to external sources of finance such as loans from banks, the financial markets and venture capital. For large public corporations the stock market provides an additional source of funds through the issuing of shares that are purchased by individuals and institutional investors, referred to as **equity finance**. In recent decades, pension funds have become increasingly important as investors in the stock market, as indicated by Table 3.1 (Clark, 1999). **Venture capital** is equity finance that investors provide to firms not quoted on the stock exchange. Such firms are generally small and have a high growth potential, attracting venture capitalists who aim to make high returns by selling their stake in the firm at a later date. Such investment is high risk with investors typically expecting returns of over 30 per cent (Tickell, 2005, p.249). The investors can be either wealthy individuals, often former entrepreneurs, known as 'business angels', or institutions in the form of professional investment companies who raise funds from financial institutions such as banks (Mason and Harrison, 1999). While the venture capital industry is relatively small compared with other financial markets, it is of considerable economic significance because of its importance in funding high-growth firms that generate employment and wealth.

Finance has tended to become more **centralized** and integrated over time with local and regional banks giving way to national and, increasingly, global systems (Dow, 1999). The UK, for example, has a highly centralized financial system, with the south-east region accounting for over half of all financial employment in the UK in 1995 (Tickell, 2005, p.249). Management and control of pension funds in the UK is overwhelmingly concentrated in the South-east, with two-thirds of domestic investment made in companies based in the same region (Tickell, 1999, p.111). 'Classic' venture capital, which concentrates on funding new businesses, is highly concentrated in south-east England and, to a lesser extent, Scotland (Mason and Harrison, 2002, p.443). In the US, New York has long been the dominant financial centre, symbolized by Wall Street in lower Manhattan (Figure 3.2) Significant regional concentrations of venture capital are also evident in Silicon Valley and New England (around Boston) as well as Texas and the Midwest (Figure 3.3). Regions that are deficient in 'classic' venture capital, including most

Figure 3.2 Wall Street.
Source: D. Mackinnon.

UK regions outside south-east England, lack access to funds for supporting dynamic new businesses, leading to low levels of entrepreneurship and innovation. Furthermore, given that venture capitalists raise funds from financial institutions such as pension funds and banks, savings are drained from such regions and invested in already prosperous locations (ibid., p.178).

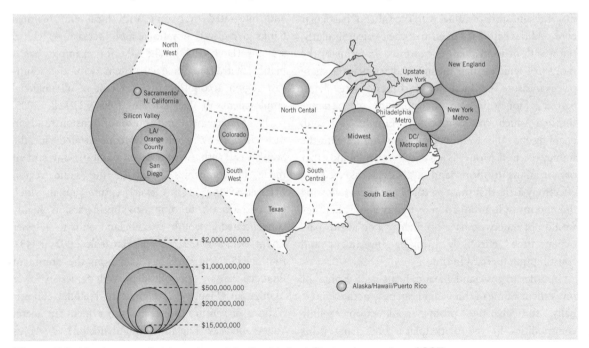

Figure 3.3 Venture capital investment in the United States by region, 1997.
Source: Mason and Harrison, 1999, p.172. PricewaterhouseCoopers LLP/National Venture Capital Association Money Tree report, based on data from Thomson Financial.

Box 3.1

Capital switching and its geographical effects: the case of the north-east of England

As the first country to industrialize, the UK enjoyed an advantage over the rest of the world in the production of heavy engineering and manufacturing goods throughout the nineteenth century. The north-east of England became a heartland region of industrial capitalism during the second half of the nineteenth century on

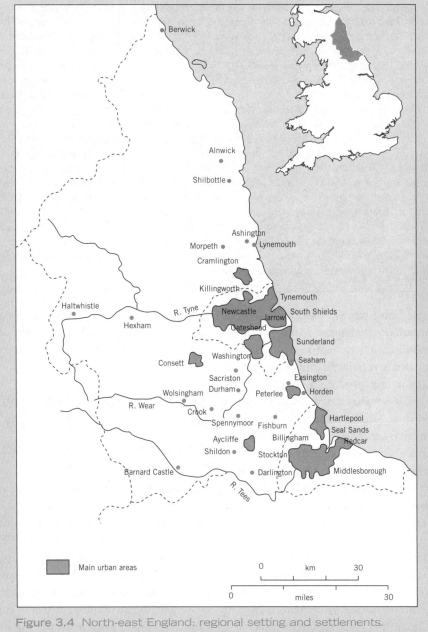

Figure 3.4 North-east England: regional setting and settlements.
Source: Hudson, 1989, p.4.

Box 3.1 (continued)

account of its iron ore and coal reserves (Figure 3.4). It became particularly dominant in the growing shipbuilding market – producing around 40 per cent of the world's ships at one point – but also became an important centre for rail and bridge engineering, and arms production. This growth was associated with the emergence of a local bourgeoisie, the 'coal combines' (capitalist class), a group of families that had originally made their fortunes through the coal trade but had diversified into the new industries to take advantage of the investment opportunities.

With the growth of competition from other countries – especially the US and Germany – and a worldwide downturn in the 1920s, the region's industries faced an economic crisis reflected in declining markets and rising unemployment. At the peak of the Depression in 1933, over 80 per cent of its shipbuilding and repair

workforce was unemployed. Faced with this situation, the region's capitalists had two alternatives: reinvest in the traditional industries, introducing new production methods and technologies, or find alternative avenues for investment. For the most part they chose the latter, switching capital from what were increasingly perceived as declining industries to new growth sectors such as car and aircraft production, the public utilities and financial services. The nationalization of the coal industry in the 1940s provided them with further opportunities when 'the coal owners' capital, locked in the fixed assets of mines and machinery, became suddenly transformed under the compensation terms into highly liquid government bonds' (Benwell Community Project, 1978, p.58).

Diversification out of the heavy industrial sector was accompanied by a geographical expansion of the coal

combines' interests at both national and international level. Through the establishment of special investment trusts in the interwar years, local capitalists were able to exploit the growing market in trading and financial transactions to the extent that by 1930 one of the largest of these trusts, the Tyneside, had only 23 per cent of its investment in north-eastern industry. Despite efforts to modernize the region by both state-led restructuring and the attempt to encourage new inward investment, the legacy of this withdrawal of capital from the North-east has been the region's transition from a core region of capitalism in the nineteenth and early twentieth century to an increasingly peripheral position since the 1930s. This is manifest in lower levels of economic development and higher levels of social deprivation than the national average.

Capital is the most mobile of the factors of production, in contrast to labour and land, which are relatively place-bound. As we have seen, it tends to become concentrated in regions that offer a high rate of return or profit on money invested, leaving underdeveloped countries and regions bereft of the capital required to invest in production facilities and development projects. But investment is not contained indefinitely within core regions: firms and investors also respond to investment opportunities in other regions. Over time, there is a 'see-saw' effect as capital flows back and forth between different sectors and regions (N. Smith, 1984). Financial deregulation in most of the major economies seems to have exacerbated this situation in recent decades, unleashing a far more rapid and volatile phase of accumulation (Harvey, 1989a). **Capital switching** has important geographical dimensions, often being transferred from regions dominated by declining sectors to 'new industrial spaces' in distant regions offering more attractive conditions for investment. An example of the

geographical effects of capital switching and uneven development is provided by the experience of the north-east of England over the twentieth century (Box 3.1).

3.2.4 Innovation and technological change

'Creative destruction' and Kondratiev cycles

Capitalism is a dynamic and unstable economic system, with periods of rapid growth followed by periods of decline and crisis. The Austrian economist Joseph Schumpeter (1943) famously described capitalism as a process of '**creative destruction**' based upon innovation and the development of new technologies. As new products and technologies emerge, they often render existing industries obsolete, unable to compete on the basis of quality or price. As part of this process, capital

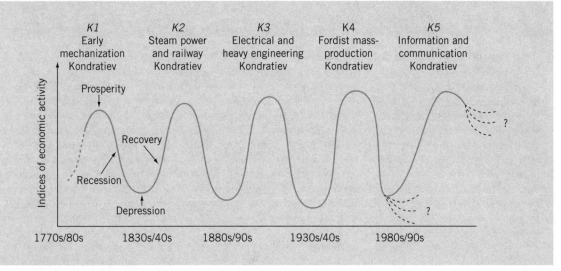

Figure 3.5 Kondratiev cycles.
Source: Dicken, 2003a, p.88.

is withdrawn from these unprofitable industries and invested in new centres of production (see Box 3.1).

The tendency for major innovations to 'swarm' or cluster together in distinct cycles or waves has been emphasized by Schumpeter and other commentators. These are sometimes known as '**Kondratiev cycles**' after the Soviet economist Kondratiev who first identified them in the 1920s. Lasting for some 50–60 years in length, each cycle is associated with a distinctive system of technology, incorporating the key propulsive industries, transport technologies and energy sources (Figure 3.5). Five Kondratiev cycles are usually distinguished since the late eighteenth century. Each cycle consists of two distinct phases: one of growth and one of stagnation (Taylor and Flint, 2000, p.14).

The notion of Kondratiev cycles is based on the analysis of price trends, which show a characteristic pattern of steady increases for about 20 years, culminating in a rapid inflationary spiral, followed by a collapse with prices reaching a trough some 50–55 years after the start of the cycle. This has sparked much debate about the specific mechanisms behind this pattern. At a basic level, though, each cycle starts with the bunching together of key inventions that create new economic opportunities for firms and entrepreneurs. At this stage, technology is expensive and demand high, resulting in rising prices; after the technology has matured, becoming routine and stan-

dardized, prices fall. An inherent problem is **overproduction**, reflecting a tendency for the volume of output to grow more rapidly than market demand. Driven by the search for profits and the pressures of competition, a large number of firms invest in new technologies and products during periods of growth. Accordingly, output increases rapidly until it reaches a point where it can no longer be absorbed by the market. As a result, prices drop and profits and wages are reduced, leading to bankruptcies and unemployment. Even during periods of growth, new technologies make established skills obsolete, leading to marginalization and unemployment for groups of workers (Box 3.2).

Successive waves of industrialization

Rather than representing a single abrupt change – the Industrial Revolution – industrialization should be seen as an uncertain and uneven process, occurring in successive waves of transformation (Figure 3.5). In the narrow sense, industrialization refers to a series of changes in manufacturing technology and organization that radically transformed the economic landscape of capitalism. It can be defined in narrow terms as the 'application of power-driven machinery to manufacturing' (Gremple, undated, p.1). While technological innovations provided a base for large rises in productivity and output, it was the development of the

Box 3.2

The decline of the handloom weavers

The changing fortunes of the handloom weavers of England in the late eighteenth and early nineteenth century illustrate how the processes of 'creative destruction' associated with industrialization actually operate, particularly in terms of the impact of new technologies on labour. In particular, the cotton textiles industry – based in Lancashire in northern England (Figure 3.6) – was transformed by a sequence of innovations in the late eighteenth century. These included Arkwright's water frame (1771), Hargreaves's spinning jenny (1778) and Crompton's mule (1779), which combined the jenny and the water frame. The effect was to revolutionize the spinning process, allowing yarn to be produced in large quantities, feeding into weaving that had been speeded up by the flying shuttle (1733).

Weaving expanded hugely in this period, absorbing the growing quantity of yarn through the multiplication of handloom weavers. In retrospect, the years from 1760 to 1810 appear as something of a 'golden age' for handloom weaving with the number of handloom weavers in cotton rising from about 30,000 to over 200,000 (Thompson, 1963, p.327). Demand for cloth was high and wages rose steadily, attracting many new entrants to the trade. Yet the general climate of prosperity disguised a loss of status as the weavers' ancient craft protections were eroded, with all restrictions on entry to the trade collapsing (ibid., pp.305–6).

The removal of such protections exposed the weavers to savage wage cuts after 1800 as markets became glutted through the expansion of output, with a surplus of cheap goods and fierce competition driving wages down. The decline actually pre-dated the spread of power looms in the cotton textile industry, but they became a key force that rendered handloom weaving uncompetitive. As a result, the average weekly wage of the handloom weaver in Bolton fell from 33s in 1795 to 14s in 1815 to 5s. 6d in 1829–34 (Hobsbawm, 1962, p.57). Production moved from the cottage to the factory as the way of life and culture of the handloom weavers were undermined.

Widespread distress and desperation saw the weavers resort to expressions of impotent 'rage against the machine' by smashing the new power looms and factories. Alongside such Luddite acts, strikes were organized and petitions to the government launched. The doctrines of political economy that held sway at the time, however, prohibited any interference

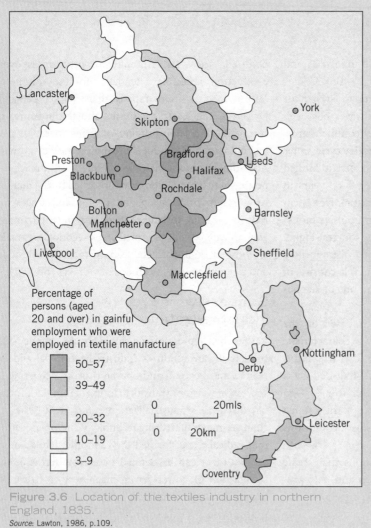

Figure 3.6 Location of the textiles industry in northern England, 1835.

Source: Lawton, 1986, p.109.

Percentage of persons (aged 20 and over) in gainful employment who were employed in textile manufacture

- 50–57
- 39–49
- 20–32
- 10–19
- 3–9

0   20mls

0   20km

Box 3.2 (continued)

with the 'natural' operation of market forces, and the government responded with repression.

By the 1820s, the great majority of weavers were living on the borders of starvation (Thompson, 1963, p.316). In the town of Blackburn, at the heart of Lancashire's textile industry, over 75 per cent of handloom weavers were unemployed in the early 1820s. In one week in April 1826, 76 people entered the Blackburn Workhouse, bringing the total 'crammed together' to 678 (Turner, undated). The labour of the handloom weavers had been rendered obsolete by the industrial revolution, demonstrating how technological innovation could displace skilled labour from established craft industries.

---

factory system that enabled this possibility to be exploited to the full (Knox *et al.*, 2003, p.244). **The factory** was at the core of industrialization, replacing the existing system of domestic labour and independent craftsmen, representing perhaps its most enduring legacy. Basically, it allowed capitalists to gather and order a large workforce under one roof, enabling them to control the labour process and to exploit the new technologies and emerging division of labour to the full (section 3.3.2).

The first Kondratiev cycle involved early mechanization from the 1770s based on water power and steam engines. This was focused on cotton textiles, coal and iron working, facilitated by the development of river systems, canals and turnpike roads for transporting raw materials and finished products. The cotton textiles industry was the main focus of the Industrial Revolution, acting as the 'pacemaker' of change and providing the basis for the first industrial regions (Hobsbawm, 1999, p.34). Output greatly increased while production costs were massively reduced. This occurred through the creation of **economies of scale**, referring to the tendency for firms' costs for each unit of output to fall when production is carried out on a large scale, reflecting greater efficiency. By 1800, for example, the number of workers needed to turn wool into yarn had been reduced by 80 per cent, and by 1812 the cost of making cotton yarn had dropped by 90 per cent since 1760 (Grempel, undated, p.2). But by the 1830s and 1840s, markets were not growing quickly enough to absorb output, leading to reduced profits, falling wages and unemployment (Box 3.2) (Hobsbawm, 1999, pp.54–5).

A second cycle of industrialization began in the 1840s, lasting until the 1890s, just as the first textile-based phase had reached its limits. It was based on the industries of coal, iron and steel, heavy engineering and shipbuilding, creating a much firmer foundation for economic growth. Advances in transportation were of central importance here, particularly the development of the railways. Again, however, a phase of expansion from the 1840s to the early 1870s gave way to depression as markets became saturated and prices fell.

The development of the internal combustion engine and electricity was central to the third wave, lasting from the 1880s to the 1920s, associated with automobiles, oil, heavy chemicals and plastics. Britain lost its lead to Germany and the US who pioneered the development of the new technologies. The characteristic pattern of a growth phase being succeeded by stagnation was again apparent with the boom years from the mid-1890s to 1914 matched by the interwar Depression.

The fourth cycle occurred between the 1930s and 1970s based upon key industries such as electronics, automobiles, aerospace engineering and petrochemicals. This is the period of Fordism, involving the mass production of consumer goods. It also saw the spread of road transport and aircraft. A fifth cycle seems to have begun in the late 1970s, centred upon the 'knowledge economy' sectors of information technology, telecommunications and biotechnology (Hall, 1985). The spread of computer networks has generated a revolution in communications while the containerization of ships' cargos and the development of jet aircraft have transformed the long-distance transport of freight and people. The service sector has become the dominant source of employment in developed economies,

Table 3.2 The growth of the service sector

	Employed in services (%)			
	1976	1986	1996	2004
France	52.2	61.4	69.4	72.6
Germany	48.7	54.7	61.6	66.6
Japan	52.0	57.1	61.2	67.1
UK	57.5	63.1	70.7	76.4
US	65.3	68.8	73.3	78.4

Source: Adapted from Labour Force Statistics, OECD, 2000 and 2005.

accompanied by a process of **deindustrialization** as manufacturing industry declines in importance (Table 3.2). The work of Manuel Castells (1989, 1996) has been particularly influential in recent years in arguing that we have moved from an industrial to an 'informational' economy and society.

Reflect

➤ Identify examples of competencies developed by specific firms.

➤ In what ways does the mobility of capital between regions contribute to the process of uneven development?

3.3 The labour process

3.3.1 The position of labour under capitalism

As we have emphasized, the **position of labour** is a central defining feature of the capitalist mode of production. Unlike previous and more primitive societies where the great mass of the population, as slaves or peasants, were in thrall and effectively owned by their masters, under capitalism workers are released from feudal ties and are free to sell their own labour. Indeed, labour has to sell its own 'labour power' to earn wages to pay for the essentials of life (e.g. food, clothing, shelter) because it does not own the means of production and therefore the means for its own sustenance.

Capital certainly needs labour in order to produce things to sell, but labour's needs are more urgent: it has to engage in production to secure a wage to sustain itself. In this sense, there is a fundamental imbalance at the heart of capitalist social relations. While labour can withdraw its labour power (i.e. strike) as a sanction against capital, its immediate needs are greater than those of capital, which can afford an empty factory/office for a certain amount of time, although not of course for a sustained period.

Labour can be regarded as a '**fictitious commodity**' that takes the appearance of a commodity but cannot be reduced to one (Polanyi, 1944). In the sense that it is purchased by capital, and combined with the means of production to produce commodities for sale, labour functions as a commodity. As such, there is a **labour market** where labour is purchased by capital, and sold by workers, for a certain price (wage). Yet, as Storper and Walker (1989) argue, echoing Marx:

> Labour differs fundamentally from real commodities because it is embodied in living, conscious human beings and because human activity (work) is an irreducible, ubiquitous feature of human existence and social life.
>
> (Storper and Walker, 1989, p.154)

Labour is not produced for sale on the market like other commodities, as Polanyi observed (Box 1.3), but emerges into the labour market from society, through the education and training system. In this sense, the supply of labour is relatively autonomous from demand. Unlike other commodities, when capital buys labour it does not buy a specified amount but rather the time and labour power of the worker. After the working day is over, workers are free to pursue their lives beyond the confines of the factory gates or office.

This brings us to another crucial point: labour must be reproduced outside the market (Peck, 1996). **Reproduction**, in this context, refers to the daily processes of feeding, clothing, sheltering and socializing that support and sustain labour, processes that rely on family, friends and the local community. As Harvey (1989, p.19) has memorably expressed it, 'unlike other commodities, labour power has to go home every night'. The notion of labour reproduction focuses attention on the connection between work, home and community. In many conventional accounts of work, the focus on full-time waged employment, usually that of men, has obscured the importance of domestic labour in the home, often performed by women (Gregson, 2000). Domestic labour includes activities such as childcare, cooking, cleaning, washing, ironing, etc. It is typically unpaid, but vital to the reproduction of the paid workforce.

The key geographical expression of the nature of labour as a 'fictitious commodity' is its **relative immobility**. Labour is reproduced in particular places, something that is determined by the need for work and home to be located in fairly close proximity for the vast majority of people. The result is a patchwork of local labour markets, the geographical range of which is determined by the distance over which people are able to commute. While this has expanded over time, with the growth of suburbs and the dominance of the private car, most people still live and work within local labour market areas. According to Storper and Walker,

> It takes time and spatial propinquity for the central institutions of daily life – family, church, clubs, schools, sports teams, union locals, etc. – to take shape … Once established, these outlive individual participants to benefit and be sustained by generations of workers. The result is a fabric of distinctive, lasting local communities and cultures woven into the landscape of labour.
>
> (Storper and Walker, 1989, p.157)

The relative immobility of labour, and its concern with sustaining and defending its communities, contrasts with the mobility of capital (although this distinction should not be taken too far as firms' investment in particular places does constrain their mobility to a certain extent). This point can be broadened to suggest that economic landscapes are formed out of the interaction between the conflicting forces of capital seeking profits and labour seeking to defend and promote its interests (Peck, 1996).

3.3.2 Divisions of labour

As we have already indicated, a key feature of the Industrial Revolution was the reorganization of production into the factory system, bringing large numbers of workers together under the control of capitalists. A central principle of the factory system and industrial society more broadly is **the division of labour,** which has technical, social and geographical dimensions (Sayer, 1995). We deal with the former two here, integrating the latter into our discussion of the geography of industrial development in Chapter 4. The technical division of labour can be defined as the process of dividing production into a large number of highly specialized parts, so that each worker concentrates on a single task rather than trying to cover several (see Box 3.3).

An increased division of labour results in the **deskilling of labour** as more rewarding aspects of work such as design, planning and variation are removed. The aim of the eighteenth-century pottery owner Josiah Wedgwood, for instance, was to increase the division of labour so as to convert his employees into 'such machines of men that cannot err' (quoted in Bryson and Henry, 2005, p.315). The subdivision and fragmentation of the labour process increases employers' control of production, reducing workers to small cogs in the system.

Following the initial development of the factory system, further developments took place in the early twentieth century. These can be described as Fordist, after the American automobile manufacturer Henry Ford who introduced the key innovations (section 1.4.4). Fordism is based on an **intensification of the labour process**, developing techniques of mass production. This was to be balanced by mass consumption with increased wages for workers giving them additional purchasing power in the market, creating the consumer demand needed to underpin mass production.

Box 3.3

Adam Smith and the division of labour

The concept of the division of labour was first formalized and developed by Adam Smith, writing in the late eighteenth century, just as the first shoots of industrialization were starting to appear. Smith argued that the division of labour in production is limited by the extent of the market. In pre-industrial times, localized markets were associated with small-scale domestic industry, employing craftsmen and artisans who undertook a number of different tasks. The rapid extension of markets on a global scale that took place from the late eighteenth century (section 3.4), by contrast, created the conditions for large-scale industrial production to flourish, employing an elaborate division of labour.

Smith famously used the example of a pin factory to demonstrate that it is far more efficient for an individual worker to concentrate on one particular task rather than to try to perform a number of activities:

> One man draws out the wire, another straightens it, a third cuts it, a fourth points it, a fifth grinds the top for receiving the head; to make the head requires two or three distinct operations . . . and the important business of making a pin is, in this manner, divided into about eighteen distinct operations.
>
> (A. Smith [1776], 1991, pp.14–15).

The key principle here is specialization. In a small factory employing 10 workers, an untrained worker could make fewer than 20 pins a day. If workers became specialized through an increased division of labour, though, Smith observed that 10 people could make 48,000 pins in a day, even when the small size of the workforce meant that one worker had to perform two or three operations. In this way, then, an increased division of labour resulted in huge rises in productivity. There are three specific reasons why an enhanced division of labour increases productivity and efficiency in the workplace (ibid., p.13):

- ➤ It improves the dexterity of workers who become highly adept in performing the same routine task thousands of time in the same day.

- ➤ Lost time, referring to the time wasted by workers moving between tasks, tools and machines, is sharply reduced.

- ➤ It facilitates the replacement of labour with machines. This is due to the fact that specialization involves breaking the labour process down into a large number of standardized and routine tasks that can be performed by machines (fixed capital).

The first key element of the Fordist organization of production occurred with the introduction of **scientific management** or **Taylorism** – after its key advocate F.W. Taylor – which involved the reorganization of work according to rational principles designed to maximize productivity. Three key elements of scientific management can be identified (Meegan, 1988, p.140):

- ➤ A greatly increased technical division of labour, based on the complete separation of the design and planning of work, undertaken by management, from its execution by workers, who became increasingly focused on simplified and repetitive tasks.

- ➤ The subdivision of operations was matched by the reintegration of the production process, involving increased coordination and control by management who were to exercise complete authority over the planning and direction of work, removing the power of foreman and workers.

- ➤ The performance and organization of workers was subject to very close monitoring and analysis by management, employing techniques such as time-and-motion studies.

This organizational revolution was matched by experimentation with new techniques to increase productivity. The most famous and widely adopted of these is the moving assembly line, first established at Ford's Highland Park automobile plant at Detroit, Michigan, between 1911 and 1913 (Figure 3.7). This revolutionized production methods in the automobile industry. Rather than the worker assembling cars by moving around a factory to pick up different parts, the parts now came to the worker, who would be placed at a fixed position with typically just one dedicated task:

Figure 3.7 The Ford assembly line.
Source: Mary Evans Picture Library.

'The man who places the part does not fasten it,' said Henry Ford, 'the man who puts in a bolt does not put on the nut; the man who puts on the nut does not tighten it.' Average chassis assembly time fell to ninety-three minutes. The lesson was obvious. Within months Highland Park was a buzzing network of belts, assembly lines, and sub-assemblies ... The entire place was whirled up into a vast, intricate and never-ending mechanical ballet.

(Lacey, 1986, cited in Meegan, 1988, p.142)

The saving of time and restructuring of work tasks had a dramatic effect on productivity; between 1911 and 1914 production of cars quadrupled from 78,000 to around 300,000, while the workforce only doubled in size over the period and even fell during 1913–14 (Meegan, 1988, p.143). Box 3.4 gives a brief flavour of the experience of 'working for Ford'.

The increased technical division of labour in the workplace is matched by what is termed the **social division of labour** in society. This refers to the vast array of specialized jobs that people perform in society, from doctors and lawyers to plumbers, painters and construction workers (Sayer and Walker, 1992). Modern, industrial societies are characterized by a highly complex division of labour in this respect. In the course of their work, individuals enter into a range of social relations with other people occupying roles such as colleagues, supervisors and clients or competitors (section 1.4). Students, for example, enter into social relations with academics in their role as university teachers during their degree programmes. The different jobs that people do have acquired different values in society, using that term in its broadest sense to incorporate the social status and prestige that particular jobs confer on people (e.g. compare an investment banker and a hairdresser). Such varying levels of status and prestige play an important part in determining pay rates for different occupations, alongside patterns of supply and demand in the labour market (Peck, 1996).

Reflect

➤ Why does labour tend to be relatively immobile geographically, compared with capital?

➤ Outline the distinction between technical and social divisions of labour.

Box 3.4

Working for Ford

The industrial sociologist Huw Beynon undertook a detailed study of working conditions at Ford's Halewood plant in Liverpool, north-west England, in the late 1960s and early 1970s, conveying workers' experience of assembly-line work:

> It's the most boring job in the world. It's the same thing over and over again. There's no change in it, it wears you out. It makes you awful tired. It slows your thinking right down. There's no need to think. It's just a formality. You must carry on. You just endure it for the money. That's what we're paid for, to endure the boredom of it all.
>
> If I had a chance to move I'd leave right way. It's the conditions here. Ford class you more as machines than men. They're on top of you all the time. They expect you to work every minute of the day
>
> (Ford workers at Halewood, quoted in Beynon, 1984, p.129).

According to another worker, while the white-collar staff that worked in the 'office' were part of Ford's, the men on the shop floor were just regarded as 'numbers'. Getting used to the incessant demands of the production process was difficult for many new workers with no experience of working in a car plant. The effects of assembly-line work could sometimes extend into domestic life as well. According to one night-shift worker:

> My wife always used to insist that I had my breakfast before I went to bed. And I would get into such a state that I would sit down to bacon and egg and the table would be moving away from me. I thought, 'crikey, how long am I going to have to put up with this?' But the pay was good. It was a case of really getting stuck in and saying 'to hell with it, get it while it's here'. And this is the way it went, but the elderly chaps couldn't stand the pace
>
> (Joe Dennis, quoted in Meegan, 1988, p.144).

It is important to recognize that the workers were not just passive, meekly accepting the dictates of management. The role of the trade unions was central and, in particular, the plant-level shop stewards' committee in representing the interest of the workers. One of the key things shaping workers' day-to-day experience was the speed of the line, with Beynon (1984, p.148) stating that 'the history of the line is a history of conflict over speed-up.' After considerable disruption and struggle at Halewood, the workers won an important victory when management conceded to the shop stewards the right to hold the key that locked the assembly line (ibid., p.149), giving them more control over their work conditions.

3.4 Consumers and consumption

3.4.1 Understanding consumption and 'the consumer'

Consumption can be defined as those processes involved in the sale, purchase and use of commodities (Mansvelt, 2005, p.6). For most people, it is a central, taken-for-granted part of life in contemporary society. Watching television, eating a burger, going to the cinema, shopping and clubbing are all acts of consumption. While such everyday activities may seem mundane and trivial, they are of considerable economic and cultural significance (P. Crang, 2005, p.360). In economic terms, the sale of commodities is crucial in enabling firms to generate revenue and profits, fuelling the process of economic growth. The importance of consumption as a driver of economic growth is emphasized through regular media references to patterns of consumer spending and levels of consumer confidence (*The Economist*, 2005a, b). In the UK, for example, consumer spending increased by 73 per cent between 1993 and 2003, outstripping the growth of consumer income (65 per cent) and underpinning the growth of the economy (*Money News*, 2006).

Developed countries, such as the UK and the US, have experienced a shift from an industrial economy to a **post-industrial** service-based **economy** in recent decades as the manufacturing sector has contracted in size and service industries such as retail and financial services have grown. As a result, the circulation of knowledge, information and images has become increasingly important relative to the production of manufactured goods (Lash and Urry, 1994). Major corporations and 'brands' such as Nike, McDonalds and Coca-Cola have focused increasingly on marketing and sales activity, outsourcing actual production to a range of suppliers. The spread and influence of such brands has become an important dimension of the process of globalization, raising concerns about the creation of a global consumer culture erasing the distinctiveness of local cultures and places (Ritzer, 2004). The retail sector is an important part of this post-industrial economy, accounting for a sizeable share of paid employment. In the US, for example, retail is the second largest industry, accounting for 12.4 per cent of all business establishments and 11.6 per cent of all employment in 2004 (Bureau of Labor Statistics, 2006). Particular retailers such as the major supermarket chains like Tesco in the UK exercise great power over manufactures and suppliers through their supply chains with efforts to reduce consumer prices translating into lower prices for suppliers.

At the same time, 'the consumer' or 'the market' is often invoked as the reason for producing goods and organizing services in particular ways. For example, bananas must be of a particular size and quality to satisfy consumer expectations while services such as banking should be provided through the Internet rather than face to face because this is what 'the consumer demands' (P. Crang, 2005, p.126). The implications of this can be serious, leading to the impoverishment of Caribbean banana growers who cannot meet these standards (Box 1.1) or the closure of bank branches. In this sense, 'the consumer' has become a kind of 'global dictator' (Miller, 1995, p.10), with the demands of affluent northern consumers in particular determining how goods are produced and services delivered throughout the world economy. Consumer demand cannot be understood in narrowly economic terms with consumption playing a central role in people's wider social and cultural lives (going shopping, visiting a bar or a restaurant). At the same time, consumption is itself shaped by 'those wider contours of society and culture' (P. Crang, 2005, p.361) with people's desire for goods bound up with their identities and positions in society.

3.4.2 Consumption, culture and identity

Consumption has become a major focus of interest for geographers and other social scientists over the last couple of decades. Much of this is a response to the previous neglect of consumption, seen as very much secondary to, and derivative of, the more fundamental process of production. Three broad premises underpin this recent concern with consumption. First, it is seen as central to the reproduction of social and cultural life, referring to people's everyday actions in supporting themselves and their families, involving feeding, clothing, sheltering, socializing, etc. Second, modern market societies are said to be characterized by a '**consumer culture**' organized around the logic of individual choice in the marketplace. Third, studying consumption enables us to better comprehend the importance of culture in shaping economic processes and institutions, representing 'the site on which culture and economy most dramatically converge' (Slater, 2003, p.149).

Two main perspectives on consumption can be identified. First, a number of influential social critics, including Karl Marx and Herbert Marcuse, have viewed it as signalling the triumph of market exchange and industrial society over deeper human qualities and meanings. As such, consumption marks the process through which culture is colonized by economic forces (ibid., p.150). In capitalist societies, needs and wants are artificially created and manufactured, inducing people to consume far more than they actually need. A number of studies have focused on the role of advertising in stimulating demand for products. While there is certainly a sense in which market demand is induced through processes like advertising, this approach invariably tends to cast the consumer in a passive role as a 'cultural dupe' or 'dope' manipulated and controlled by corporations and the media. Some versions of postmodernism (Box 2.5) have reproduced this theme of the **passive consumer**, emphasizing their powerlessness in the face of an infinite universe of abstract signs and meanings

A second view emphasizes the **active role of consumers** in utilizing things for their own ends. Earlier work in this vein viewed consumption and leisure practices as sources of social status and distinction, as captured in the institutional economist Thorstein Veblen's notion of 'conspicuous consumption' and the sociologist Pierre Bourdieu's more recent work on cultural capital and taste. Contemporary studies have moved away from this concern with social status and distinction to stress how consumers creatively rework the products they buy, generating new meanings in the process (P. Crang, 2005, p.363). Rather than reading off the process of consumption from production and corporate strategies, as critical approaches have tended to do, one has to understand the social and cultural relations in which it is entangled. From this perspective, consumption is seen as a relatively fertile arena for the expression of individuality and creativity, compared with the world of work, which involves considerable drudgery and monotony for many people.

An example of the wider social and cultural relations in which active consumption is rooted are those of family and friendship. According to the anthropologist Danny Miller, instead of being driven by individual greed and hedonism, consumption is based on acts of love and devotion involving the purchase of commodities for partners, children and friends (Miller, 1998b). Of particular significance here is gift-buying, focused around rituals such as Christmas and birthdays (Box 2.7). Much of this work on active and creative consumption has been ethnographic in nature, attempting to understand through detailed fieldwork and observation the meaning and significance that people attach to consumption.

3.4.3 Changing patterns of consumption

One of the key expressions of **modern consumer culture** is the department store, which came to prominence in the second half of the nineteenth century. Described as 'the quintessential consumption site of the late nineteenth and early twentieth century' (Lowe and Wrigley, 1996, p.18), department stores presented 'the most visible, urban manifestation of consumer culture and the economics of mass production and selling'

Figure 3.8 Macy's: A famous New York department store with nineteenth-century origins.
Source: D. Mackinnon.

Box 3.5

Gender and consumption in nineteenth-century New York City

While the rapid and massive expansion of New York's retail district in the nineteenth century reflected the spectacular growth of the city's economy, its form was shaped by the shopping habits of the middle classes, particularly women. Shopping in the city's burgeoning department stores was an important activity for middle-class women in New York, becoming almost a daily ritual for some (Domosh, 1996). The very wealthy defined the styles and fashions that middle-class women sought to emulate, allowing the latter to define their own tastes and, by extension, their social status (ibid., p.261). Rather than being entirely confined to the domestic sphere, women made frequent trips from their suburban homes to downtown Manhattan, meaning that they played an important role in shaping this and other American city centres. While shopping was a frequent activity, they also met for church prayer groups, for lectures and concerts, and to pay bills or make social calls (ibid., p.258).

A distinct retailing area developed in late nineteenth-century New York, focused on Fifth Avenue between Union and Madison Squares, extending west to Sixth Avenue and east to Broadway (Figure 3.9). This was an 'urban landscape designed specifically for consumption', made up of 'ornamental architecture and grand boulevards, of restaurants and bars, and of small boutiques and large department stores' (ibid., pp.263–4). Improvements in urban transportation through the development of a rapid

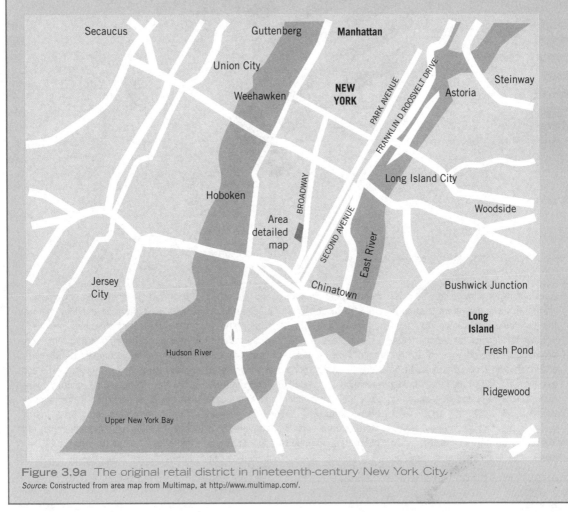

Figure 3.9a The original retail district in nineteenth-century New York City.
Source: Constructed from area map from Multimap, at http://www.multimap.com/.

Box 3.5 (continued)

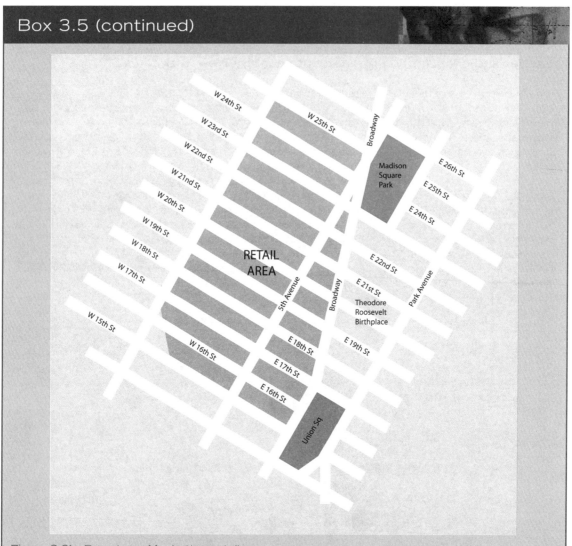

W 24th St
W 23rd St
W 22nd St
W 21st St
W 20th St
W 19th St
W 18th St
W 17th St
W 15th St
W 16th St
W 25th St
Broadway
Madison Square Park
E 26th St
E 25th St
E 24th St
E 22nd St
E 21st St
RETAIL AREA
5th Avenue
Broadway
Park Avenue
Theodore Roosevelt Birthplace
E 18th St
E 17th St
E 16th St
E 19th St
Union Sq

Figure 3.9b Downtown Manhattan retail area.

transit network allowed middle-class women to travel downtown to shop in the morning, returning to their uptown residences for lunch before proceeding back downtown in the afternoon. Wide, paved streets, lit by gas and electricity, made the retailing district feel safe and congenial for women.

The city's first department store, Stewart's, opened on Broadway in 1846. Its four storeys, devoted entirely to retailing, and its white marble façade were unprecedented in the city (ibid., p.264). The interior of the store was arranged in order to maximize the display of goods and to create an appropriate atmosphere for women. It included large mirrors, chandeliers, a gallery and a parlour. A later store was opened by Stewart further uptown on Broadway in 1862, becoming an important tourist attraction. Its six storeys contained a main

floor and five encircling balconies, lit by a skylight that provided natural light. Reflecting the analogies drawn between consumption and religion, a commentator compared Stewart's with nearby Grace Church. By the end of the nineteenth century, the domestic sphere had become more fully incorporated into stores in the form of tea rooms, restaurants, art galleries and grand architectural displays.

(Domosh, 1996, p.257) (Figure 3.8). A vast array of goods was placed on public display in the store and prices fixed for standardized goods. While we take this for granted today, prior to the development of modern stores, goods were not on display; customers had to ask to view them and prices were negotiated between the customer and shop owner or clerk (ibid., p.264). In the new department stores, shoppers were able to directly compare the prices and qualities of different goods as shopping became a knowledgeable and skilled activity, encouraged by the retailers and advertisers.

Shopping also became highly **gendered** as the department stores targeted middle-class women in particular (Box 3.5). In this way, a crucial set of links was forged between gender, class and culture in the late nineteenth century that continue to shape retail consumption today. Shopping became a major part of women's work, with the purchase of goods replacing the domestic production of food and clothing, creating a market for manufactures and retailers. The realm of consumption was defined as female, associated with leisure and self-indulgence, in contrast to the masculine domain of production, governed by the work ethic and associated notions of self-denial and self-discipline (ibid., p.262). Department stores provided spaces in which women could be taught to shop through the display of goods as spectacle, the use of advertising and demonstration and the assistance of specialist staff (Hudson, 2005, p.147). The introduction and manipulation of fashion was a key mechanism for increasing demand, meaning that frequent changes in style were required to keep up. Shopping became an important female duty, with store owners cultivating associations between women, fashion and religion, referring to their stores as 'cathedrals' and goods as 'objects of devotion' (Domosh, 1996, p.266).

While mass consumption became established in the nineteenth century, it was consolidated and reinforced during the period of Fordism from the 1940s to the 1970s. As we have emphasized (sections 3.2.4, 3.3.2), Fordism was a system of industrial organization based on a balance between mass production and **mass consumption**. The key link here was higher wages for workers, received in exchange for increased productivity. Fordism involved the mass production of consumer durables, such as automobiles, fridges and washing machines, produced in standard forms. Rising wages meant that more and more workers were able to afford such goods, ensuring a larger market for manufacturers and retailers. A key trend in the geographical organization of society was the growth of suburbs, particularly in North America, facilitated by state investment in infrastructure such as roads and electricity. Suburban lifestyles become closely associated with mass-consumption patterns, with every household requiring its car, washing machine and lawnmower (Goss, 2005, p.261).

Mass markets for standardized goods became increasingly saturated from the late 1960s, however, as economic growth slowed and the Fordist system experienced growing problems. As a result, mass consumption has been eclipsed by the rise of **post-Fordist patterns of consumption** since the 1970s. These are defined by flexibility, reflecting the fragmentation of markets into distinct segments and niches. Accordingly, patterns of consumption are defined individually rather than collectively. Consumer choice and identity has become increasingly important, with individual consumers regarding the purchase and consumption of commodities as expressions of their lifestyles and aspirations (Mansvelt, 2005, p.44). Individualized patterns of consumption oriented towards identity and lifestyles are a key component of postmodern culture, characterized by an emphasis on flexibility, difference and diversity (Box 2.6). The rapid circulation of ideas, images and signs, fuelled by the advertising industry and the media, is another central aspect of postmodern consumer culture.

Producers and retailers have become increasingly **consumer-oriented**, striving to tailor goods and services to the demands of individuals and specific groups of consumers. It is in this sense that the consumer has become a kind of 'global dictator' (Miller, 1995, p.10), determining how production is organized across a wide range of economic sectors and geographical locations. The growth of information and communication technologies has allowed retailers to store, process and convey data about changing patterns of consumer demand. The introduction of point-of-sales terminals in the 1980s allowed retailers to rapidly transmit information about consumption trends from shops to company headquarters, from where it was passed on to designers and suppliers. The clothing company Benetton is one of the best examples of a company that has grown through a strategy of niche marketing and product differentiation, employing advanced communication technologies to the fullest possible extent in the spheres of design, manufacturing, procurement and marketing (Knox *et al.*, 2003, pp.188–9).

> ## Reflect
>
> ➤ How does the idea of the active consumer differ from previous theories adopted by social scientists?
>
> ➤ To what extent do you view consumption as an important means of expressing your individuality and identity?

3.5 The state

In this book (section 2.4.3), we adopt a broadly regulationist perspective, rejecting the notion of an autonomous self-regulating economy. Rather, the economy is regulated through a wide range of forms of social regulation, including social habits, administrative rules and cultural norms, with **the state** playing a key role in harnessing and coordinating these different mechanisms (Aglietta, 1979). By 'state', we are referring to a set of public institutions that exercise authority over a particular territory, including government, parliament, civil service, judiciary, police, security services and local authorities. As this suggests, the state is a complex entity, encompassing a number of institutions beyond what is normally referred to as government. The role of the state in social regulation is directed towards the objective of ensuring social and economic stability, thereby creating the conditions that allow the national economy to expand.

The state assumes two key functions in this respect (Johnston, 1996). First, the **'accumulation' function** means that a key task of the state is supporting and promoting economic development within its territory, ensuring that business can accumulate capital for investment and growth. Such an objective underpins many of the activities listed in Table 3.3, involving the state in designing economic strategies, promoting innovation and 'competitiveness' in certain strategic industries and administering education and training programmes for workers. In recent decades, governments have viewed their performance in managing and overseeing the economy as vital to the retention of power, striving to build a reputation for economic competence. In order to gain power in Britain in 1997, for instance, the leaders of the (new) Labour Party had to drop its old 'tax and spend' policies, fostering a new image of themselves as responsible and prudent custodians of the public finances. This required them to develop good relations with business and the City of London.

The second key function is **'legitimation'**, referring to the range of activities undertaken to maintain social order, ensuring that the capitalist system and the associated social order are regarded as legitimate and 'natural' by the majority of citizens. As well as managing the legal system, this has seen states construct elaborate welfare systems over the course of the twentieth century to try to spread the benefits of growth and offer social protection to its citizens against the vagaries of the market, including unemployment and ill health. This **welfare state** has come under attack from neoliberal commentators and politicians since the 1970s, but much of its basic architecture remains in place in the majority of advanced industrial countries. While the 'legitimation' function requires the state to sponsor collective projects aimed at increasing the wellbeing of its citizens, it remains dependent on the expansion of the capitalist economy (accumulation) to ensure its continued financial viability through the revenue provided by taxation.

The key regulatory activities of the state are listed in Table 3.3. These include ensuring macroeconomic stability through fiscal and monetary policy, helping to secure the reproduction of labour through, among other things, education and training programmes, health and safety legislation and unemployment benefit, and the provision of basic infrastructure. More broadly, states are charged with safeguarding property relations, regulating trade and overseeing the operation of financial markets. The state is also of course, an important economic entity in its own right, employing large numbers of people, purchasing a wide range of goods and services and harnessing public and private

Table 3.3 The economic roles of the state

Economic role of the state	Details
1. Maintenance of property rights and governance frameworks	Maintenance of law and order Maintenance of private property rights Recognition of institutional property rights Rules of ownership and use of productive assets Rules for the exploitation of natural resources Rules for the transfer of property rights Protection of intellectual property rights The governance of economic relations between: family members, employers and workers, landlords and tenants, buyers and sellers.
2. Management of territorial boundaries	Provision of military force Regulation of money flows, goods flows, service flows, labour flows, knowledge flows.
3. Control of macroeconomic trends	Fiscal policy Monetary policy
4. Governance of product markets	Regulation of the market power of firms The selection and regulation of natural monopolies The provision of public goods and goods unlikely to be supplied fairly
5. Governance of financial markets	Rules for the establishment and operation of financial institutions Designation of the means of economic payment Rules for the use of credit Maintenance of the lender of last resort
6. Provision of basic infrastructure	Includes transportation and communication systems, energy and water supply, waste disposal systems Assembly and conduct of communications media Assembly and conduct of public information Land use planning and regulation
7. Selection and development of economic growth strategy	Promotion and maintenance of strategic industries Development of science and technology Urban and regional development Identification of key outcomes and targets (employment, growth, innovation, etc) Selection and implementation of financial incentives
8. Production and reproduction of labour	Demographic planning Provision of universal education and training Governance of workplace conditions Wages policies Social wage provision Supply and governance of childcare Governance of retirement pensions
9. Other legitimation activities	Maintenance of public health Citizenship rights Income and wealth distribution Reduction of poverty Cultural development Socialization Environmental protection/enhancement

Source: Adapted from O'Neill, 1997, p.295.

resources behind major strategic initiatives and projects (Painter, 2000).

The state is explicitly geographical in form, having clearly marked borders that separate the territory of one state from that of others. Within these territorial limits, states have sought to define and promote national economies, building integrated national markets through the creation of common legal standards and financial rules, the expansion of transport and communication systems and the careful regulation of flows of goods, money and people across their borders. At the same time, states contain considerable internal diversity in terms of economic conditions, cultural values and political allegiances. The problems of territorial management created by such internal diversity have generally been recognized through the creation of a tier of local government that administers state programmes and represents local interests (Duncan and Goodwin, 1988).

The notion of a coherent national economy has been seriously disrupted by the process of economic globalization in recent years as flows of goods, money and information across national borders have grown rapidly (section 1.2), making it more difficult for governments to regulate their economies in the conventional fashion. Despite this, states remain important actors within globalization (Dicken *et al.*, 1997), not least in prosecuting the interests of their own multinationals in overseas markets and in international trade forums such as the World Trade Organization (WTO).

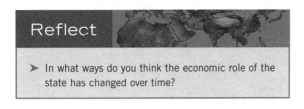

Reflect

➤ In what ways do you think the economic role of the state has changed over time?

3.6 Summary

This chapter has examined the role of the main economic actors and outlined the basic underlying processes that have shaped the development of capitalism since the late eighteenth century. The basic process of production is described in terms of the circuit of capital, in which capital, labour and the means of production are combined to generate commodities that are sold in competitive markets for a profit. The firm can be seen as a distinctive bundle of competencies and assets, expressed in terms of skills, practices and knowledge. While capital tends to be concentrated in prosperous core areas such as south-east England, it remains highly mobile, subject to being 'switched' from declining regions to new growth regions. The process of innovation and technological change is an inherent feature of capitalism, reflecting the pressures of profit-seeking and competition between firms. Economic historians have observed the tendency of major innovations to 'swarm' together, underpinning 'long waves' or Kondratiev cycles of economic development. This chapter has also considered the role of labour, noting that, although workers are required to sell their labour to capitalists for a wage, labour should be viewed as a 'fictitious commodity' on account of its irreducible human and social qualities (Polanyi, 1944). At the same time, the evolving technical division of labour in production has shaped the development of industrial capitalism, finding perhaps its fullest expression in the Fordist mass-production industries that typified the middle decades of the twentieth century.

The role of consumers and consumption is central to modern economic life, providing the market demand that underpins economic growth. The retail sector is an important part of the economy in developed countries, for example accounting for 11.5 per cent of employment in the US. Recent work on consumption has been based on notions of the 'active consumer', rejecting the view of the consumer as a 'dupe' manipulated by corporations and the media. Modern mass consumption grew in the nineteenth century, finding particular expression in the department store, described as 'the quintessential consumption site of the late nineteenth and early twentieth century' (Lowe and Wrigley, 1996, p.18). Middle-class women were targeted as a key market by the retailers, forging a set of links between gender, class and consumer culture that remains influential today. Mass-consumption practices were consolidated and reinforced under Fordism in the twentieth century. Since the 1970s, a new post-Fordist mode of consumption has emerged, characterized by a focus on different market segments and niches with consumption defined individually rather than collectively. Companies have become increasingly consumer-oriented, with the focus on meeting consumers' demands and expectations leading to the Western consumer being described as a 'global dictator' (Miller, 1995, p.10). We also considered the role of the state in regulating the economy, focusing on its 'accumulation' and 'legitimation' functions. While the role of the state in the economy has been attacked by Conservative politicians since the late 1970s, much of the welfare state remains in place and the state is an important actor within contemporary processes of globalization (Chapter 6).

Exercise

Select a major innovation introduced during the twentieth century (e.g. motor car, moving assembly line, commercial jet aircraft, television, personal computer) and assess whether the actual benefits of this innovation outweighed the negative effects.

1. Consider the following: the growth and spread of the product or method; who owned or controlled the technology; its appeal to consumers; how production was organized; the impact on existing industries and workers; and its effects on experiences of time and space.

Key reading

Gregory, D. (2000) 'Capitalism', in Johnston, R.J., Gregory, D. Pratt, G. and Watts, M. (eds) (2000) *The Dictionary of Human Geography*, 4th edn, Oxford: Blackwell, pp.56–9.
An introduction to capitalism as a distinctive mode of production based on the private ownership of the means of production and the commodification (commercialization) of labour power. Discusses different approaches to the analysis of capitalism and outlines the circuit of capital.

Hobsbawm, E.J. (1999) *Industry and Empire*, new edn, Harmondsworth: Penguin, pp.34–56, 87–111.
A highly readable account of the development of the British economy since the Industrial Revolution by the renowned Marxist historian. Particularly informative on the Industrial Revolution and links with empire.

Harvey, D. (1985) 'The geopolitics of capitalism', in Gregory, D. and Urry, J. (eds) *Social Relations and Spatial Structures*, London: Macmillan, pp.128-163.
A forceful and stimulating essay on the key processes of capital accumulation by the leading Marxist geographer. Difficult but worth persevering with for the valuable insights it offers into the role of capitalism in shaping the geographical landscape.

Meegan, R. (1988) 'A crisis of mass production', in Allen, J. and Massey, D. (eds) *The Economy in Question*, London: Sage, pp.136-83.
Probably the best geographical introduction to Fordism as an economic system based on mass production and mass consumption. Examines the growth of Fordism and the experience of working in Fordist industries.

Peck, J. (2000) 'Places of work', in Sheppard, E. and Barnes, T.J. (eds) *A Companion to Economic Geography*, Oxford: Blackwell, pp.133-48.
A very useful summary of the geography of labour and labour markets. Relates the changing organization of labour to wider processes of industrial restructuring, emphasizing the distinctiveness of local labour markets and the shift towards flexible production.

Slater, D. (2003) 'Cultures of consumption', in Anderson, K., Domosh, M., Pile, S. and Thrift, N. (eds) *Handbook of Cultural Geography*, London: Sage, pp.146-63.
A wide-ranging survey of recent work on consumption that stresses the importance of consumption as a crucial link between the economy and culture. Identifies different approaches to the study of consumption and discusses recent debates about new shopping patterns and the emergence of a global consumer culture.

Useful websites

http://en.wikipedia.org/wiki/Capitalism
A description of the basic principles and operations of capitalism in accessible terms from Wikipedia, the free online encyclopedia. Defines and discusses key elements of capitalism such as private ownership of the means of production, the pursuit of self-interest and the operation of free markets in an engaging manner.

http://en.wikipedia.org/wiki/Industrial_Revolution
An account of the Industrial Revolution from the same

source. Identifies the main causes of the Industrial Revolution and the key innovations as well as outlining how these were applied in key industries such as coal mining and textiles. The development of factories and machines is emphasized alongside the changing organization of labour.

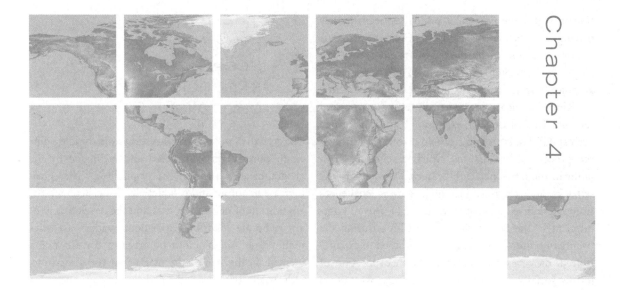

Spaces of production and consumption

Chapter map

In the introduction to this chapter, we emphasize the geographical expansion and historical evolution of the capitalist economy. We then outline the relationship between the Industrial Revolution and the geographical expansion of capitalism, facilitated by the development of the 'space shrinking' technologies of transport and communications (Leyshon, 1995). In section 4.3, the uneven *geography* of production is examined in relation to successive waves of industrialization from the late eighteenth century (section 3.2.4). We focus particular attention on the growth of 'new industrial spaces' since the 1970s. We turn our attention to consumption in section 4.4, examining key spaces of consumption at the global and local scales. The notion of a single global consumer culture is assessed and the role of specific sites of consumption such as mega-malls, streets, car boot sales and the home is highlighted.

4.1 Introduction

As we indicated in section 3.2, the contemporary economy is dominated by a capitalist mode of production in which the pursuit of profit has been the key driving force behind development. Over the last four hundred years or so, this has led to the emergence of a single world economy as capitalism has spread geographically from its origins in Western Europe across the globe to incorporate virtually all countries and regions. This process of expansion has transformed pre-capitalist societies and systems, not least through the process of colonialism, which reached its high point in the late nineteenth century. Despite periodic setbacks and reverses, including serious challenges from alternative social systems such as forms of communism in the Soviet Union and China during the twentieth century, capitalism has continued to expand. Indeed, the latest stage involves the incorporation of former socialist countries through China's 'open-door' policies and following the collapse of Communist regimes in the former Soviet Union and Eastern Europe.

The growth of capitalism has been not only uneven geographically but also historically, with periodic crises like the Great Depression of the 1930s and the 'stag-flation' (a combination of economic stagnation and rising inflation) of the 1970s posing particularly serious problems. Such crises seem to reflect, among other things, a tendency towards overexpansion as favourable economic conditions encourage firms to produce more and more goods, leading ultimately to a slump when the market cannot absorb all this output. Such internal contradictions of capitalism were recognized by commentators as diverse as Karl Marx, Joseph Schumpeter and John Maynard Keynes, leading them to forecast or fear its demise. Adopting a longer-term view, however, capitalism has proved remarkably durable and adaptable in finding short- and medium-term solutions to its internal crises, although it has never been able to completely solve the underlying problem of overproduction. As we have seen, the regulation approach explains this in terms of the role of historically specific **regimes of accumulation** like Fordism in sustaining and stabilizing capitalism (section 2.4.3). Since the early 1980s, the spread of neoliberal policies has shaped the development of a new phase of capitalist globalization (Harvey, 2005).

4.2 The Industrial Revolution and the geographical expansion of capitalism

The growth of industrial production from the late eighteenth century was supported and facilitated by the geographical expansion of capitalism. The great increase in output and productivity associated with the application of new technology and the development of an elaborate division of labour within factories required, as Adam Smith observed, a corresponding increase in the size of the market. At the same time, industrialization was dependent on an increased volume and range of raw materials, drawn from different parts of the globe. Through the growth of capitalist production and trade, an **international division of labour** was created during the nineteenth century. This involved the developed countries of Europe and North America producing manufactured goods while the underdeveloped world specialised in the production of raw materials and foodstuffs. This global trading system was supported and justified by the doctrine of **comparative advantage**, classically expressed by David Ricardo, the great English economist, in 1817. It emphasizes the benefits of international trade, stating that countries should export the goods that they can produce with greater relative efficiency. In recent years, the theory of comparative advantage has provided an important intellectual foundation for the dominant neoliberal ideology of **globalization**.

While the growth of a world economy can be traced back to the voyages of exploration and discovery in the sixteenth century, **geographical expansion** gained new momentum as the Industrial Revolution took off, culminating in the 'age of empire' between 1875 and 1914 (Hobsbawm, 1987). The inherently expansive nature of capitalism as an economic system has underpinned a search for new markets, new sources of raw materials and new supplies of labour, forging economic relationships between territories on a global

Box 4.1

Trade and comparative advantage

Comparative advantage is the principle that a country should specialize in producing and exporting goods in which it has a comparative or relative cost advantage over others and import goods in which it has a cost disadvantage. For example, while the developed country in Table 4.1 has an absolute advantage in producing both wheat and cloth (it can produce them more efficiently), the developing country has a comparative advantage in wheat, and the developed country in cloth. Since wheat is relatively cheaper to produce than cloth in the developing country, it only needs to sacrifice 1 metre of cloth to produce 2 kilos of wheat. Conversely, cloth is relatively cheaper to produce than wheat in the developed country, with the production of 8 metres entailing the sacrifice of 4 kilos of wheat. In the developing country, by contrast, producing one metre of cloth involves giving up 2 kilos of wheat. Thus, the principle of comparative advantage states that countries should specialize in the goods that lead them to give up least in terms of the production of other goods. Through specialization, both countries gain by focusing on the good that they can produce most efficiently and importing the other.

One key question concerns the sources of comparative advantage. How is it that certain countries can produce some goods more efficiently than others? Ricardo explained this in terms of countries' different endowments of the factors of production: land, labour and capital. For example, Canada has a lot of land, China a lot of labour and the US is rich in capital. This means that relative costs of producing goods vary between countries, providing a basis for trade. Basically, countries should specialize in producing goods that use the factors that they have in abundance, enabling them to produce these goods cheaply (grain in Canada, textiles and footwear in China, pharmaceuticals in the US). In this sense, Ricardo believed that trade patterns had a natural basis. The doctrine of comparative advantage supported the colonial trading system where the European countries exported capital-intensive manufactured goods and the colonies produced raw materials and agricultural goods that were labour- (mines, plantations, etc.) and land-intensive.

In recent decades, however, it has become apparent that most trade takes place between developed countries with similar factor endowments. This has helped to stimulate the development of a **new trade theory** by the eminent economist Paul Krugman (Box 2.2) and others that recognizes that comparative advantage does not simply reflect pre-existing factor endowments. Rather, it is actively created by firms through the development of technology, human skills and economies of scale, something that is often referred to in terms of **competitive advantage**. This helps to account for patterns of trade and regional specialization at a more detailed level, for example aircraft in Seattle, cars in southern Germany and finance and business services in London or New York.

Table 4.1 Quantities of wheat and cloth production

	Kilos of wheat	Metres of cloth
Developing country	2	1
Developed country	4	8

Source: Sloman, 1999, pp.659–60.

scale. As Marx and Engels [1848], 1967, pp.83–4) wrote:

The need of a constantly expanding market for its products chases the bourgeoisie over the whole surface of the globe. It must nestle everywhere, settle everywhere, establish connections everywhere ... All old-established national industries ... are dislodged by new industries ... that no longer work up indigenous raw material, but raw material drawn from the remotest zones; industries whose products are consumed not only at home, but in every corner of the globe ... In place of the old local and national seclusion and self-sufficiency, we have intercourse in every direction, universal interdependence of nations.

It is this continuing search for new markets, raw materials and labour supplies that underpins recent discussions of globalization (see Chapter 5), which can be viewed as the latest chapter in the ongoing geographical expansion of capitalism.

The process of **colonialism** in the eighteenth and nineteenth centuries entailed a forcible transformation of pre-capitalist societies in Asia, Africa and Latin America, which 'were no longer locally oriented but had now to focus on the production of raw materials and foodstuffs for the "core" economies' (Knox *et al.*, 2003, p.250). The raw cotton for the mills of Lancashire was supplied by slave plantations in the West Indies and, from the 1790s, southern states of the US, ensuring that the 'most modern centre of production thus preserved and extended the most primitive form of exploitation' (Hobsbawm, 1999, p.36).

At the same time, the significance of the colonies as outlets for manufactured goods increased greatly. Cotton exports, for instance, multiplied by ten times between 1750 and 1770 (ibid., pp.35–6), with the vast majority exported to colonial markets, originally in Africa before India and the Far East took over from the middle of the nineteenth century. The early development of the modern cotton industry relied on Britain's effective monopoly of colonial markets in this period, reflecting both its commercial superiority and naval supremacy. In many cases, the export of manufactured goods from the core countries resulted in active deindustrialization in the periphery as traditional local industries were undercut and destroyed by modern factory-based production. Thus, the Indian textile industry was decimated by the much cheaper products of the Lancashire mills. The resultant distress among the native hand-loom weavers prompted Marx's ([1867], 1976, p.555) comment that 'the bones of the cotton-weavers are bleaching the plains of India'.

The geographical expansion of capitalism was facilitated by new transport and communications technologies during the nineteenth century. On the transport side, the development of railways and steamships resulted in a dramatic **'shrinking' of space**, transporting a growing volume of goods and people over long distances. In 1870, for instance, 336.5 million journeys by rail were made in Britain (Leyshon, 1995, p.23). The rapid construction of the English railway network in the 1840s had far-reaching consequences:

> In every respect this was a revolutionary transformation ... it reached into some of the remotest areas of the countryside and the centres of the greatest cities. It transformed the speed of movement – indeed of human life – from one measured in single miles per hour to one measured in scores of miles per hour, and introduced the

Figure 4.1 A freight train passing through Laramie, Wyoming.
Source: D. Mackinnon.

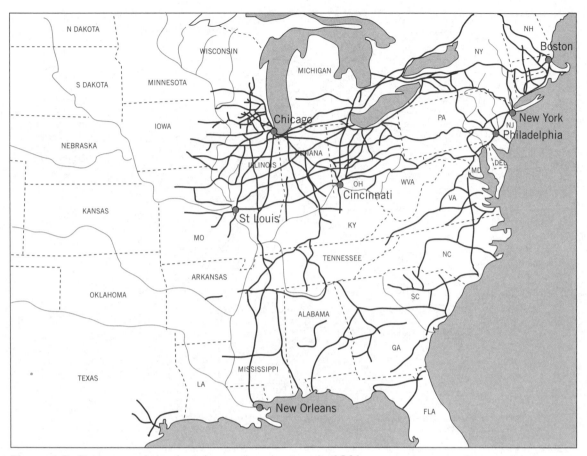

Figure 4.2 Chicago and the American railroad network, 1861.

Source: 'Map: Nature's Metropolis with American Railroads, 1861', from *Nature's Metropolis: Chicago and the Great West* by William Cronon. Used by permission of W.W. Norton & Company, Inc.

notion of a gigantic, nation-wide, complex and exact interlocking routine symbolised by the railway timetable ... it revealed the possibilities of technical progress as nothing else had done.

(Hobsbawm, 1999, p.88)

Railways were subsequently built across much of continental Europe and North America as well as in European colonies overseas (Figure 4.1), opening up these territories to large-scale investment and trade and creating strong demand for coal, iron and steel. In North America, the railways were crucial to the development of the economy during the nineteenth century, linking the Great Plains to the ports of the Great Lakes and the markets of the East Coast and Europe (Leyshon, 1995, p.28). Huge cities such as Chicago grew as transportation hubs and agricultural markets and processing centres, linking the resources of the

American interior to the wider world economy (Figure 4.2) (Cronon, 1991). The steamship also played an important role in both transforming the speed of movement between continents and providing a market for the heavy engineering and shipbuilding industries.

On the communications side, the late nineteenth century saw the invention of the telegram (the 1850s), the telephone (1870s) and the radio (1890s). In terms of facilitating communication over distance and connecting up the realm of everyday life with a geographically dispersed 'out there', these developments were of great significance, paving the way for later developments such as television, the Internet and satellite communications. As the German philosopher Martin Heidegger noted in 1916:

I live in a dull, drab colliery village ... a bus ride from third-rate environments and a considerable

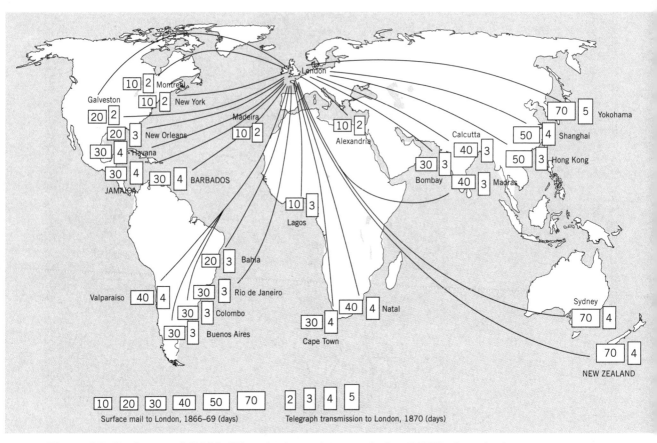

Figure 4.3 Surface mail (1866–69) and telegraph transmission (1870), times in days.
Source: 'Annihilating space? The speed-up of communications', in *A Shrinking World? Global Unevenness and Inequality*, edited by Allen, J. and Hammett, C.
(Leyshon, A. 1995). By permission of Oxford University Press.

journey from any educational, musical or social advantages of a first-class sort. In such an atmosphere life becomes rusty and apathetic. Into this monotony comes a good radio set and my little world is transformed'.

(quoted in Urry, 2004, p.125)

A telegraph cable was laid across the Atlantic Ocean in 1858, enabling rapid communication between North America and Europe for the first time (Leyshon, 1995, p.24). By 1900, the global telegraph system was complete, creating a communications system that dramatically reduced the time and cost of sending information between places (Figure 4.3). The telegraph was particularly important, supporting the growth of

trade and the growing integration of financial markets in particular. For example, foreign exchange could now be easily and rapidly exchanged between markets in London and New York as the telegraph enabled traders to inform one another of the rate at which they would sell pounds for dollars and vice versa (leading to the sterling–dollar exchange rate becoming known as 'cable') (ibid., p.25).

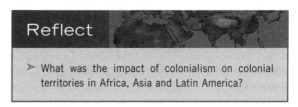

Reflect

➤ What was the impact of colonialism on colonial territories in Africa, Asia and Latin America?

Box 4.2

Myrdal's model of cumulative causation

A useful model of the process of uneven regional development is that of **cumulative causation**, derived from the work of the Swedish economist Gunnar Myrdal (1957). This explains the spatial concentration of industry in terms of a spiral of self-reinforcing advantages that build up in a particular area (Figure 4.4) and the adverse effect this has on other regions, creating a core–periphery pattern. Once an industry starts to grow in a region, for whatever reason (section 1.2.2), it will tend to attract ancillary (supporting) industry made up of firms supplying it with various inputs and services. At the same time, the expansion of employment and population created by the growth of industry creates a large market that draws in further capital and enterprise. The expansion of industry

and the growth of population, moreover, create increased revenues for local government, resulting in the provision of an improved infrastructure for industrial development (Figure 4.4).

The process of cumulative causation in a growing region is linked to the fate of surrounding ones through flows of capital and labour. Myrdal identified two contrasting types of effects. The first he termed **'backwash' effects** when investment and people are sucked out of surrounding regions into the growth region, which offers higher profits and wages. In this situation, the virtuous circle of growth in the latter is matched by a vicious circle of decline in the former, leaving such regions suffering from classic symptoms of underdevelopment such as a lack of capital and

depopulation. The prevalence of 'backwash' effects, then, means that industry becomes concentrated in growth regions, creating an entrenched pattern of core–periphery differentials.

The second set of effects, however, are **'spread' effects** where surrounding regions benefit from increased growth in the core region. One important mechanism here is increased demand in the core region for food, consumer goods and other products, creating opportunities for firms in peripheral regions to supply this growing market, (Knox *et al.*, 2003, p.243). At the same time, rising costs of land, labour and capital in the core region, together with associated problems like congestion, can push investment out into the surrounding regions. Rising costs reflect increased

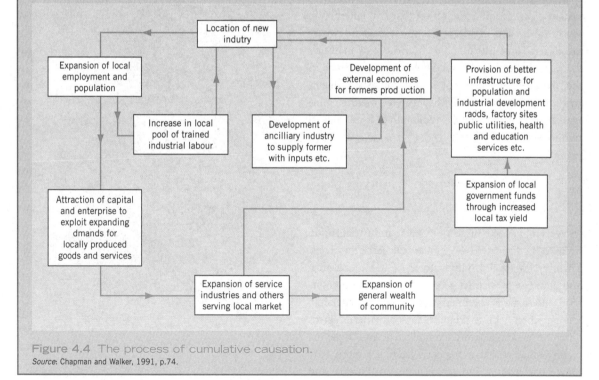

Figure 4.4 The process of cumulative causation.
Source: Chapman and Walker, 1991, p.74.

demand and the tendency for growth to outstrip the capacity of the underlying infrastructure to support it. This problem, which has periodically affected the economies of major core regions like south-east England, is often referred to as **'over-heating'**. As a result, capital flows out into lower-cost regions, followed by labour. This is a process of **spatial dispersal**, where industry moves out of existing centres of production into new regions. It will create a geographically balanced economy if it is the dominant process over a long period of time.

4.3 The rise and fall of industrial regions

Industrialization was a regional phenomenon during the nineteenth century, originating in areas of northern and central England before spreading to parts of continental Europe and North America (Pollard, 1981). It gave rise to a distinct pattern of **regional sectoral specialisation**, involving particular regions becoming specialised in certain sectors of industry. Characteristically, all the main stages of production from resource extraction to final manufacture were carried out within the same region. This pattern of **spatial agglomeration** (Box 4.2) contrasted with the system of proto-industrialization that had emerged in seventeenth- and eighteenth-century England where local merchants distributed or 'put out' raw materials to be manufactured by smallholders and artisans in their cottages and workshops. Such domestic industry was small-scale and widely dispersed across the countryside, often based on the exploitation of local mineral resources and water power.

4.3.1 Nineteenth-century industrialization

As we emphasized in section 3.3.4, industrialization occurred in distinct waves, known as **Kondratiev cycles**. These cycles are not only historical phenomena, they also gave rise to distinct geographies as certain countries and regions assumed technological leadership, leaving others behind. When we look at patterns across different cycles, a more complicated and dynamic picture emerges. Many formerly leading regions (e.g. nineteenth-century industrial areas in Europe and

North America) have been challenged and eclipsed by 'rising' regions such as the US 'sunbelt' or East Asia since the 1970s. At the same time, a select number have been able to maintain their position – particularly

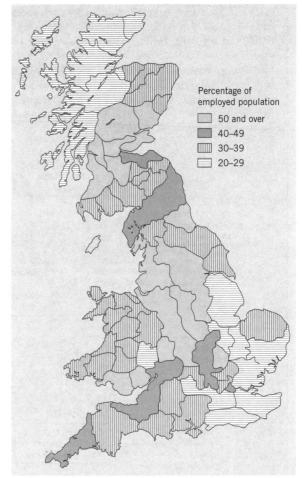

Figure 4.5 UK manufacturing employment, 1851.
Source: Lee, 1986, p.32.

Percentage of employed population
- 50 and over
- 40–49
- 30–39
- 20–29

Figure 4.6 Europe in 1875.
Source: Peaceful Conquest: The Industrialisation of Europe 1760–1870 (Pollard, S. 1981). By permission of Oxford University Press.

metropolitan core regions around cities such as New York and London – while much of the developing world outside East Asia has remained peripheral to the world economy. In broad terms, then, the economic geography of the world can be explained in terms of the interaction between the patterns of investment associated with successive Kondratiev cycles (Massey, 1984).

The first wave of industrialization, based on cotton, iron smelting and coal, took off in certain regions of Britain, principally Lancashire, the West Riding of Yorkshire, the West Midlands, north-east England and west central Scotland (Figure 4.5). Possession of substantial coal reserves gave these regions important advantages as industry became increasingly dependent on coal for energy from the 1820s. The development and improvement of canal systems was crucial in enabling raw materials and finished goods to be transported economically.

The textiles industry – at the heart of the industrial revolution – was concentrated in two main regions:

Lancashire, which specialized in cotton products, and the West Riding of Yorkshire, which focused on woollens. The city of Manchester, for example, experienced explosive growth as the industrial metropolis of the early nineteenth century ('cottonopolis'), with its population multiplying tenfold between 1760 and 1830 (Hobsbawm, 1999, p.34).

These industrial regions of Britain maintained their success during the second Kondratiev cycle from the 1840s to the 1890s, based on the railways, iron and steel and heavy engineering. At the same time, other regions of continental Europe such as southern Belgium, the German Ruhr, parts of northern, eastern and southern France, experienced rapid industrialization (Figure 4.6), as did also the US Northeast. The third wave saw industrialization spread to 'intermediate Europe', parts of Britain, France, Germany and Belgium not directly affected by the first two waves as well as northern Italy, the Netherlands, southern Scandinavia, eastern Austria and Catalonia (Pollard, 1981). The position of the US

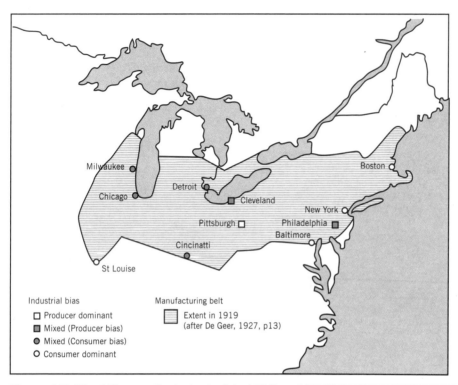

Figure 4.7 The US manufacturing belt in 1919.
Source: Knox *et al.*, 2003, p.156.

manufacturing belt in the Northeast and Midwest was reinforced by new rounds of cumulative causation (Figure 4.7) while Japan also began to experience industrial growth.

Peripheral Europe, on the other hand – encompassing most of Spain and Portugal, northern Scandinavia, Ireland, southern Italy, east central Europe and the Balkans – was left behind, becoming specialized in a subordinate role supplying agricultural products and labour to the core regions (Knox *et al.*, 2003 p.148). The relationship between these industrial cores and the surrounding territories was defined by 'backwash' rather than 'spread' effects as industrialization sucked in capital and labour, creating an uneven economic landscape of urban-industrial cores surrounded by extensive rural peripheries (Box 4.2).

By the early twentieth century, the spatial concentration of industry in a small number of specialized industrial regions in Britain was readily apparent. These regions were built on a coalfield base, supplying the energy for the heavy industries that emerged from the 1840s. In addition to textiles, the iron and steel and ship-

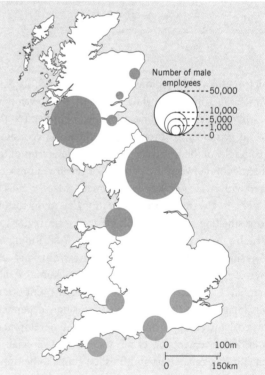

Figure 4.8 Shipbuilding employment in Britain, 1911.
Source: Slaven, 1986, p.133.

building industries became highly concentrated. In shipbuilding, for instance, north-east England and west central Scotland accounted for 94 per cent of employment in Britain in 1911 (Figure 4.8). South Wales was another leading centre of heavy industry by this stage, focused around coal, iron and steel, while the West Midlands became specialized in engineering and metal industries. These industrial regions developed a certain '**structured coherence**' as centres of heavy industry (Harvey, 1982), becoming working-class regions with strong Labour Party and trade union traditions.

4.3.2 Fordism and mass production

Under the fourth cycle based on the mass production of consumer durables, certain established manufacturing regions such as the US manufacturing belt and the English Midlands further consolidated their position. Proximity to the main centres of population became important for these market-oriented Fordist industries. In Britain, for example, the new mass-production industries were drawn to the Midlands and south-east

England. In the late 1920s and 1930s particularly, the Soviet Union experienced very rapid state-led industrialization, although this was very much focused on heavy capital goods like iron and steel and heavy engineering rather than consumer goods. This was focused on a core manufacturing belt stretching south and east from St Petersburg and the eastern Ukraine through the Moscow and Volga regions to the Urals (Knox *et al.*, 2003, p.164) (Figure 4.9). The Japanese economy experienced very rapid growth, averaging around 10 per cent a year, in the 1950s and 1960s, driven by automobiles and electronics as well as more traditional heavy industries like shipbuilding and iron and steel.

The pattern of regional sectoral specialization created by successive waves of industrialization in Europe and North America began to break down from the 1920s. As the economic difficulties of the 1920s crystallized into a major depression in the 1930s, the output of industries such as coal, cotton and shipbuilding slumped. The effects were devastating in terms of unemployment, dereliction and poverty. At the peak of the slump in 1931-2, 34.5 per cent of coalminers, 43.2 per cent of cotton operatives, 43.8 per cent of pig

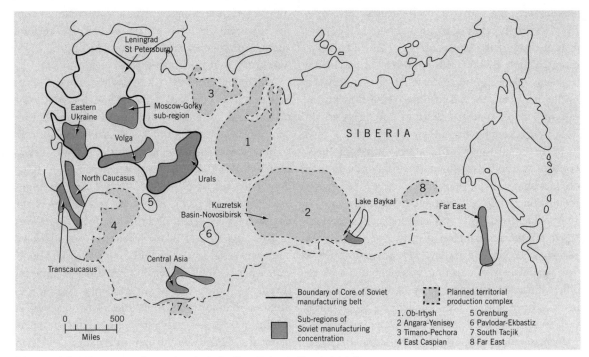

Figure 4.9 The manufacturing belt in the former Soviet Union.
Source: Knox *et al.*, 2003, p.164.

Table 4.2 The spatial division of labour in manufacturing

Functions	Characteristics of the division of labour	Location
Research and development	Conceptualization/mental labour; high level of job control	South-east England
Complex manufacturing and engineering	Mixed; some control over own labour process.	Established manufacturing regions like West Midlands
Assembly	Execution, repetition, manual labour, no job control.	Peripheral regions such as Cornwall or north-east England

Source: Adapted from Massey, D., *Spatial Divisions of Labour: Social Structures and the Geography of Production*, 1994, Macmillan, reproduced with permission of Palgrave Macmillan.

iron workers, 47.9 per cent of steelworkers and 62 per cent of shipbuilders and ship repairers in Britain were out of work (Hobsbawm, 1999, p.187).

Such unemployment was concentrated in the traditional industrial regions of northern England, Scotland and Wales, prompting the official recognition of the 'regional problem' by government in 1928 as central Scotland, north-east England, Lancashire and South Wales were designated as 'depressed regions' requiring special assistance (Hudson, 2003). As such, a growing North–South divide in economic and social conditions in Britain has been apparent since the 1930s, although its roots stretch back further (Box 4.3). Rearmament brought a gradual economic recovery from the late 1930s, reinforced by war and subsequent reconstruction. By the 1960s, however, the underlying problems of nineteenth-century industrial regions in northern Britain, the US manufacturing belt, north-eastern France, southern Belgium and the German Ruhr were becoming increasingly severe.

By the 1960s and 1970s, a new phase of 'neo-Fordism' was apparent – broadly corresponding to the decline phase of this Kondratiev cycle – as mass-production technologies became increasingly routine and standardized. This was creating a new pattern based on the geographical dispersal of industry to peripheral regions. As Massey (1984) demonstrated, a new 'spatial division of labour' was emerging in which different parts of the production process are carried out in different regions, reflecting underlying geographical variations in the cost and qualities of labour.

Increasingly, corporations were separating the higher-level jobs, in areas such as senior management and research and development, from the lower-level

and more routine jobs, such as the processing or final assembly of products (Table 4.2). Through the emergence of large, multi-plant corporations, this division of labour takes on an explicitly spatial form, with companies locating the higher-order functions in cities and regions where there are large pools of highly educated and well-qualified workers, with lower-order functions locating increasingly in those regions and places where costs (especially wage rates) are lowest. As a result, the organization of particular industries becomes spatially 'stretched' through corporate hierarchies with different stages of production carried out in different regions (Massey, 1988). This contrasts with the nineteenth-century pattern of **regional sectoral specialization** where all the main stages of production were concentrated within the same region.

When applied to the UK, Massey's analysis reveals a sharp divide between London and the south-east of England, where a disproportionate amount of company headquarters and R & D facilities are located, compared with the outlying regions of the UK, such as Scotland, Wales and the north of England, which are dominated by 'branch plant' activities. For example, over 70 per cent of the headquarters locations of the largest 100 manufacturing firms are in the South-east (Hayter 1997, p.210) Concentrations of professional and managerial labour could be found in the South-east while peripheral regions contained significant surpluses of lower-cost labour, particularly women, to perform routine work in factories. This dispersal of routine production has also occurred on an international scale through the 'new international division of labour' (Froebel *et al.*, 1980) as MNCs based in Western countries have shifted low-status assembly and processing operations to developing countries where costs are much lower (Chapter 7). Partly as a result, the newly industrialized countries of East and South-east Asia have become important centres of industrial production.

4.3.3 Deindustrialization and 'new industrial spaces'

The geography of the 'fifth Kondratiev cycle' is based on the rise of new 'sunrise' industries such as advanced electronics, computers, financial and business services and biotechnology, employing **flexible production** methods. At the same time, many traditional manufacturing regions have experienced serious deindustrialization since the late 1960s as manufacturing industry has declined in the face of competition, overproduction and reduced demand. The legacy for old industrial regions has been one of high unemployment, poverty, industrial dereliction and decay. The West Midlands of England, for example, part of the industrial heartland of the UK in the 1950s and 1960s, lost over half a million manufacturing jobs between 1971 and 1993, 50 per cent of its total manufacturing employment (Bryson and Henry, 2005, p.358).

The process of '**creative destruction**' associated with the rise of a new technology system based upon information technology has led to some dramatic geographical shifts, in terms of, for example, the US 'rustbelt' (the North-east and the Midwest) and

'sunbelt' (the South and the West) and north-western and southern Germany (Box 4.3). For example, while the mid-Atlantic region of the US (New Jersey, New York, Pennsylvania) experienced a net loss of over 175,000 jobs between 1969 and 1976, the South Atlantic region experienced a net gain of over 2 million jobs in the same period (Knox *et al.*, 2003, p.230). New investment in high-technology industries has been attracted to '**new industrial spaces**' distinct from the old industrial cores, which offered attractive environments and a high quality of life for managerial and professional workers. This is part of a broader spatial division of labour with industries like computers and semi-conductors organized on a global basis. Typically, for instance, R & D functions might be based in Silicon Valley, skilled production carried out in the central belt of Scotland (the so-called 'Silicon Glen'), assembly and testing in Hong Kong and Singapore and routine assembly in low-cost locations in the Philippines, Malaysia and Indonesia (ibid., pp.235–6).

Three different kinds of 'new industrial spaces' have been identified in Europe and North America (Knox *et al.*, 2003, p.237):

➤ Craft-based industrial districts that contained clusters of small and medium-sized firms producing products such as textiles, jewellery, shoes, ceramics, machinery, machine tools and furniture (Box 1.4). Examples include the districts of central and north-eastern Italy, Jura in Switzerland, parts of southern Germany and Jutland in Denmark. This production system was based on high levels of subcontracting

---

Box 4.3

Britain's North–South divide

Concerns about a North–South divide in levels of wealth and prosperity in Britain have been periodically expressed since the 1930s (Massey, 2001). This became a topic of intense political debate in the 1980s and again in the late 1990s, reflecting concerns about the social and spatial impacts of the policies of the Conservative and New Labour govern-

ments, respectively. In the 1980s, the combined impact of the neoliberal reforms of the Thatcher government and wider processes of deindustrialization seemed to have resulted in the emergence of 'two nations': the prosperous and dynamic South in which most of the growth industries were located and the stagnant and impoverished North, scarred by industrial

dereliction, poverty and unemployment (Martin, 1988). The late 1990s saw evidence emerge that the divide had widened under New Labour (Massey, 2001, pp.5–6), creating political problems for a government that drew many of its Ministers and Members of Parliament (MPs) from the North.

As Martin (1988) argues, the roots of the North–South divide in Britain

Box 4.3 (continued)

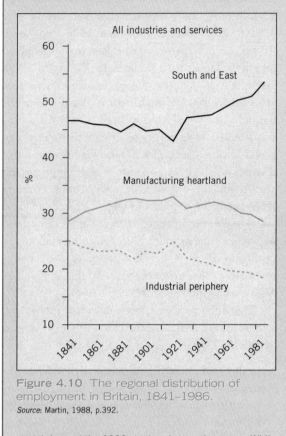

Figure 4.10 The regional distribution of employment in Britain, 1841–1986.
Source: Martin, 1988, p.392.

stretch beyond the 1930s to the nineteenth century. While the image of a dynamic, industrial North and sleepy agricul-

tural South in this period remains powerful, the real situation was more complex. For one thing, the South East was actually the most affluent region throughout the nineteenth century, reflecting the diversified nature of its economy, and the gap with other regions widened from the 1850s. At the same time, the economies of many northern regions remained somewhat precarious, exhibiting a heavy dependence on a narrow range of export-oriented staple industries. This was particularly the case for what Martin calls the 'industrial periphery' of Scotland, the North and Wales (Figure 4.10). These industries collapsed in the difficult economic climate of the 1920s, plunging the North into an economic and social crisis. As we have indicated, moreover, the new consumer industries of the interwar years were drawn to the large market of the South East and the Midlands.

The post-war period saw the economic dominance of the South East consolidated through the expansion of modern, lighter industries and the service sector, compared with the sluggish growth of much of the North. The deindustrialization of the 1970s and 1980s decimated the northern regions whereas the financial and business service and high-tech manufacturing sectors that grew rapidly in the 1980s and 1990s were predominantly located in the south and east of the country. While the recession of the early 1990s was most severe in the South East, leading to a temporary narrowing of the divide, this was soon overtaken by the geography of economic recovery after 1993, with the South East gaining the most new jobs between 1993 and 1997 (Peck and Tickell, 2000, p.157). Since 1997, over a million manufacturing jobs have been lost, many of them in the North (ONS, 2005).

and outsourcing, often relying on family labour and artisan skills.

➤ Centres of high-technology industries such as advanced electronics, computer design and manufacturing, pharmaceuticals and biotechnology. Examples include Silicon Valley in California (see Box 4.4) and Route 128 around Boston; in the US; the M4 corridor and the Cambridge region in the UK; and Grenoble and Sophia-Anatolis in France. Such areas have been characterized by rapid growth and high levels of innovation, often based around networks of small firms, although they have spawned some important MNCs. They are often close to major cities but offer a high quality of life for workers with an integrated local labour market

that enables workers to switch jobs without leaving the area. Links with universities are often significant in terms of providing the research and development infrastructure that supports innovation and learning.

➤ Clusters of advanced financial and producer services, often in central districts of large world cities such as London, New York and Tokyo. The City of London is a good example. The concentration of corporate headquarters functions and major financial institutions found in such metropolitan regions is supported by specialized networks of firms in activities such as accountancy, legal services, management consultancy and advertising. These regions are characterized by high levels of

Box 4.4

The development of 'Silicon Valley'

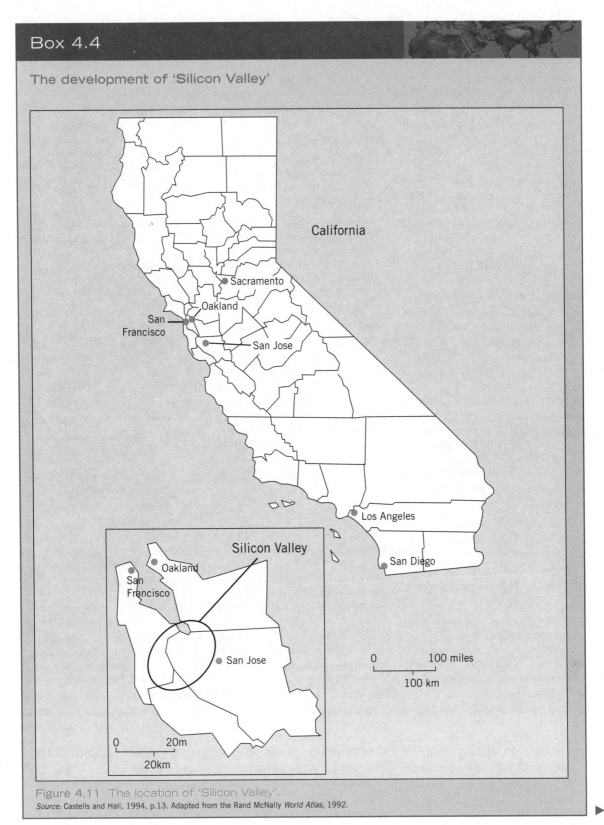

Figure 4.11 The location of 'Silicon Valley'.
Source: Castells and Hall, 1994, p.13. Adapted from the Rand McNally *World Atlas*, 1992.

Box 4.4 (continued)

Over the last half century or so, Silicon Valley in California has become probably the most renowned centre of high-tech industry in the world. As such, it has attracted attention from a range of government and development agencies seeking to emulate its success. While the development of Silicon Valley reflects some specific historical circumstances, its experience does highlight some of the factors that explain why particular types of high-tech industry tend to become concentrated in particular places, including links with universities, the availability of skilled labour and the crucial role of social networks (Saxenian, 1994).

Silicon Valley is a strip of land in Santa Clara County to the south of San Francisco, stretching from Palo Alto to San Jose (Figure 4.11). In the 1940s and 1950s, this was a sparsely populated agricultural area focused on fruit production. Writing in the early 1990s, Castells and Hall (1994, p.15) summarize its subsequent development and transformation in terms of five main stages:

➤ The historical roots of technological innovation, dating back to the early twentieth century.

➤ The growth of high-tech industry in the 1950s around the Stanford Industrial Park.

➤ The growth of innovative electronics companies in the 1960s through spin-offs from the original firms, underpinned by contracts from the US Department of Defense.

➤ The consolidation of semi-conductor firms and the launching of the personal computer era in the 1970s.

➤ The growing dominance of the computer industry, the internationalization of production and a new round of innovative spin-offs in the 1980s and early 1990s.

➤ We can add another stage to this that has been based upon a new generation of Internet technology from the mid-1990s with a phase of rapid growth followed by recession and retrenchment since early 2001.

The links with Stanford University were crucial. An engineering professor, Frederick Terman, was instrumental in the creation in the early 1950s of the Stanford Industrial Park, which attracted a growing number of innovative firms (Saxenian, 1994). The establishment of Fairchild Semiconductors in 1957 was particularly significant as a base for the growth of a number of spin-off firms, including Intel and National Semiconductors. This phase of growth was underpinned by a strong demand from the military for electronics devices.

The invention of the personal computer generated a new round of growth in the 1980s, spawning firms such as Apple and Sun Microsystems as the Valley became increasingly specialized in advanced microelectronics and computers. The Internet facilitated a further round of growth in the 1990s through firms such as Google and Yahoo. Since 2001, the region has been hard hit by the collapse of the Internet economy, losing a fifth of its jobs in three years (*The Economist*, 2004). The key question is whether the region has the capacity to reinvent itself once more to generate a new round of innovation in the face of increasing competition.

In explaining the success of Silicon Valley, Saxenian has drawn attention to its distinctive **social networks**. These have facilitated high levels of informal communication and cooperation between individual engineers and entrepreneurs, often working in different firms, allowing technical information and ideas to be rapidly circulated and shared. High levels of labour mobility represent a key mechanism for the diffusion of technology as people move between firms or leave to establish their own spin-off ventures. The success of the early pioneers provided a ready source of venture capital and created a culture of entrepreneurialism and individualism that has spurred subsequent generations of innovators. The great advantage of this industrial system is that it enabled the region to maintain competitive advantage through continuous innovation, something that, somewhat paradoxically, required cooperation between firms (Saxenian, 1994, p.46). In particular, close geographical proximity fostered face-to-face communication and trust between key individuals, providing the region with the adaptability and responsiveness that has allowed it to maintain its competitive advantage.

specialization and large pools of skilled, white-collar labour, although low-wage female and ethnic minority labour is also important. Geographical proximity is important for specialized service firms that are often putting together packages for firms, requiring face-to-face communication and trust (Thrift, 1994).

4.4 Spaces of consumption

Consumption is also implicated in the production and reproduction of geographic space at a range of scales, from the local to the global (P. Crang, 2005, p.360). It shapes the local spaces in which everyday life takes place, with contemporary urban landscapes, for instance, organized to facilitate consumption through the construction of shopping centres, retail parks and the like, often in out-of-town locations accessible by car. Consumption has been viewed as increasingly important to the fashioning of global space too through the creation of a **global consumer culture** centred upon brands like McDonalds, Coca-Cola and Nike. For many commentators, this is erasing the dis-

tinctiveness of local places and cultures, heralding the 'end of geography' (Ritzer, 2004).

4.5.1 A global consumer culture?

The notion of a homogeneous global consumer culture has become popularized through the media. The central image here is of the erasure and dissolution of distinctive local cultures in the face of global corporations and brands (Figure 4.12) (Slater, 2003, p.157). International tourism is seen as a key agent of this kind of cultural imperialism, subordinating local cultures to the dominance of Western consumer norms (Chapter 12). As an increasing number of studies have shown, however, this cultural homogenization argument is highly simplistic, resting on a number of problematic assumptions (P. Crang, 2005, p.367). Not least among these is that it reintroduces the discredited notion of the passive consumer, with non-Western populations powerless to resist Western consumer norms propagated by powerful corporate interests (Slater, 2003, p.158). At the same time, the authenticity of non-Western cultures is seen as dependent on being pure

Figure 4.12 McDonald's in Beijing.
Source: D. Mackinnon.

and uncontaminated by external forces and influences. In reality, however, cultures are a product of the relationships and connections between places, blending elements from different sources (Massey, 1994). Think, for example, of the importance of drinking tea or eating a curry to contemporary British culture.

A number of recent studies have shown that global consumer cultures assume locally specific forms, blending with pre-existing local cultures in particular ways. Research on how McDonald's is consumed in East Asian countries, for example, found that the chain has been localized and incorporated into local practices, with restaurants functioning also as 'leisure centres, where people can retreat from the stresses of urban life' (Watson, 1997, quoted in P. Crang, 2005, p.368). The consumption of such non-local products can actually be seen as part of the production of identifiable local cultures, with people appropriating such goods for

their own ends, as shown by Miller's research on the consumption of Coca-Cola in Trinidad (Box 4.5).

There is a clear sense in which modern consumer culture actively embraces cultural and geographical difference, with the globalization of food, for instance, presenting consumers in Western countries such as the UK and US with the 'world on a plate' through a choice of ethnic cuisine from different cultural regions (Cook and Crang, 1996). In any large British city, for example, consumers are presented with a wide choice of ethnic restaurants, stretching beyond the popular Indian, Chinese, Mexican and Italian to include specialities such as Lebanese, Thai and Vietnamese, and an array of goods of diverse geographical origins in supermarkets. Rather than eradicating geographical difference, consumption produces new geographies, presenting us with particular representations of the global and the local, the foreign and the domestic, etc. (P. Crang, 2005,

Box 4.5

Coca-Cola: 'a black sweet drink from Trinidad'

Coca-Cola is usually regarded as one of the pre-eminent global brands, central to the creation of a global consumer culture that is actively marginalizing and subordinating more authentic local cultures. As several commentators have observed, however, this view is highly simplistic (P. Crang, 2005). One study that demonstrates this is the economic anthropologist Danny Miller's work on the consumption of Coca-Cola on the Caribbean island of Trinidad. Rather than representing the dominance of Western consumer culture, Miller shows that Coca-Cola is consumed in locally specific ways in Trinidad, having become absorbed into local cultures and traditions. This leads him to term it 'a black sweet drink from Trinidad'.

Instead of being viewed as an imported Western drink, Coke is seen as authentically Trinidadian. It is

regarded as a basic necessity and the common person's drink (Miller, 1998a, pp.177–8). For local consumers, the categories that frame and inform their choice of product are not those used by the producer and advertisers, but the distinctly local concepts of the 'black' sweet drink and the 'red' sweet drink. The latter is a traditional category particularly associated with the Indian population. The former is summarized in the centrality of a 'rum and coke' as the most popular alcoholic drink on the island, although the consumption of 'black' sweet drinks without alcohol is equally common. Coke is probably the most popular of these 'black drinks', becoming associated with the black African population. These links are historical associations rather than describing actual patterns of consumption with possibly a higher proportion of Indians than blacks

drinking Coke, identifying with its modern image while many Africans consume red drinks, associating it with an image of Indianness that is an essential part of their Trinidadian identity.

These local cultural specificities and complexities impose real limits on the marketing strategies of the producers, indicating that consumption is shaped by a wide range of locally specific factors beyond their control. More broadly, the case study shows how the consumption of a prominent global brand is dependent on locally specific cultural practices and traditions. In this way, mass-consumption practices effectively transform global products into locally specific forms, suggesting that capitalism should be viewed as a diverse collection of practices rather than as a set of overarching economic imperatives.

pp.371, 372). As emphasized in section 1.2, however, the geographies of production and distribution that actually generate the goods on sale in Western shops are typically obscured by the emphasis we attach to the physical appearance and price of goods. This process of **commodity fetishism** is inherent to capitalism (section 1.2.1).

While much work on consumption has been cultural in orientation, and sometimes rather celebratory in tone, some political economists have sought to overcome commodity fetishism by reasserting the relations of production. This involves uncovering the working conditions and regimes through which particular commodities are produced and distributed before being sold to consumers (Goss, 2005, p.257). The downside of this is the tendency to reduce consumption to production. As we saw in section 1.2, one framework that helps to overcome the opposite extremes of commodity fetishism and productionism is the **commodity-chain** approach. This traces how commodities link together diverse actors performing various roles such as farmers, labourers, haulage companies, shipping companies, retailers and consumers. The increasing globalization of the economy means that such actors are often located in separate countries and continents, emphasizing how commodity flows create a range of linkages between people, things and places (Mansvelt, 2005).

4.4.2 Places of consumption

Another major focus of attention has been particular **places of consumption**. Key sites include the department store, the mall, the street, the market and the home as well as a host of more inconspicuous sites of consumption (e.g. charity shops, car boot sales) (Mansvelt, 2005). Tourist regions and heritage parks are also sites of consumption, albeit of landscapes and experiences rather than material goods and services. The department store is seen as the classic consumption site of the late nineteenth and early twentieth century, particularly in the major cities of North America and Western Europe such as New York, London and Paris (section 3.4.3). More recently, such stores have been reinvented in the form of the 'flagship store', which incorporates its own labels, building on concepts introduced by chains

like Habitat. Harrods and Harvey Nichols are good examples of such chains, which claim to be selling 'lifestyles rather than simply goods, combining "designer interiors, rituals of display and leisure, sexuality and food"' (Lowe and Wrigley, 1996, p.25).

The **mall** or **shopping centre** has attracted a lot of interest from geographers and other consumption researchers, representing perhaps the most visible and spectacular kind of retail environment. It consists of a range of shops and entertainment facilities within an enclosed space that is usually privately owned and managed (Mansvelt, 2005, p.61). Malls are widely viewed as the iconic space of contemporary retailing, representing the 'urban cathedrals' of contemporary capitalism (Goss, 1993). The world's first fully enclosed mall was opened in Southdale, Minneapolis, in 1956, becoming the prototype for thousands of others over the succeeding decades.

Shopping centres are designed in order to maximize the exposure of consumers to goods, with those of the 1960s and 1970s designed as 'machines for shopping'. More recently, planners and developers have sought to provide spectacular places that people want to spend time in, thus maximizing spend, incorporating entertainment facilities, food courts and visual features as well as shops (P. Crang, 2005, p.373). A number of regional malls, such as the Metro Centre in Gateshead and Meadowhall in Sheffield, were opened in Britain in the 1980s. Further developments occurred through the development of mega malls, such as the West Edmonton Mall and the Mall of America, in North America in the 1980s and 1990s (see Box 4.6). Again, people's use of such spaces is not wholly determined by the intentions of developers and chains. They provide certain groups with a place to socialize or 'hang out' instead of shop, with teenagers, for instance, often coming into conflict with centre management (ibid., pp.374–5).

More recently, the focus of attention has moved away from malls as representing the grand and spectacular towards more mundane and everyday sites of consumption such as the street, the home and car boot sales. Streets provide a great variety of environments for consumption, with particular types of shop often concentrated in particular districts. Inconspicuous consumption spaces such as car boot sales, charity

Box 4.6

The Mall of America

The Mall of America was opened in 1992 in Bloomington, Minnesota. It is probably the most spectacular example of the modern mall in the world today, representing 'the largest fully enclosed retail and family entertainment complex in America' (Mall of America, 1997, quoted in Goss, 1999, p.45). It receives between 35 and 40 million visitors a year, more than the Grand Canyon, Disneyland and Graceland combined. The Mall contains over 520 stores, over 50 restaurants, 14 movie screens, the largest indoor family theme park in the US, a 1.2 million gallon aquarium and a range of other attractions (Mall of America, 2005a). According to the owners, 'first in the industry to mix retail and entertainment, Mall of America has become the model for combining signature attractions with retail to create an outstanding entertainment venue' (Mall of America, 2005b).

The account presented here is based largely on Goss's study (Goss, 1999). In contrast to other studies that focus on how consumers attach meaning and significance to consumption – utilizing methods such as interviews and focus groups – Goss's approach emphasizes the symbolic construction of the retail environment, 'reading' it in terms of the images and themes that the developers and owners seek to convey. It is important to appreciate that Goss does not consider how consumers actually receive and view the Mall since other work on consumption suggests that they are likely to attach meanings to it that may be quite distinct from the intentions of the developers (P. Crang, 2005). While elements of his analysis may be disputed, Goss's research is indicative of how geographers have sought to understand the mall as a key site of consumption.

A key narrative (story) is that of authenticity, stressing the Mall's rootedness in the local environment and culture. An important source of this is the site itself, previously the Metropolitan Stadium: home of local sports teams the Minnesota Twins (baseball) and the Minnesota Vikings (American football). The connections with nearby Southdale are also stressed. As such, the site is a strong symbol of local identity, representing a natural place of congregation and conveying a local 'sense of place' as rooted and authentic. Notions of travel and tourism are incorporated into the fabric of the Mall itself, divided into districts based on imagined tourist-retail destinations (West Market, North Garden, etc.).

A close inspection of certain displays of goods provided some glaring examples of the process of commodity fetishism in terms of how the descriptions of goods served to obscure their actual origins. Most notably perhaps, a stuffed bear in the shop 'Love From Minnesota' had a tag in its right ear saying 'Minnesotans who live deep in the northwoods among the loons, wolves and scented pine trees, listen to the gentle lapping of the waves of the shoreline while they handicraft unique memories of our homeland, like this one, to share with you'; a tag on the other ear said, 'Bear made by the Mary Meyer Corp., Townsend, Vermont ... Made in Indonesia' (Goss, 1999, pp.54–5).

In terms of time, the developers of the Mall sought to mobilize meaning and a sense of magic through the four key themes of nature, primitivism, childhood and heritage. In this way, a strong sense of nostalgia is evoked, building a collective dream of authenticity that becomes attached to the products sold within the Mall. According to Goss (ibid., p.72), 'it [the Mall] must promote the spontaneity of crowds in order to evoke [the] natural commerce of the marketplace'. The obvious point underpinning all of this is that elaborate narratives of authenticity are constructed to sell products, creating an aura of mystery in order to overcome the perceived meaningless and superficiality of modern life.

shops and retro-vintage clothes shops involve the valuing and purchase of second-hand commodities (Crewe and Gregson, 1998). Domestic space has also been re-examined as a site of consumption with research examining how a range of consumer goods are utilized within the home. The role of home-based shopping has also been examined in terms of catalogues, classified adverts and Tupperware, for example.

Food and cooking represent another strand of research, with studies focusing on this as an expression of social relations, particularly those of gender, within the household. The body can also be seen as an important site of consumption, being central to the creation of identity through appearance and image, underpinning the consumption of items such as clothes and cosmetics.

The growth of information and communication technologies (ICTs) has had a profound impact on retailing and consumption, opening up a new sphere of **electronic commerce**. Major retailers were routinely using electronic point-of-sale data to automate and control flows of commodities between stores, warehouse and distribution networks by the late 1980s (Hudson, 2005, p.160). More recently, ICT has facilitated new and expanded forms of home-based consumption through the telephone, Internet and cable networks. There has been significant growth in Internet auction sites, for example, with the best known of these, eBay, established in the USA in 1995, now overtaking Amazon.com, the online bookstore, as the world's most popular shopping site (Mansvelt, 2005, p.3). More generally, the Internet is often used in conjunction with more conventional sites such as shops, providing consumers with the ability to scan websites, identify items of interest and gather information before visiting shops to purchase them. 'Thus the Internet becomes a new search technology rather than an electronic space of sale per se' (Hudson, 2005, p.162).

> ## Reflect
>
> ➤ Do you agree that the notion of a single global consumer culture is crude and simplistic? Justify your answer.

4.5 Summary

In geographical terms, the capitalist system has expanded progressively over time from its roots in Western Europe, encompassing much of the globe by the late twentieth century. While facilitated by successive revolutions in the 'space-shrinking' technologies of transport and communications, this process has been fundamentally driven by a quest for new markets, fresh sources of raw materials and unexploited reservoirs of labour. The geography of economic activity has shifted from the pattern of **regional sectoral specialization** that characterized the nineteenth century, supported by colonial markets and raw material supplies, to the **spatial division of labour** of today. The former pattern

involves all the main stages of production being concentrated within a particular region, while the latter is based on the spatial 'stretching' of production with different parts of the overall process carried out in different regions. The period since the 1960s has seen the dispersal of routine, low-value activities such as the assembly and processing of goods to low-cost locations in peripheral regions of developed countries and the developing world, creating a 'new international division of labour' (Froebel *et al.*, 1980). At the same time, higher-value activities like research and development and financial services have become increasingly concentrated in metropolitan regions and world cities.

Spaces of consumption were assessed at both the global and local scales, highlighting the links between economy and culture (Slater, 2003). The notion of a single global consumer culture defined by major brands such as McDonald's, Coca-Cola and Nike is hugely influential, but rests on a limited and simplistic view of cultures as separate and pure. In reality, even highly symbolic 'global' products are consumed in locally specific ways, as Miller's research on Coca-Cola in Trinidad demonstrates (see Box 4.5). At the same time, different cultures have become increasingly mixed and entangled, with different elements defined as global and local, foreign and domestic (Crang, 2005). Geographers have also focused on local sites of consumption, with research on large malls and shopping centres giving way to work on mundane and everyday spaces such as car boot sales, urban markets and charity shops. The broader point to take from this chapter is that of the continuing importance of place and locality within a globalized economy, focusing attention on the interaction between local conditions and global processes.

Exercises

Concentrate on the city, region or state in which you live. Investigate the restructuring of its economy since the early nineteenth century. Prepare an essay that addresses the following questions.

1. What have been the main industries located there?

2. What markets did these industries serve and where

were raw materials and components supplied from?

3. In what ways has this area been affected by successive waves of industrialization?

4. How have the economic fortunes of the place changed over time.

5. What key institutions have shaped the process of economic development?

6. Can this be understood in relation to changing spatial divisions of labour?

7. Has the area been affected by deindustrialization since the 1970s?

8. Have any new industries or new forms of investment grown in the same period?

9. What are the economic prospects of the area in the early twenty-first century?

Key reading

Bryson, J. and Henry, N. (2005) 'The global production system: from Fordism to post-Fordism', in Daniels *et al.* (2005), pp.313–24.
A useful introductory summary of the changing geographies of production at a global scale, emphasizing particularly the transition from Fordism to post-Fordism.

Crang, P. (2005) 'Consumption and its geographies', in Daniels *et al.* (2005), pp.359–79.
A good introduction to contemporary debates on consumption. Focuses particularly on the notion of a global consumer culture and studies of local places of consumption.

Knox, P. and Agnew, J. (2003) *The Geography of the World Economy*, 4th edn, London: Arnold, pp.143–156.
An informative summary of the changing geography of production in developed countries in terms of successive waves of industrialization and the development and restructuring of particular core regions.

Leyshon, A. (1995) 'Annihilating space?: the speed-up of communications', in Allen, J. and Hamnett, C. (eds) *A Shrinking world?: Global Unevenness and Inequality*, Oxford: Oxford University Press, pp.11–54.
Worth revisiting (Chapter 1 reading) for the historical perspective on processes of time–space compression during the nineteenth century, driven by developments such as railways and the telegraph.

Massey, D. (1988) 'Uneven development: social change and spatial divisions of labour', in Allen, J. and Massey, D. (eds) *Uneven Redevelopment: Cities and Regions in Transition*, London: Hodder & Stoughton, pp.250–76.
An introduction to the concept of spatial divisions of labour, illustrated with reference to the changing geography of the UK.

Useful websites

www.oecd.org
The Organization for Economic Cooperation and Development mainly represents the developed countries. Offers access to a wealth of statistics, documents and surveys on trends and effects of economic restructuring.

http://en.wikipedia.org/wiki/Consumption
A brief definition of consumption and summary of contemporary understandings of the term that links to related concepts.

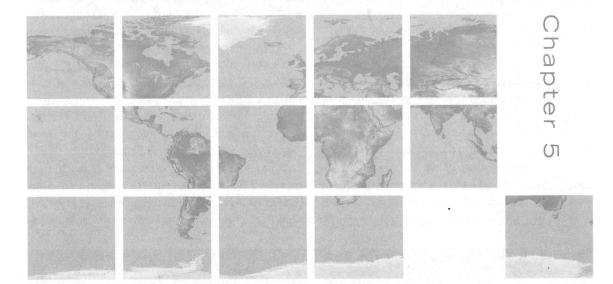

Chapter 5

The uneven geographies of globalization

Topics covered in this chapter

- Different perspectives on globalization.
- The usefulness of globalization as a concept for understanding the changing geography of the world economy.
- The novelty and significance of contemporary globalization compared with earlier periods of economic integration.
- Globalization, uneven development and changing economic relationships between people and places.
- Winners and losers resulting from globalization.

Chapter map

In this chapter, our purpose is to consider the significance of **globalization** in relation to recent changes in

the world economy and the changing geography of economic activity in particular. We start by considering three different perspectives on globalization: the 'hyperglobalists', 'sceptics', and 'transformationalists' (see Held *et al.*, 1999). We then proceed to provide a brief history of globalization, charting its rise with the expansion of the world economy from the mid-sixteenth to the nineteenth centuries. Section 5.4 identifies the key contemporary processes of globalization and Section 5.5 examines the geographically uneven effects of globalization. The following section identifies some of the main 'winners' and 'losers' of globalization reflecting its uneven nature, and the penultimate section briefly charts the emergence of the **anti-globalization movement**.

5.1 Introduction

The concept of globalization has taken centre stage in recent discussions about the world economy. Countless

3

words have been written on the subject by academics, business commentators and journalists. Why all the fuss? In beginning to provide an answer, what unites most accounts is an attempt to get to grips with changes in the geography of the world economy since the late 1960s/early 1970s (section 1.2.1). Five initial points can be made. First, globalization is associated with the dramatic technological changes that have occurred through the advent of the Internet and global communications, which have made connections between people and places more rapid and regular. Second, for many people in the developed economies of the world, globalization symbolizes the increasing competition from low-wage economies in distant places such as India and China, and associated processes of deindustrialization at home and the transfer of jobs abroad. Third, it reflects the emergence of a global financial system that moves billions of dollars around the world electronically in an instant (section 8.4.3). Fourth, it is associated with the attempt to construct new global governance systems and rules of international trade and development. Fifth, and finally, it is associated with what has been termed the anti-globalization movement; a growing coalition of trade unions, environmental groups and other campaigners who oppose the direction that globalization is taking, dramatically symbolized by the protests at the World Trade Organization meeting in Seattle in 1999.

Beyond these general points, there is much argument and debate. At the heart of most debates are disagreements over the nature and extent of globalization. To what extent has the economy shifted from one

Table 5.1 Three perspectives on economic globalization

	Hyperglobalists	Sceptics	Transformationalists
Extent and intensity	Existence of single global economy. Growth of MNCs and global finance supersede national level economic organization.	Globalization is a myth. Serves to mask neoliberal project. Firms still predominantly nationally embedded and trade and investment flows concentrated among advanced states.	Globalization involves changes, increased flows and connections.
Timescale	Globalization has emerged since 1970s as governments follow neoliberal policies.	World economy actually more globalized 1890–1914.	Globalization not entirely new. Globalizing forces evident from 16th century.
Costs/benefits to society	Globalization is good for us, reducing government interference. Free trade and free market will benefit everyone in long term.	'Globalization' is bad for us. Unrestricted free markets and international trade (i.e. 'globalization') result in increasing inequality and poverty.	Globalization has good and bad aspects. Increasing connections, but inequality.
Geographical implications	'End of geography' in the sense that distance and location no longer matter, but geographical clusters at sub-national scales rise and fall according to market forces	Persistence of existing geographical patterns. Core and periphery at both international and national levels.	Produces new geography – new forms of unevenness, exclusion and difference.
Political dimension/ implications	Globalization is inevitable. Governments powerless to intervene except to provide supportive business environment	Markets can be controlled and regulated. Governments have key role. Need for new international institutions and renewed cooperation between nation states	Globalization can be shaped and influenced by governments though new forms of policy and scales of intervention (e.g. EU, WTO, Tobin Tax) may be required

Source: adapted from Held *et al.*, 1999

centred predominantly at the national scale, but with some overseas trading connections, to one that is increasingly integrated at the global scale?

5.2 Three perspectives on globalization

Broadly speaking, and following Held *et al.* (1999), we can recognize three different types of perspective (Table 5.1). The first of these has been termed the 'hyperglobalist' perspective and is associated with enthusiasts for globalization such as the management theorist Kenichi Ohmae (1990, 1995), for whom a fully integrated global economy is becoming a reality. A second perspective, that of the 'sceptics', argues that globalization is a myth, or at best extremely overstated (e.g. D. Gordon, 1988; Hirst and Thompson, 1999; Weiss, 1997), while a third perspective, that of the 'transformationalists', sits somewhere between the other two.

5.2.1 The hyperglobalist perspective

The **hyperglobalist position** contends that nation states are being bypassed by transnational business networks of production, trade and finance, leaving their governments unable to regulate and control market forces. Globalization is traced to the late 1970s when the US and UK governments became the first major powers to embrace free market policies, abandoning the interventionist doctrines that had governed economic policy since the 1940s. As a result, restrictions on international investment and trade were reduced. The subsequent growth of multinational corporations, allied to the increasing mobility of finance, requires states to follow orthodox pro-business policies.

For hyperglobalists, we have reached an 'end of geography' position, whereby economic activity will constantly shift from place to place, locating wherever the infrastructure, costs and regulatory environment are most beneficial for business. Critically, from this perspective, globalization is irreversible. The genie cannot be put back in the bottle and neither should it be. Globalization is viewed as positive since it increases economic well-being by enabling a more efficient allocation of resources through the market and free trade. At the same time, it represents reduced government interference in our lives, leading to greater individual freedom. The role of government should be minimized to overseeing competition, although a more moderate version emphasizes the importance of investing in skills and knowledge-intensive activities, such as research and development, product design and high-tech activities, asserting that it is pointless to try to protect lesser-skilled activities from low-wage competition overseas (Reich, 1991).

5.2.2 The sceptic perspective

The **sceptics stand** in opposition to the hyperglobalists and use various pieces of evidence in their argument that globalization is overstated. Two key claims are the argument that the world economy was more globalized, in terms of trade, between 1890 and 1914 (Kitson and Michie, 2000) and that even the most transnational of corporations are still firmly embedded in particular nation states (see section 7.4) where headquarters are located, taxes are paid and the highest share of assets and operations found (Allen and Thompson, 1997). Most investment and trade still takes place between the advanced industrial states or within macro-regions (e.g. EU, North America, South-east Asia, etc.).

Globalization is seen as a useful political myth to support the arguments of free marketeers (neoliberals) over those who argue for government intervention. It provides an appealing and progressive story of increased economic prosperity through international integration and competition, masking the reality of an increased gulf between rich and poor as cuts in state expenditure, resulting from neoliberal economic polices, make it harder for governments to redistribute income. For sceptics, state intervention is not only still possible, but increasingly necessary to control the excesses of financial markets (section 8.5.3) and regulate multinationals. Any increase in global economic integration, without state intervention or international cooperation, will lead to a widening divide between rich and poor. Rather than reducing the gap between the global North and South, globalization will accentuate uneven development.

5.2.3 The transformationalist perspective

Like the sceptics, the **transformationalists** realize that globalization is a longer-term historical process, with its origins in the colonial expansion of Western European states in the sixteenth century. Where this approach departs from the sceptical view, however, is in its acceptance that the world economy is currently going through some important qualitative changes (section 1.2.1), reshaping relations between governments, corporations and communities (e.g. Castells, 1996; Giddens, 1990). But transformationalists also recognize, counter to the arguments of the hyperglobalists, that the effects of globalization cannot be quite so easily determined. Globalization has good and bad aspects, increasing trade and connections between people and bringing wealth to some places, but also creating new forms of inequality. A **new international division of labour** is emerging in which some formerly undeveloped countries (e.g. China) are experiencing rapid development, but at the same time other places are being left behind (e.g. large parts of Africa and Latin America) or losing their position through de-industrialization (old industrial cities in North America and Western Europe). As this suggests, transformationalists reject the 'end of geography' conclusion drawn by the hyperglobalists. Globalization does not make places all the same, but leads to new forms of difference as global flows of commodities, people and finance interact with distinctive and specialized local patterns of economic activity.

So which of these perspectives offers the most convincing explanation? In providing an answer, we need a more careful assessment of the evolution of the world economy, alongside an appreciation of the geography of uneven development.

Reflect

➤ What are the main points of difference in the three perspectives on globalization?

5.3 Globalization and the development of the world economy

The actual process of economic globalization is rooted historically in the early **geographical expansion** of capitalism, as the search for ever greater profits led capitalists to seek investment opportunities further and further afield (sections 3.4, 3.5). The world economy originated with the rise of capitalism in the mid-sixteenth century in north-west Europe, expanding through the growth of the Dutch and British Empires in particular, which were founded first and foremost on trade, unlike their earlier Spanish and Portuguese counterparts, which developed through plunder and conquest. International economic integration increased hugely in the nineteenth century, following the Industrial Revolution, through the creation of a world market for manufactured goods, initially cotton and textiles.

Globalization found its fullest expression (prior to the present epoch), during the 'age of empire' in 1875–1914 (Hobsbawm, 1987), underpinned by the key transport and communication technologies of railways, steamships and the telegraph. International trade was based on an international division of labour in which industrial countries exported manufactured goods and underdeveloped countries in Africa, Asia and Latin American exported natural resources such as rubber, copper and tea. By 1913 it is possible to talk of an integrated global economy involving all continents in world trade flows, although even at this stage the dominance of Europe and North America is evident (see Table 5.2). Alongside increased economic integration, the nineteenth century also witnessed an unprecedented level of international migration with around 50 million Europeans emigrating between 1850 and 1914 (King, 1995), spurred on by the promise of better economic opportunities in the new colonies. Of these, 70 per cent moved to North America, mostly the United States, with others travelling to Africa and Australasia.

Despite the considerable globalization processes at work, the world economy that emerged in the late nine-

Table 5.2 Exports and imports by geographical region, 1913 (as percentage of world total)		
	Exports %	Imports %
US and Canada	14.8	11.5
UK and Ireland	13.1	15.2
North-west Europe	33.4	36.5
Other Europe	12.4	13.4
Oceania	2.5	2.4
Latin America	8.3	7.0
Africa	3.7	3.6
Asia	11.8	10.4

Source: derived from Knox and Agnew, 1989, p.252

Table 5.3 Shares of world GDP, 1820 and 1913		
	1820 %	1913 %
Western Europe	23.6	33.5
Western offshoots*	1.9	21.7
Japan	3.0	2.6
Asia (excl. Japan)	56.2	21.9
Latin America	2.0	4.5
Eastern Europe/former USSR	8.8	13.1
Africa	4.5	2.7
World	100	100

*Australia, Canada, New Zealand and the US.
Source: Adapted from Maddison, Table 3-1c. Shares of world GDP, 1000–1998, *The World Economy: A Millennial Perspective*, © OECD 2001.

teenth century was still a relatively shallow one economically and geographically. Levels of trade and migration were certainly unprecedented, but, for the vast majority of people and firms, global connections still played a relatively small part in their lives. While every continent was becoming integrated into a world economy, the process remained extremely selective and partial. The extent of multinational production remained limited, with most manufacturing processes still organized on a national basis. With the exception of the UK – which by 1914 had over £6 billion, 45.5 per cent, of the world total for foreign direct investment (Knox and Agnew, 1989, p.256) – few countries had significant overseas investment beyond the development of infrastructure to extract raw materials.

For most African, Asian and Latin American economies, external relations with the European powers were dictated by colonialism and characterized by economic and political dependency. World industrial production – where most of the value was added and therefore wealth concentrated – remained the preserve of a small number of developed economies in Europe and North America. The result was an increasingly uneven distribution of wealth between world regions, with European and North American countries absorbing an increasing share at the expense of the colonies (see Table 5.3). For the colonized territories, external **dependency** was reflected in their internal geographies of development with transportation and settlement patterns being dominated by the exploitation of raw materials (whether foodstuffs such as coffee and sugar or minerals such as iron ore or coal). Development was restricted to coastal areas and port cities with interior communications restricted to links to important centres for resource extraction (i.e. mining or agriculture) (see Figure 5.1).

Within the developed economies of Western Europe, North America and Japan, the period from the 1920s through to the 1970s was characterized by a phase of industrial capitalism associated with Fordist mass-production techniques, the growth of large corporations and the emergence of a spatial division of labour (see Chapters 3 and 4). Although there were a growing number of US and European multinationals operating overseas after 1945, most manufacturing activity remained nationally oriented, with a number of core industrial regions in each country supplying predominantly domestic markets, and it was not until the 1960s that a more global production system and international division of labour emerged (see Chapter Seven).

Reflect

➤ What were the main processes behind the globalization of the world economy in the years up to the First World War?

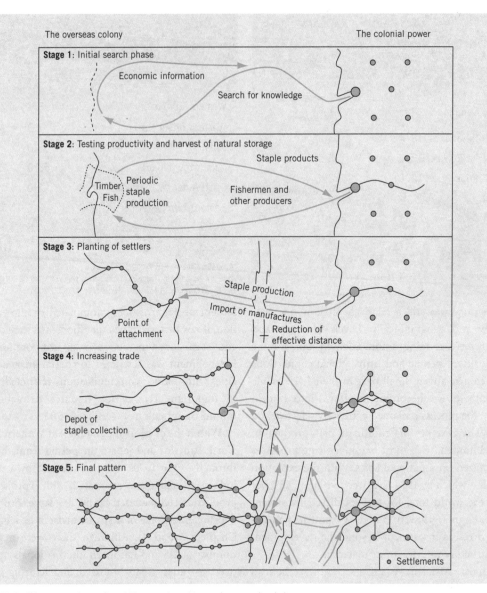

The overseas colony

The colonial power

Stage 1: Initial search phase

Economic information

Search for knowledge

Stage 2: Testing productivity and harvest of natural storage

Staple products

Timber

Fish

Periodic staple production

Fishermen and other producers

Stage 3: Planting of settlers

Staple production

Import of manufactures

Point of attachment

Reduction of effective distance

Stage 4: Increasing trade

Depot of staple collection

Stage 5: Final pattern

○ Settlements

Figure 5.1 Transport and settlement pattern in a colonial economy.
Source: Geographies of Development, 2nd edn, Potter, R.B., Binns, T., Elliott, J.A. and Smith, D., 2004, p.102. Pearson Education Limited.

5.4 Contemporary processes of economic globalization

Although the world economy experienced considerable processes of integration in the past, the contemporary period – since the 1970s in particular – has seen a dramatic increase in the volume and intensity of global flows and connections (Held and McGrew, 2002). Admittedly, the situation is complicated by the fact that some trends date back to the 1940s and 1950s while others are more recent and stem from the 1970s and in some cases the 1980s or 1990s. A summary of the processes at work, and the timescales over which they have operated, can be outlined as follows.

5.4.1 Increasing levels of trade

In the first instance, there has been an exponential increase in levels of trade *across* and between all the world's major regions since 1950. After a decline in world trade from 1914 to 1945 (see above), post-war reconstruction and peace between the major economic and political powers of Western Europe and North America was followed by a dramatic recovery and then an unsurpassed period of growth in world trade from 1950 to the early 1970s. After a period of economic slowdown in the 1970s, world trade again increased dramatically in the 1980s and 1990s, only being halted by the Asian financial crisis of 1998 (see Box 5.3). As a basic indicator of the growing 'interconnectedness' of the world economy, while world production of merchandise goods increased six-fold between 1950 and the mid-1990s, world trade increased by a factor of twenty (see Figure 5.2) (Dicken, 2003a, p.35).

As a consequence of the above, world trade has become more important and more integral to processes of national economic development. After falling back, or stagnating, between 1913 and 1950, exports (of merchandise) have increased significantly as a proportion of Gross Domestic Product (national wealth) in every region of the world, except Africa, between 1950 and 1998 (Table 5.4). Having said this, levels of globalization continue to vary widely by country. For the more developed economies, generally speaking, the smaller the economy, the more it will be oriented towards international trade. Thus, for the world's largest economy, the United States, merchandise exports (traded goods rather than services) represent only around 10 per cent of the value of the country's GDP, while for a smaller country such as the Netherlands, they account for over 60 per cent (Maddison, 2001, p.363). However, while a country the size of the US will have a relatively low international orientation and a high level of self-sufficiency, its overseas impact as the largest economy in the world is considerable, through both its domestic demand for foreign goods and the external impact of its multinational corporations, banks and investment houses.

5.4.2 The rise of foreign direct investment and the multinational corporation

Alongside the increase in trade in commodities, there has been a dramatic rise in **foreign direct investment** (FDI) – overseas investment in production by firms based in other countries, including the establishment or acquisition of plants, factories and offices – at a global scale since 1945. Increases in FDI have been particularly significant in two periods: between 1945 and 1960, and since 1985 when FDI has far outstripped exports to the extent that 'the primary mechanism of integration has shifted from trade to FDI' (Dicken, 2003b, p.52) (see Figure 5.2). The growing importance

Table 5.4 Merchandise exports as percentage of GDP, 1913–1998 (1990 prices)

	1913 %	1950 %	1973 %	1998 %
Western Europe	14.1	8.7	18.7	35.8
Western offshoots	4.7	3.8	6.3	12.7
Eastern Europe/former Soviet Union	2.5	2.1	6.2	13.2
Latin America	9.0	6.0	4.7	9.7
Asia	3.4	4.2	9.6	12.6
Africa	20.0	15.1	18.4	14.8
World	7.9	5.5	10.5	17.2

Source: Adapted from Maddison, Table 3-2b. Merchandise Exports as Per Cent of GDP in 1990 Prices, World and Major Regions, 1870–1998, *The World Economy: A Millennial Perspective*, © OECD 2001.

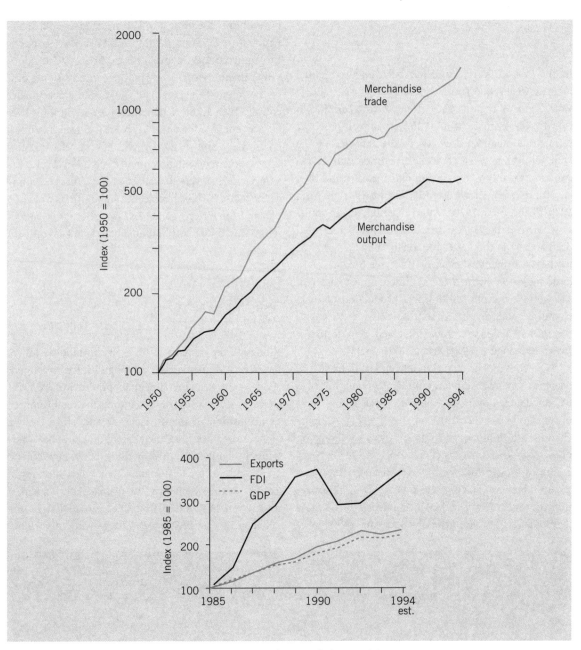

Figure 5.2 a and b The growing interconnectedness of the world economy.
Source: Dicken, 2003a, pp.35, 53.

of FDI reflects the increased significance of the **multinational corporation** as a key actor in the global economy with 'the power to coordinate and control operations in a large number of countries (even if it does not own them), but whose geographically dispersed operations are functionally integrated, and not merely a diverse portfolio of activities' (ibid., p.30) (see Chapter 7). By 1997 it is estimated that there were 53,000 MNCs operating worldwide with 450,000 foreign subsidiaries (http://www.polity.co.uk/global/summary.htm, last accessed 10 July 2006).

5.4.3 Global consumption patterns

Linked to the emergence of global systems of production has been a growth in global consumption patterns (section 4.3). The importance of global consumption is reflected in the declining consumption of local products and the growing significance of imported commodities. During the 1980s and 1990s this process accelerated as many growing economies in the global South, particularly in Asia and Latin America, experienced an unprecedented increase in imports. The UN's Human Development Report of 1998 estimated that at the global scale merchandise imports grew from $2 trillion in 1980 to over $5 trillion by 1995. The share of manufactured goods in total imports rose in almost every country although there were wide variations. For example, between 1980 and 1995 the share of manufacturing goods in total imports increased from 19 per cent to 54 per cent in Japan, from 40 per cent to 71 per cent in Brazil, from 51 per cent to 81 per cent in Thailand and from 50 per cent to 79 per cent in the US (UNDP, 1998, p.46). More generally, such developments indicate that everyday consumption patterns and habits involve increasingly dense and often complex sets of global connections and linkages, as we witnessed in Chapter 1 with the example of the banana (Box 1.2).

5.4.4 Global financial flows

There has also been a massive increase in the volume of money circulating across the globe in financial markets (see Lee, 2003) and in the range of financial products that are traded (see section 8.4.3). For example, daily turnover on the major foreign exchange markets grew from over $10–20 billion in 1973, about twice the value of trade, to an average of $1.5 trillion in 1998, about 70 times the value of trade (Figure 5.3) (Pollard, 2005, p.347). In this way, it is possible to talk of a virtual economy, or what geographer Nigel Thrift (2002) refers to as a 'phantom state' operating alongside the more 'real' economy in which products are manufactured in specific locations. This **hypermobility of capital** – made possible in part by advances in electronic technology and the ability to shift billions of dollars from one part of the globe to another at the touch of a button – is seen as having its own disciplining effect on

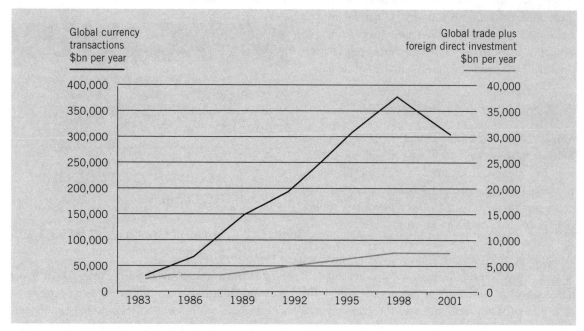

Figure 5.3 The growth in global currency transactions relative to global trade plus foreign investment.
Source: Lee, 2003, p.29.

Box 5.1

International economic organizations

These are the key organizations created at the end of the Second World War as part of the Bretton Woods system, alongside the regime of fixed exchange rates. They play an important role in managing the world economy. Since the early 1980s, the policies of these organizations have been shaped by neoliberalism (Box 6.5).

➤ The International Monetary Fund (IMF). The role of the IMF is to promote monetary cooperation between countries, and to support economic stability and trade. The provision of financial assistance to

countries experiencing budgetary problems allows the IMF to set conditions requiring countries to reform their economies. See http://www.imf.org/.

➤ The World Bank (officially the International Bank for Reconstruction and Development). Its role is to provide development assistance to countries, mainly in the less-developed world. The Bank runs a range of programmes and initiatives aimed at reducing poverty and narrowing the gap between rich and poor countries. See http://www.worldbank.org/.

➤ The World Trade Organization (WTO), established in 1995, taking over from GATT (the General Agreement on Trade and Tariffs). The role of the WTO is to ensure a free and open trading system, working through successive 'rounds' or conferences where member countries come together to negotiate agreements. See http://www.wto.org/.

(See sections 6.5.2, 11.3.3, 11.6 particularly for further discussion of the role of these organizations.)

national economies and states. To attract and keep mobile and footloose investment, governments are likely to pursue policies that are pro-business and fiscally conservative (e.g. low taxes, low inflation). In this context, Lee talks of the new power of financial markets and those who make decisions within them, leading to a situation in which everywhere becomes in effect 'marginalized' by these global flows or circuits of capital.

5.4.5 Neoliberalism

A final feature of the period since the 1970s has been a changing economic policy context, with the emergence of a dominant regime known as **neoliberalism** and **global governance institutions** (Box 5.1) involved in attempting to regulate the economy along neoliberal lines. Neoliberalism in simple terms is the commitment of governments and policymakers to the principle of 'free markets' and involves the advocacy of trade **liberalization**, financial market **deregulation**, the elimination of barriers to foreign investment, the reduction of state involvement in the economy and the protection and encouragement of private property rights. Since the 1970s, it has replaced **Keynesianism** – based on a commitment to full

employment and state intervention to stimulate growth during periods of recession – as the dominant mode of thought governing global economic policy-making (section 6.5).

Reflect

➤ How does contemporary globalization differ from earlier periods of global economic integration?

➤ To what extent has globalization been produced by national governments?

5.5 Patterns of global inequality

5.5.1 The regionalization of world trade

While there is considerable evidence of increased global economic integration, it is important to recognize that globalization continues to be a highly uneven phenomenon in geographical terms. In the first instance, while there has been an increase in world trade, a closer look at the figures raises questions about the 'globalness' of

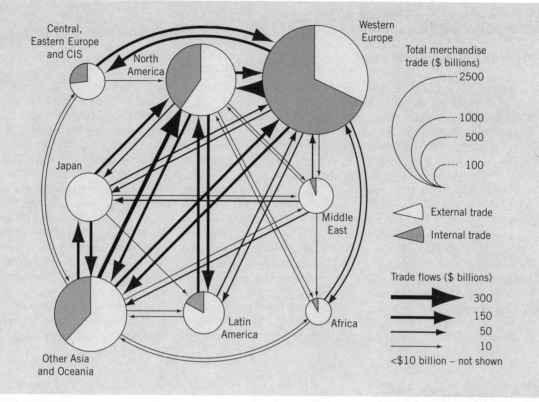

Figure 5.4 The world trade network.
Source: Dicken, 2003a, p.41.

this process. In fact, world trade has, if anything, become highly '**regionalized**' in recent years (see Box 5.2) in the sense that for the majority of countries the most important trading partners are neighbouring states. For example, Dicken (2003a, p.41) notes that for the world's most significant trading region, Western Europe, over two-thirds of trade takes place internally between European states. Similarly, for Asia, as the second largest trading region, half of its trade is internal while for North America, as the other major trading region, the figure is 40 per cent. The connections between these three world regions are also stronger than connections to other parts of the world (see Figure 5.4). Trading connections involving Africa, the Middle East and the former communist countries of Eastern Europe and the Soviet Union are far fewer in comparison.

5.5.2 Concentration of power

By the same token, while world financial markets may be global in their scope and hypermobile in their effects, they are underpinned by the decisions of a small number of market traders and analysts, operating out of a select group of cities such as London, New York and Tokyo. As Thrift (2002, p.39) puts it: 'what we find is a system where people are often in interaction with only four or five other people at a time – on a trading floor or a bank branch'. The same also applies to key 'global' institutions that in reality operate out of particular local contexts and revolve around networks of a small elite cadre. For example, the International Monetary Fund and the World Bank, the key global institutions in funding international development, are both physically located in the US in Washington DC. More importantly, the elite decision-makers of these

Box 5.2

The rise of the world's macro regions

The increase in trade within 'macro-regions' – large world regions, such as North America or Western Europe, incorporating a number of nation states – is associated with the emergence of new supranational bodies designed to enhance trade and economic growth. Every world region now has some form of organization dedicated to the task of regional economic integration (see Figure 5.5), but the most advanced is the European Union where an element of political integration has accompanied economic integration. The other most significant development has been the North American Free Trade Agreement (NAFTA), creating a single, tariff-free market between Canada, the US and Mexico. This is an event that has been accompanied by considerable controversy across the three countries as the removal of barriers to trade brings the threat of increased competition to a diverse range of groups from peasant farmers in Mexico to trade unionists from the US and Canada. There is now therefore considerable opposition to NAFTA as part of a growing anti-globalization movement (see Box 5.4).

The EU – or European Economic Community (EEC) as it was then called – was formed by six nation states through the Treaty of Rome in 1957. While integration was driven by the need to avoid future conflict, in the light of the two world wars, concerns about Europe's competitiveness in relation to the US and Japan were also important. Competitiveness would be enhanced through economic integration based on the removal of barriers to trade between members, including tariffs and various regulations and restrictions. The EEC developed as a customs union, with internal barriers to trade eliminated and a common external tariff against imports from other countries established by the late 1960s. The latter was what distinguished the EEC from a simple free-trade area. Common policies on

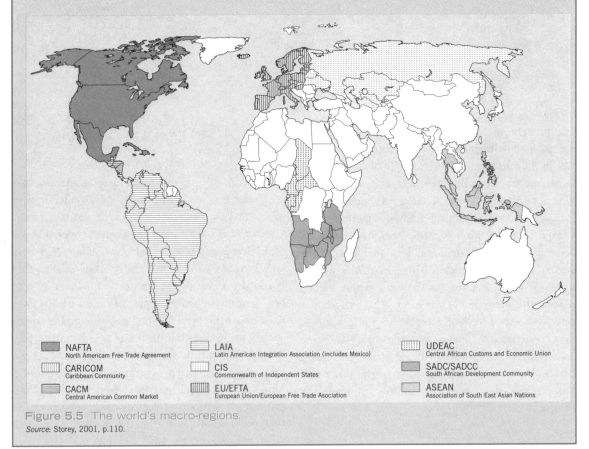

Figure 5.5 The world's macro-regions.
Source: Storey, 2001, p.110.

NAFTA
North American Free Trade Agreement

CARICOM
Caribbean Community

CACM
Central American Common Market

LAIA
Latin American Integration Association (includes Mexico)

CIS
Commonwealth of Independent States

EU/EFTA
European Union/European Free Trade Association

UDEAC
Central African Customs and Economic Union

SADC/SADCC
South African Development Community

ASEAN
Association of South East Asian Nations

Box 5.2 (continued)

agriculture and fisheries were also introduced in the 1960s. Successive enlargements in 1973, 1981, 1986, 1994 and 2004 have expanded the EU to 25 members, incorporating former communist countries in Central and Eastern Europe.

Economic integration gained momentum again in the 1980s when EU states agreed a programme to establish a Single European Market (SEM) based on the free movement of goods, services, labour and capital by January 1993. Again, this was presented as a response to Europe's perceived lack of economic competitiveness compared with the US and Japan (Williams, 1995), gaining the support of Britain's largely Eurosceptic Thatcher government for this reason. The next major step in integration was the Treaty on European Union, agreed in Maastricht in December 1991, when member states agreed to create a single currency to replace national currencies.

After a period of conversion, when participating members were to meet strict 'convergence criteria' such as low inflation and the reduction of budget deficits, the euro was introduced in January 2002. However, it remains controversial and only 12 of the 15 EU countries (of the time) signed up to it, with the UK, Denmark and Sweden retaining their own national currencies.

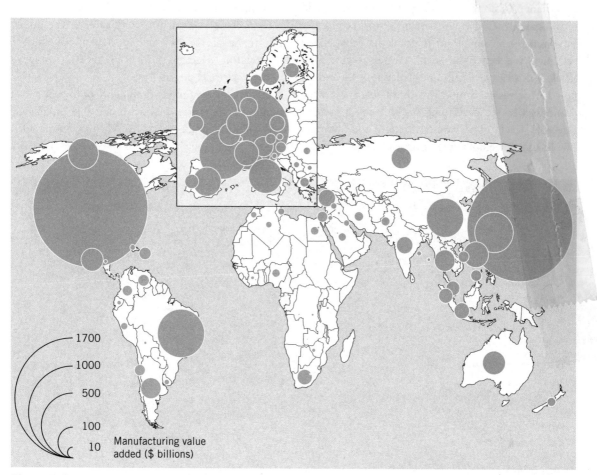

Figure 5.6 World map of manufacturing production.
Source: Dicken, 2003a, p.36.

agencies are dominated by US and European economists and policy-makers who tend to regard free market policies – developed in the US and the UK in the late 1970s and early 1980s – as universal solutions to economic problems (Stiglitz, 2002). The implication is that the vast majority of people – particularly in the global South – are geographically remote from the key decisions that affect their lives and are heavily dependent upon a few distant 'others'.

5.5.3 Uneven development at the sectoral level

There is a highly uneven geography of activity in both the manufacturing and services sectors. A glance at the world map of manufacturing production shows that although some developing countries have become important producers, particularly in Latin America and Asia (especially China), the developed world continues to command an important share of the value added activities, and hence the wealth created (Figure 5.6). While industries and firms in the developed world do face an increasing competitive threat from low-wage economies in the developing world, leading to considerable levels of deindustrialization and job loss (Table 8.1), this should not be overstated. Rates of dein-

dustrialization vary widely, being most severe in the UK and the US, although both retain a significant world market share, and less so in countries such as Germany. Indeed, if we examine the breakdown of the value added in manufacturing at the country level (Table 5.5), the dominance of the developed world becomes even more evident.

The geography of services shows a slightly different picture with a few countries once again dominant at a global scale, but with the US and the UK continuing to be the leading export countries (Figure 5.7). While much service activity, by its very nature, has to be localized due to the need for face-to-face contact with customers, there is a growing amount of export activity in areas such as management consultancy, accounting, legal services, banking and finance, telecommunications and information technology. In this sense, there is a growing amount of trade in the transfer of knowledge, an issue we return to in Chapter 9. The tourist industry has also become an important global industry, in attracting overseas visitors to boost a country's economy (see Chapter 12).

Despite the continuing dominance of the developed economies, there has been a considerable shift in the geography of manufacturing over the past few decades. The world economy in 1960 was still characterized by a

Table 5.5 World rankings for manufacturing value added, 2000

	Ranking	Share of world total %	MVA per capita $
United States	1	24.1	5,306
Japan	2	14.0	6,865
Germany	3	8.5	6,414
China	4	7.0	350
Italy	5	4.6	4,951
France	6	4.5	4,732
UK	7	3.5	3,696
South Korea	8	2.6	3,434
Spain	9	2.0	3,194
Canada	10	2.0	4,040
Top 10 share		72.8	

Source: UNIDO, 2004, pp.183, 195–7

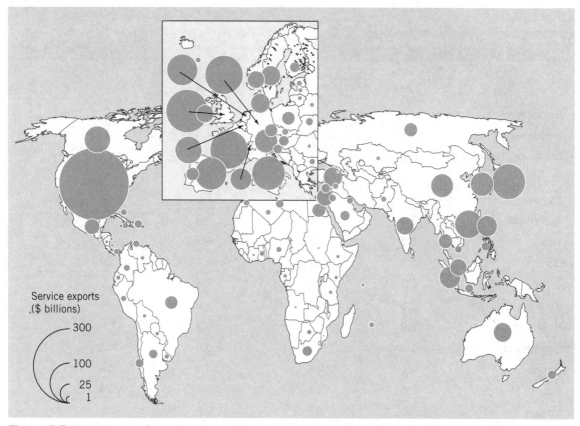

Figure 5.7 World map of services exports.
Source: Dicken, 2003a, p. 43.

relatively simple geographical division of labour with almost all the manufacturing being carried out by the core economies of Europe and North America, exploiting the natural resources and raw materials of the rest of the world. In this sense, with the exception of Japan and the 'Old Commonwealth' countries of Australia and New Zealand, most of the rest of the world found itself in dependent relations as part of the periphery. This situation has changed over the last fifty years with developing countries increasing their share of world manufacturing from 5 per cent in 1953 to 23 per cent in the late 1990s (Dicken, 2003a, pp.37–8). If we consider the world rankings for manufacturing exports, there has been a marked change from the situation in the 1960s, when export trade was dominated by Europe and North America, to the situation today where newly industrializing countries have captured a growing share of world markets (see Table 5.6).

In capturing the significance of this development, many scholars in the 1980s talked about a '**New International Division of Labour**' emerging (e.g. Froebel *et al.*, 1980; Lipietz, 1986) (section 7.3.2) in which multinational corporations from the 'First World' were increasingly locating production facilities in less-developed countries. While the more technical and skilled aspects of production remained in the developed economies, the less-skilled activities could be relocated to Third World locations to take advantage of lower wage costs and social regulations. This applied particularly to those industries that could be broken up most easily into their constituent parts, allowing the more routine tasks to be dispersed to low-wage countries. Clothing and electronics were two cases in point (see Dicken, 2003a).

Geographically, this shift in production activities has been a highly uneven process, dominated in particular

Table 5.6 World rankings for manufacturing exports, 1963–2004, as percentage of world totals

	2004		1963	
	%	Rank	%	Rank
Germany	10.0	1	15.6	2
United States	8.9	2	17.4	1
China (exc. Hong Kong)	6.5	3	N/A	N/A
Japan	6.2	4	6.1	5
France	4.9	5	7.0	4
Netherlands	3.9	6	3.3	9
Italy	3.8	7	4.7	6
UK	3.8	8	11.4	3
Canada	3.5	9	2.6	12
Belgium	3.3	10	4.3	7
Hong Kong	2.9	11	0.9	15
South Korea	2.8	12	0.0	N/A
Mexico	2.1	13	N/A	N/A
Russian Federation	2.0	14	N/A	N/A
Taiwan	2.0	15	0.2	N/A

N/A = not available
Source: derived from Dicken, 2003a, p.40, and WTO, 2005, Table I.5.

by East Asia, with a few Latin American countries, notably Brazil and Mexico, also developing a presence in world export markets. Recent Mexican growth has been almost entirely due to the increase in trade with the US through NAFTA and the transfer of jobs by US corporations across the border to lower-wage locations. However, the most remarkable development in the last 30 years has been the emergence of East Asia as a core region of the capitalist economy. The rise of the region was associated, initially, with Japan from the 1950s onwards. Japan was followed by the four 'Tiger' economies of Hong Kong, South Korea, Singapore and Taiwan (see Box 5.3), which grew dramatically from the 1960s but first received serious attention during the 1980s and, more recently, the emergence of China and, on a smaller scale, Malaysia, Thailand and Indonesia. The share of world value added in manufacturing accounted for by East Asia has risen from 4 per cent in 1980 to around 14 per cent in 2000 (UNIDO, 2004, p.84). China has, however, grown fastest in the shortest time period, increasing its share from 1.5 per cent to 7.1 per cent between 1980 and 2000, making it the fourth largest manufacturer in the world (on the basis of value added, although it is now the largest in terms of numbers employed).

Reflect

➤ In what ways can contemporary globalization be described as an uneven process?
➤ Is further global economic integration inevitable?

Box 5.3

The rise of the newly industrialized countries of Asia

The **newly industrialized countries (NICs)** of East Asia (Hong Kong, Singapore, South Korea and Taiwan) have moved from the periphery to the core of the world economy over the past forty years. Although China, as the world's most populated country, has long had the potential to play a much greater role on the world economic stage, the emergence of the four comparatively small NICs – the so-called Asian 'tiger' economies – is equally, if not more, impressive. As a result of rapid rates of economic growth across the East Asian region over the past three decades (see Table 5.7), the NICs have moved from undeveloped low-income economies to being among the more advanced industrialized economies. Hong Kong and Singapore in particular have levels of GDP per capita on a par with some of the richest European countries.

There has been considerable debate about the factors behind the success of the 'East Asian miracle' (Henderson, 1993). Neoclassical economists have explained it as an example of the positive impact of market forces, with the countries being characterized by limited government intervention and high exposure to international competition and trade, which allowed them to develop efficient industries and low wage rates to compete in global markets rather than being protected by tariff barriers and government subsidy. However, more detailed empirical investigation (e.g. Amsden, 1989; Wade 1990) suggests that this is only part of the story and that forms of state involvement have been key both to initial industrialization and to successful export strategies (section 6.4).

Another factor, often neglected by mainstream economists, was the impact of the Cold War. During the 1950s and 1960s, at a time when communism was perceived as a threat to the capitalist world economy, the NICs were seen as a bulwark against the neighbouring communist states of China, North Korea and North Vietnam. South Korea in particular received considerable political and economic assistance from the US, while all four states were given carte blanche to develop in their chosen fashion, which usually involved the suppression of labour unions and democratic forces. High growth, rising living standards and relatively low levels of inequality – by the standards of many other capitalist and developing world economies – were used to retain public support.

The region suffered an economic crisis in 1998 when the NICs, along with other economies in the region, experienced a dramatic outflow of foreign investment. The reasons behind the crisis were complex but many observers believe the relaxation of government controls on financial activities, particularly the amount of money flowing in from overseas, allowing more speculative investment in property and the stock market, may have played a part. Notably, the crisis was felt unevenly across the region, affecting South Korea heavily but having less impact in Singapore and Taiwan. Worst hit were the new 'tigers' of Thailand, Malaysia and Indonesia where companies' overexposure in financial and property markets led to a banking collapse. South Korea's problems have been blamed on the overborrowing by its major industrial conglomerates, the 'Chaebol', in order to compete internationally (Henderson *et al.*, 1998).

Table 5.7 Growth rates by categories of state in the developing world and former communist world, 1975–2001

	GDP per capita annual growth rates	
	1975–2001	1990–2001
Arab States	0.3	0.7
East Asia and the Pacific	5.9	5.5
Latin America and the Caribbean	0.7	1.5
South Asia	2.4	3.2
Sub-Saharan Africa	−0.9	−0.1
Central and Eastern Europe and CIS	N/A	−1.6

N/A = not available
Source: UNDP, 2003, p.352

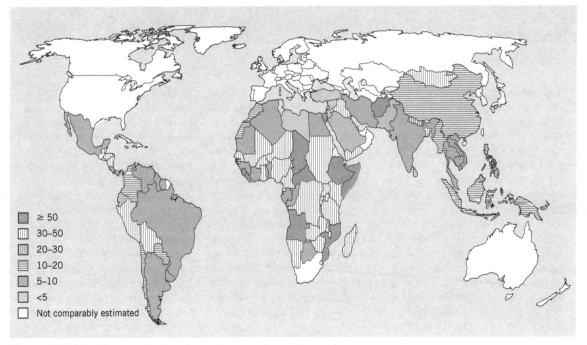

Figure 5.8 World map of chronic undernourishment.
Source: Atkins and Bowler, 2001, p.131.

5.6 Winners and losers in the global economy

While globalization has brought benefits to some countries, large parts of the world have experienced either economic stagnation or decline. The period since the late 1970s has seen a widening of the gap between the world's richest and poorest countries. While the ratio of per capita income between the richest and poorest world regions fell from 15:1 to 13:1 between 1950 and 1973, it increased to 19:1 by the end of the twentieth century (Maddison, 2001, p.125). At the level of individual countries, there are even starker contrasts in income. The income gap between the world's richest and poorest countries increased from 44:1 to 72:1 between 1973 and 1992. Prior to industrialization in 1820, the ratio has been estimated at 3:1 (Dicken, 2003a, p.513).

According to sociologist Manuel Castells (2000), in the period between the 1960s and 1990s the poorest 20 per cent of the world's population saw their share of global income decline from 2.3 per cent to 1.4 per cent of the global total, while the richest 20 per cent saw

their share rise from 70 to 85 per cent. Another startling figure revealed by Castells, vividly capturing the scale of global inequalities, is that the assets of the world's 358 billionaires ($US) add up to the combined incomes of the poorest 45 per cent of the world's population. The sad reality is that one in five people in the world are still living on less than $1 per day, with around 70 per cent of these people concentrated in sub-Saharan Africa or South Asia (Dicken, 2003a, p.513).

5.6.1 The effects of inequality

The effects of such glaring inequalities are brought home by trends in consumption. In 1998 the UN revealed that 20 per cent of the world's people in the highest-income countries account for 86 per cent of total private consumption expenditure while the poorest 20 per cent accounted for only 1.3 per cent (UNDP, 1998, p.2). In terms of individual products, the top 20 per cent consumed 45 per cent of all meat and fish while the poorest 20 per cent consumed 5 per cent; the comparative figures for energy use were 58 per cent

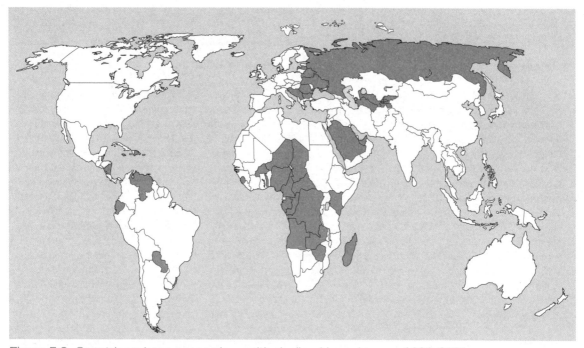

Figure 5.9 Countries whose economic wealth declined in real terms, 1990–2001.
Source: UNDP, 2003, pp.278–81.

and 4 per cent. The inability of many people in the less-developed world to meet their basic needs are revealed by figures on malnutrition: in sub-Saharan Africa, for example, the number of undernourished people – defined as an 'inadequate intake of calories' (Atkins and Bowler 2001, p.325) more than doubled between 1970 and 1990 from 103 million to over 215 million. Once again, there are wide discrepancies in the experience of the global South where undernourishment and starvation are rife compared with the North (Figure 5.8)

Underpinning these trends in global inequality have been widely varying rates of economic growth both between developing countries and between developed and less-developed economies in the period since the late 1970s (see Table 5.7). In particular, large parts of Latin America and sub-Saharan Africa have seen their economies either stagnate or decline, while the countries of Eastern Europe and the former Soviet Union have also been adversely affected by the transition to a capitalist economy during the 1990s (see Figure 5.9).

5.6.2 The effects of financial hypermobility

Additionally, the increasing degree of financial mobility in the world economy and the deregulation of economies by national governments pursuing neoliberal policies seems to be generating greater instability in the world economy. One consequence is that international investors are now able to shift funds very rapidly from one country to another in search of higher returns. This has had devastating effects in particular places. The opening up of an economy to greater foreign trade can lead to a gradual influx of speculative investment that, in the short term, can bring prosperity. But, when international investors lose confidence in an economy, there is a rapid outflow of such 'hot money', leading to financial crisis, collapse of the banking system and knock-on effects for the rest of the economy (Box 5.4).

A case in point is Mexico, which, as a major oil exporter, suffered an economic crisis in 1981 due to the collapse in oil prices. In the second half of the 1980s, and under pressure from the International Monetary

Box 5.4

Financial crisis and the collapse of the Argentine economy

In 2002, the collapse of the Argentine economy was referred to by the *New York Times* as 'the largest financial collapse of any country in history' (Lopez-Levy, 2004, p.46). In what had formerly been one of Latin America's more prosperous and developed economies, GDP fell by 16.3 per cent and manufacturing output by 20 per cent, while unemployment climbed to 20 per cent (Rock, 2002, pp.55–6). The economic collapse was accompanied by social and political upheaval as all classes of people took to the streets in protest. A major source of unrest was the attempt by the government to freeze bank accounts to prevent a run on the banking system in order to continue paying back its massive debts while allowing the mass exodus of international financial capital, with foreign banks transferring up to US$130 billion out of Argentina at the same time that small local investors were unable to access their own savings (Lopez-Levy, 2004). The result was a massive demonstration in Buenos Aires, a violent response from the police and the death of 36 people, leading to the resignation of President de La Rua, followed by three other presidents, before some semblance of stability was restored.

For the world's economic and political elite, the collapse was particularly shocking given the country's status as a model of 'neoliberal' economic reform in the 1990s.

From the mid-1970s up to 2002 successive governments had steadfastly followed the advice of the International Monetary Fund (IMF) in privatizing its industries, deregulating and opening up the economy to foreign investment. This set in train an increasing degree of international involvement in the economy so that by the time of the current crisis 50 per cent of all financial deposits in the country were held by foreign investors; foreign control of the banking sector increased from 14 per cent in 1989 to 75 per cent by 2000 (Pettifor et al., 2001), and foreign ownership of industrial production increased from 46 per cent in 1963 to 75 per cent by 2002 (Lopez-Levy, 2004). After a brief boom in the early 1990s as investment flowed in, conditions worsened in the latter half of the decade. A rising currency (pegged to the US dollar under the IMF's policy advice) made the country's exports expensive at a time when the Asian financial crisis of 1997 triggered a world economic downturn. Coupled with a government austerity programme to reduce public spending, thereby cutting back on public services, a loss of competitiveness among domestic firms sent unemployment spiralling and precipitated the current crisis.

The Argentinian crisis clearly illustrates the type of global–local connections that are at work in the contemporary world economy. A key set of connections, for example, is

between Argentina's ruling business and political elite and the international financial establishment in the form of the IMF, the World Bank and the US Treasury, who increasingly set the institutional rules for trade and investment in the world economy. For example, the Argentinian finance minister responsible for the nationalization of foreign debt in 1982, Domingo Cavallo, was himself a Harvard-educated economist, a local actor who is also embedded in a wider and more global economic and policy-making community. It is important to emphasize that these economic relationships do not just work in one direction, with local actors completely at the mercy of broader global actors and processes. For while there had been mass capital flight from the country, the continuing scale of the debt held by international financial institutions was such that the refusal by the new populist government of Nestor Kirchner to continue repayments in September 2003 threatened to plunge the whole world economy into crisis. Ultimately, Kirchner was able to strike a much better deal with the IMF over repayments than many in the international financial community had believed possible. This was achieved by reminding the international financial community in a speech to the United Nations General Assembly that 'nobody has ever been known to collect any debt from the dead' (ibid., p.135).

Fund to address its massive debts, the Mexican government cut its tariff barriers to overseas trade and privatized 90 per cent of its state-owned enterprises (Dicken, 2003a, p.192). The result was a dramatic influx of foreign investment, which had the effect of driving up the value of the peso to levels that made much of Mexican industry uncompetitive in export markets. Eventually the government was forced to devalue the currency, leading to an outflow of funds and a new financial crisis in 1995. Although the country's economy has subsequently recovered, partly as a result of the NAFTA agreement, questions remain about the nature of Mexico's growth trajectory (see section 7.6.2).

5.6.3 The gap between rich and poor

The uneven nature of economic development is not just felt at the global level between countries; it is also experienced at lower geographical scales, within countries and even within cities at the very local scale. Since 1980, the gap between the rich and poor in some of the world's most developed countries has increased. An article in *The Guardian*, reflecting on economic trends in the UK, noted that the gap between rich and poor in the early 2000s was greater than at any point since the nineteenth century (*The Guardian*, 10 May 2002). In the US too, the gap between the wealthy and the rest of the population has widened enormously in recent years. A recent survey of the salaries of company chief executives compared with the wage of the average employee by *Business Week* revealed that the gap had risen from a ratio of 42:1 in 1982 to 300:1 in 2003 (www.faireconomy.org/press/2004/CEOPayRatio_pr.html, last accessed 1 November 2005).

5.6.4 Globalization in 'world cities'

In 'world cities' such as London and New York, some of the very richest people in the country live and work side by side with some of the most impoverished neighbourhoods. In London, for example, the financial district of the City of London, where investment brokers can earn annual bonuses running into millions of pounds, is only a few hundred metres away from the borough of Tower Hamlets where 40 per cent of people of working age are out of work. Rather than functioning as part of an integrated national economy, such world cities are increasingly shaped by global flows and trends. Globalization is associated with a polarization of employment and income within world cities,

providing a range of high-paying jobs in financial and business services, alongside other low-status employment in routine services such as cleaning, retail, bars and restaurants (Sassen, 1991).

5.6.5 Levels of inequality within countries

Levels of inequality tend to vary widely between countries (Table 5.8). Within the developed world, there is a considerable difference between some Anglo-American economies that typically have favoured a more **market-oriented** approach to development in recent years and some continental European and Nordic countries that continue to be characterized by higher levels of state intervention and support for the less affluent. But some of the highest levels of inequality are in the less-developed or even newly industrializing economies. In countries such as Brazil and Mexico, the benefits from economic growth have

Table 5.8 Income inequality indicators for selected countries, 2005

	GINI coefficient
Namibia	70.7
Brazil	59.3
South Africa	57.8
Mexico	54.6
Argentina	52.2
Malaysia	49.2
China	44.7
Kenya	42.5
United States	40.8
Vietnam	37.0
United Kingdom	36.0
France	32.7
Russian Federation	31.0
Hungary	26.9
Sweden	25.0

NB: The closer to 100, the greater the level of inequality.
Source: UNDP, 2005, p.55

largely been absorbed by a rich minority who have become part of what the UNDP (1998, p.62) terms a 'global elite' or 'global middle class' who share the same high-consumption lifestyles of those in the developed economies. In contrast, some of the former socialist countries, such as the Russian Federation and Hungary, have relatively low levels of inequality, although the opening up to a market-based economy is leading to greater levels than hitherto.

Reflect

➤ Why does Lee use the phrase the 'marginalization of everywhere' in his analysis of the impact of global finance?

➤ What groups of people and countries have benefited most from economic globalization?

5.7 The growth of the Global Justice Movement

Growing resentment at global inequalities and the way the world economy is being organized and controlled has given rise to a new global resistance movement, sometimes referred to – rather erroneously – as the '**anti-globalization movement**'. While many have viewed the movement as being a negative response to globalization, this is something of a misnomer as many in the movement are challenging the form that globalization is taking, rather than globalization itself. In particular, it is the neoliberal or free market agenda driving globalization (section 6.5) that campaigners oppose, associating it with increased corporate control, policies of privatization and liberalization, and inequality and poverty in the developed world particularly.

5.7.1 The battle in Seattle

For most people, what we prefer to call the **Global Justice Movement (GJM)** – concerned with addressing the material inequalities and injustices produced by neoliberal globalization – first came to prominence in the American city of Seattle in December 1999. It was

here that the World Trade Organization – the international body set up to promote free trade and market liberalization – held its annual meeting, involving the leaders and finance ministers of the most powerful countries. The WTO has therefore come to be seen as the symbol of neoliberal economic globalization, and has become a focus for what the US and former World Bank economist Joseph Stiglitz has called globalization's 'discontents' (Stiglitz, 2002).

Around 70,000 protesters (Tormey, 2004, p.39) came together at Seattle to oppose the direction being taken in the global economy and succeeded in shutting down the conference before a trade deal had been reached, offering a wake-up call to the world's economic and political elite. At the same time, Seattle was just a manifestation or 'moment' in the emergence of a broader movement. While often associated with key celebrity spokespeople such as the journalists Naomi Klein and George Monbiot, and intellectuals such as Noam Chomsky, Walden Bello and Susan Strange, the real significance of the GJM lies in the coming together of a huge range of different movements and campaigns previously devoted to single-issue politics. There have been mass protests and movements in the recent past – think of the anti-nuclear campaign or the environmental movement – but Seattle represented a convergence of disparate and hitherto often opposing groups against a 'common enemy'. These included trade unionists and environmentalists, NGOs and direct-action networks, street dwellers and peasant movements from the developing world, and middle-class consumer activists from the developed world, united by their opposition to the direction of the world economy. The movement has subsequently developed through other 'global days of action' in Prague, Genoa, Quebec City and elsewhere. Wherever there is a major meeting of world leaders or economic officials through institutions such as the WTO, World Bank, IMF and G8, protesters will converge to try to disrupt or contest their legitimacy.

5.7.2 The future of the Movement

Despite its success in marshalling opposition to 'corporate globalization', however, the Global Justice

Box 5.5

The emergence of the World Social Forum

Since the turn of the millennium, debate within the Global Justice Movement has moved from opposition to neoliberalism to an increasing demand for political and social alternatives that seek to democratize economic decision-making. The most significant initiative in this respect has been the establishment of the World Social Forum (WSF), which was deliberately established in opposition to the World Economic Forum, a high-powered and exclusive meeting of world leaders and corporate chief executives that takes place every year in January in the Swiss resort town of Davos. The decision to set up the WSF was itself an interesting example of geographically dispersed movements coming together, following an exchange of ideas between the Brazilian Workers Party and the French group ATTAC.

From the perspective of economic globalization, **ATTAC (Association pour la taxation des transactions financières pour l'aide aux citoyens**; roughly translated, Association supporting a tax on financial transactions to help citizens) is a particularly interesting movement. It was created in 1998 by journalists from the French political magazine *Le Monde Diplomatique* to campaign for a global tax on international currency speculation aimed at putting 'sand in the wheels' of global capital (known as the Tobin Tax after the US economist James Tobin). With its catchphrase, 'The world is not for sale', ATTAC has subsequently become an international social movement committed to reforming the global economy (see http://www.attac.org/), with tens of thousands of members in 33 countries mainly in Europe, Africa and the Americas.

The first forum was held in the Brazilian city of Porto Alegre in 2001, spawning a regular annual meeting for the discussion of alternative policies and agendas on a diverse range of environmental, anti-war, feminist and social issues as well as economic matters. The 2005 WSF was the largest yet, with over 140,000 activists present, and was important enough to attract two heads of state as speakers, the Brazilian President 'Lula' da Silva and the revolutionary Venezuelan leader, President Hugo Chávez.

Movement faces some important challenges in the years ahead. Not least among them is the need to move from global protest to forging alternatives to the current world economic order. Its diversity could prove to be a weakness in this respect. While most actors within the Movement are opposed to neoliberal economic policies, there are a varied and often competing range of ideological and tactical positions. A key distinction is that between reformists – which covers a spectrum from relatively moderate liberal positions to radical social democratic positions – and genuine anti-capitalists, which include a plethora of competing Marxisms, different strands of anarchism and 'deep greens' (Tormey, 2004). This suggests that only a minority have as their aim the overthrow of capitalism with the majority looking for greater social regulation and control of markets. Similarly, very different positions on globalization are evident in terms of the reform/abolition divide and the geographical scale at which the economy should be regulated. Localists advocate a relocalization of the economy and the dis-

mantling of global neoliberal institutions such as the IMF and World Bank (e.g. Hines, 2000) – the most radical option – while others call for greater intervention at the nation level (e.g. George, 1999), and globalists favour new forms of global governance and regulation (e.g. Monbiot, 2003).

5.7.3 The movement for 'fair trade'

Alongside the emergence of the Global Justice Movement there has been an increasing concern to promote **ethical** or 'fair trade'. Since the 1970s, concerned consumers in the more developed economies of Western Europe and North America have been campaigning to improve the conditions in which farmers and commodity producers in the global South sell their food on world markets. Because many developing countries remain dependent upon one or two commodities for export, and are therefore susceptible to fluctuations in global commodity markets, producers can be hit hard by price reductions over which they have little control. In many cases, farmers are in dependent relationships with foreign MNCs who are able to dictate terms of supply. In contrast:

> Fair Trade is a trading partnership, based on dialogue, transparency and respect, which seeks greater equity in international trade. It contributes to sustainable development by offering better trading conditions to securing the rights of marginalized producers and workers especially in the South. Fair trade organizations (backed by consumers) are engaged actively in supporting producers, awareness raising and in campaigning for changes in the rules and practice of conventional international trade (FINE, 2001, cited in Moore, 2004, pp.73–4).

Fair trade organizations have two main goals, which are to provide a model of equitable trade that links consumers and producers and guarantees the latter a basic price, while at the same time challenging the current system of 'free trade' (Moore, 2004, p.74). Typically, fair trade initiatives will also invest in the communities that they support in education and in the provision of services and market information. Fair trade is dominated by food products, such as coffee, tea and sugar, although there are also a growing number of ethical trading initiatives in the clothing industry. It has been estimated that by 2003 fair trade products were globally worth around $500 million, with products being on sale in 17 countries and supporting over 900,000 families of producers in the developed world (ibid., pp.74–6). However, the sector is still small. A sobering fact is that the largest British retailer, Tesco, with a turnover of £28.6 billion, has 90 times the global turnover of free trade (ibid., p.75).

Reflect

> How far does the Global Justice Movement represent an effective counterweight to the power of MNCs?
> What are the key challenges it faces in the years ahead?

5.8 Summary

Our purpose in this chapter has been to explore the main features of economic globalization. It is important to note that globalization is an ongoing process, not a final outcome or 'end state', shaped by the decisions of governments and firms, and operating unevenly in time and space (Dicken et al., 1997). In adopting a geographical perspective on globalization, we emphasize that the 'local' and the 'global' are not separate bounded geographical scales but are 'co-dependent'. Local economic processes are shaped by global connections, but global actors and institutions are also located within and shaped by particular contexts at both the local and national levels. In this sense, globalization is about new and increasingly dense connections, networks and flows of people, organizations and materials that are located in different places.

A second point is that the nature of the relationships underpinning these connections and networks is uneven, not just in terms of differences in the economic and political power of different groups and organizations, but also because of the wide variation in the geographical scales at which people, firms or other economic actors operate. The World Bank is a powerful institution when dealing with an impoverished African country, not just because it is an institution with global reach, but also because its policies are backed by (and, many argue, favour) the most powerful national state governments.

As the latter part of the chapter has indicated, these relationships are not only economic, but social and political too. The emerging global economy is, in this regard, a contested one, made up of a new set of relationships between multinational corporations, nation states, global institutions, labour unions, and NGOs and grass-roots activists. This brings us on to a final point: in conceptual terms it is important to recognize that geographical scales do not exist as fixed, pre-given levels of economic organization (local, regional, national, supranational, global) but are themselves created by human activity to meet particular needs (Smith, 1990; Swyngedouw, 1997). The emergence of a global economy has resulted in a 'scaling up' of some forms of economic organization and governance from nation states (e.g. over monetary policy and trade) to supranational organizations such as the EU's Central Bank (in the case of monetary policy for member states) or the World Trade Organization, but also, in some cases, a scaling down as states transfer some powers to local and regional authorities (section 6.5).

Exercises

1. Returning to the three different positions set out in Table 5.1, review the chapter and list the evidence for and against each position.

2. In your opinion, which of three positions is most accurate in representing globalization and its effects upon the world economy? Why?

3. On the basis of recent trends, what are the likely changes to the geography of the world economy over the next decade?

Key reading

Dicken, P. (2003a) *Global Shift: Reshaping the Global Economic Map in the Twenty-First Century*, 4th edn, London: Sage, Chapter 3, The Changing Global Economic Map.
This is an excellent introduction to and summary of key globalization trends.

Held, D., McGrew, A., Goldblatt, D. and Perraton, J. (1999) *Global Transformations: Politics, Economics and Culture*, Cambridge: Polity, introductory chapter.
This chapter sets out in greater detail the three different perspectives on globalization that we review here.

Hirst, P. and Thompson, G. (1999) *Globalization in Question: the International Economy and the Possibilities of Governance*, 2nd edn, Cambridge: Polity.
An excellent review of the empirical evidence for globalization that is a landmark text for sceptics.

Lee, R. (2003) 'The marginalization of everywhere?: emerging geographies of emerging markets', in Peck, J. and Yeung, H.W. (eds) *Remaking the Global Economy*, London: Sage, pp.61–82.
An excellent overview of the workings of the financial markets and their geographical effects, highlighting the frightening vulnerability of places to unregulated flows of capital.

Tormey, S. (2004) *Anti-capitalism: A Beginner's Guide*, Oxford: Oneworld.
A useful summary of the different resistance movements that have developed to challenge neoliberal globalization.

Useful websites

http://www.polity.co.uk/global
Useful website, providing brief summaries of different globalization topics.

http://hdr.undp.org/
United Nations site for Human Development Reports, which provide comparative data on the economic and social performance of each country.

http://www.forumsocialmundial.org.br
World Social Forum: key organization for the anti-globalization movement.

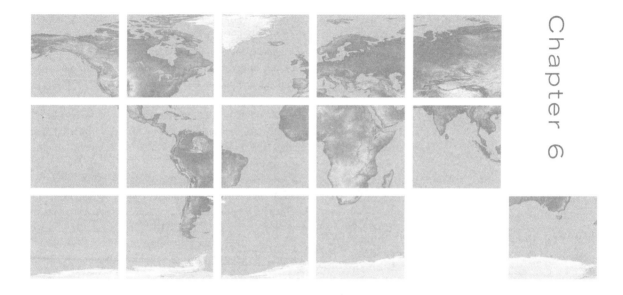

Chapter 6

The state and the economy

Topics covered in this chapter

➤ The nature of the state and its role in the economy.

➤ The concept of the 'qualitative state'.

➤ The development of the Keynesian Welfare State.

➤ The 'developmental states' of East Asia.

➤ The reorganization of states since the 1970s, including:

- the evolution of neoliberalism;
- the effects of globalization, emphasizing the growth of supranational institutions such as the European Union;
- the 'competition state' and efforts to reduce welfare spending and to introduce workfare schemes;
- changing forms of regional economic development.

Chapter map

In the introduction to the chapter, we define the state and examine its role in the economy. Section 6.2 focuses on the general concept of the state, emphasizing that states are the results of historical processes rather than existing as natural entities, and highlighting the key notion of the 'qualitative state' (O'Neill, 1997). Section 6.3 reviews the development of the Keynesian Welfare State, dominant in developed countries from the 1930s to the 1970s. This is followed by an examination of the notion of a 'developmental state', focusing particularly on the experience of the East Asian 'tiger' economies. We then turn to consider contemporary changes in state–economy relations, assessing how the state has been reformed since the late 1970s in response to processes of globalization and the spread of neoliberal or free market policy. Key aspects of change such as the internationalization of the state through the expansion of bodies such as the European Union (EU), the shift from welfare to workfare and the introduction of new forms of regional development policy are discussed.

6.1 Introduction

The role of the state in the economy is wide-ranging, but often invisible to individual consumers, shaping the provision of goods and services in ways that are not always immediately apparent. Whenever you go to the pub for a drink, for example, you will find that the state decides: how long the pub can stay open; the size of the measures of beer, wine or spirits offered; how much of the price is taken in tax; how the drinks are labelled; the standards of hygiene governing the kitchen; and the minimum wages paid to the staff (Painter, 2006, p.753). This example highlights the role of the state in regulating economic activities, although the nature of the products on offer perhaps dictates the extent of the state's involvement in the drinks industry. A key underlying argument made in this chapter is that the state should be viewed as a dynamic process rather than a fixed 'thing' or object (Peck, 2001). Instead of focusing solely on the size of the state, expressed in terms of levels of taxation or expenditure, for example, we should examine how states intervene in economic life, the economic policies that the state pursues and the effects of these on different social groups and regions (O'Neill, 1997).

Our analysis of the geography of state intervention is informed by our **political economy** approach, incorporating a regulationist position that rejects the notion of an autonomous, self-regulating economy. The idea of a self-regulating economy is central to mainstream, neo-classical economics, emphasizing the role of the market in ensuring that supply and demand are balanced through the price mechanism, consigning the state to a limited role of upholding property rights and enforcing business contracts. Instead, we believe that the economy is regulated through a wide range of political, social and cultural mechanisms in addition to market forces (Aglietta, 1979). The state plays a key role in harnessing and coordinating these different mechanisms, formulating a wide range of rules and laws covering matters such as business taxation, trade policies, employment standards and financial markets (Table 3.2). The role of the state in the economy is generally directed towards the promotion of economic growth, attempting to create the conditions that allow businesses to make profits and workers to find employment, thereby generating revenue through various forms of taxation. From a geographical perspective, states play a key role in regulating wider processes of **uneven development**, sometimes introducing policies that are focused on particular types of place (e.g. depressed regions). Beyond these very general dimensions of state regulation, the specific forms and functions of the state change over time, as highlighted by the regulationist notion of **modes of regulation**, referring to specific institutional arrangements that help to create relatively stable periods of economic growth known as **regimes of accumulation** (section 2.4.3).

6.2 Understanding the 'qualitative state'

6.2.1 Defining the state

The **state** is the basic organizing unit of political life (Figure 6.1). Before proceeding further, it is important to distinguish the state from **the nation**. By the state, we are referring to a set of institutions for the protection and maintenance of society (Dear, 2000, p.789). These institutions include parliament, the civil service, the judiciary, the police, the armed forces, the security services, local authorities, etc. As this suggests, the state is a complex entity stretching beyond what is normally referred to as government (the national executive of ministers and civil servants) (section 3.5). States exercise legal authority over a particular territory, holding a monopoly of legitimate force and law-making ability (Mann, 1984). The nation, by contrast, refers to a group of people who feel themselves to be distinctive, on the basis of a shared historical experience and cultural identity, which may be expressed in terms of ethnicity, language or religion. The two come together to form nation states in cases where the state territory contains a single nation. This is often presumed to be the norm, but there are many examples of multinational states that contain different national groups (the UK for one

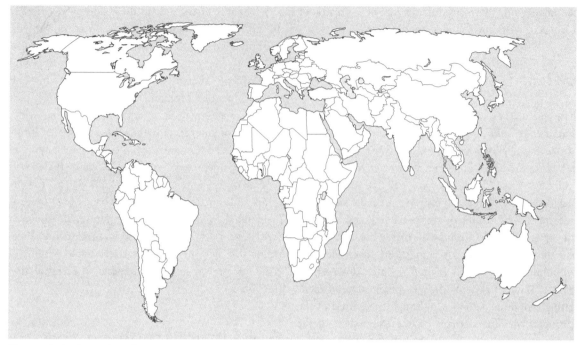

Figure 6.1 A world of states.

is made up of English, Scots, Welsh and Northern Irish). The corollary of this is that not all national groups have their own states (e.g. the Kurds in Iraq and Turkey, the Basques in Spain and France).

6.2.2 The growth of the state

The modern state is often regarded as a natural phenomenon that has always existed, in much the same way as many economists have naturalized the modern capitalist economy. In reality, however, states are a product of historical processes operating in the modern era (Painter, 1995, p.31), gradually taking shape in Western Europe following the Treaty of Westphalia in 1648. The growth of the state was closely associated with the rise of absolute monarchies with centralized bureaucracies, taxation systems and large armies. The state has played a crucial military role, concentrating resources to attack rival states or to defend its territory against potential threats from outside its borders. It has also accumulated administrative power, gathering information about its population and territory through

various forms of surveillance and information storage, from the invention of writing to the advanced computer systems of today (Giddens, 1985). Such technologies have underpinned the development of information-gathering devices such as registration systems, population censuses and surveys. Through such devices, information is collected, recorded within the state bureaucracy and used to govern the population, a process that is crucial to the consolidation and perpetuation of state power.

6.2.3 The changing nature of the state

As emphasized in the introduction to this chapter, it useful to view the state as a dynamic process rather than regarding it as a fixed object or thing. This is consistent with our regulationist perspective, which emphasizes the role of the state in stabilizing and sustaining capitalist forms of development in terms of both its '**accumulation**' and '**legitimation**' functions (section 3.5), focusing attention on specific institutional arrangements and processes. Rather than

Box 6.1

The 'qualitative state'

O'Neill's conception of the qualitative state is based on three main points. First, he rejects the notion of the state as a single, unified entity with a highly centralized structure. Instead, it is structured by a continual process of interaction between state agencies, such as the UK Treasury, and non-state actors and forces, such as business organizations like the Confederation of British Industry (CBI). Second, the state always plays a crucial role in the construction and operation of markets, including those operating at the international scale. Two key ways in which states have actively constructed markets in recent years is through the privatization of formerly state-owned industries such as electricity or telecommunications, and the fostering of globalization through policies that have sought to lower trade barriers and promote competition, with organizations such as the World Trade Organization (WTO) playing a key role. An appreciation that the state is always involved in the operation of markets focuses attention

on 'the nature, purpose and consequences of state action' (O'Neill, 1997, p.290). Third, a qualitative view of the state overcomes the politically disabling argument that the powers of the nation state are being eroded by globalization and neo-liberalism, emphasizing, instead, that they are being transformed in particular ways. Rather than being powerless in the face of globalizing processes, governments can still regulate markets to achieve broader social goals such as full employment or universal health care.

This emphasis on the importance of state action in supporting the operation of markets echoes the writings of Karl Polanyi in the 1940s and 1950s. Polanyi argued (see Box 1.3) that the state played a crucial role in matching the supply and demand of the 'fictitious commodities' of land, labour and capital during the great transformation wrought by industrialization in the nineteenth century. The underlying point of Polanyi's work is that

markets do not operate in a social and political vacuum. The state intervenes in the economy in a wide variety of ways, implementing a number of measures and programmes to support its goal of promoting economic development (Table 3.2). This involves it engaging with a wider range of non-state interests such as MNCs, financial institutions, business organizations, trade unions and consumer groups. Rather than being a single, fixed entity, the state is made up of a large number of different institutions and agencies, with the relationship between these often characterized by tension and conflict. Getting these diverse agencies to work in harmony is always a real challenge for governments and politicians. Successive attempts to reform the British National Health Service, for example, highlight the difficulties of implementing specific policy programmes, causing much frustration for politicians such as Tony Blair over the limited progress of his public sector 'reform agenda'.

focusing on the size of the state and the extent of its intervention, measured in terms of its share of national wealth, the level of taxation and the volume of welfare payments, recent work has emphasized the nature of such intervention and the social and economic goals towards which it is directed. This can be seen as a shift of emphasis from a concern with quantitative aspects of state intervention to an interest in its qualitative characteristics (Painter, 2000, p.363). This new approach examines the specific roles that the state plays in the economy, the key policies that it introduces and the geographical effects of these. It is well expressed by the Australian economic geographer Philip O'Neill (1997) in his concept of the **'qualitative state'** (Box 6.1).

The notion of the 'qualitative state' can be used to inform analyses of contemporary processes of state restructuring. It helps to focus attention on the changing forms and functions of the state in relation to the interrelated processes of globalization and neoliberal reform. The qualitative perspective highlights the importance of the state itself as an actor in processes of globalization and economic restructuring, rather than assuming that the state is always acted upon by other more powerful forces, such as MNCs and financial markets. States have contributed to the process of globalization through measures such as the reduction of trade barriers and the abolition of controls on the movement of capital.

> How does the qualitative concept of the state as a process differ from traditional understandings of the state as a fixed object with widely accepted functions and a highly centralized structure?

6.3 The Keynesian Welfare State

6.3.1 Origins and development

Comprehensive welfare states were established in the developed countries of Europe and North America in the 1940s. In Britain, the famous Beveridge Report of 1942 laid the basis of the post-war welfare state, proposing a comprehensive national insurance system to fund a range of services such as retirement pensions, widows' benefits, sickness and unemployment benefits, and invalidity allowance (Johnston, 1993, p.158). The Keynesian element of the welfare state refers to the methods of economic management and planning developed by the state in the post-war period, guided by the theories of the British economist John Maynard

Keynes. While **Keynesian economic policies** and the welfare state are distinct entities, they emerged together and became closely interlinked. The former ensured full employment to fund welfare payments and the latter supported mass consumption and helped to reproduce the labour force through the provision of social services such as health, education and housing (Martin and Sunley, 1997, p.287). The growth of the welfare state in the post-war period is evident from Table 6.1 which shows the coverage of unemployment schemes among a group of industrialized countries in 1925 and 1975 (Painter, 2002, p. 163).

6.3.2 Key features

The main features of the **Keynesian Welfare State** are summarized in Table 6.2. It was closely bound up with the consolidation of Fordism as a regime of accumulation or mode of growth based on mass production and consumption, providing the system of state regulation to support and underpin it (Table 2.2). The primary geographical scale at which the Keynesian Welfare State operated was that of the nation state, regarded as the key unit of economic organization. Keynesianism assumed a high degree of closure of national economies to international trade and financial flows, enabling increased demand for goods to be met by increased supply from within the domestic economy.

Keynesian economic theory emphasized the need for the state to take an active role in managing the national economy through fiscal policies, referring to taxation and government expenditure. The key focus here was on managing the aggregate level of demand for goods and services in the economy as a whole or the total spending of firms and households on purchasing goods and services. Maintaining high levels of spending and consumption is crucial. This involved stimulating demand in downturns by increased government expenditure – supporting public works and employment schemes, for example road building projects – or reducing taxes (Box 6.2). Such policies would create additional demand in the economy, leading to increased employment and income. In periods of economic growth, the state should dampen down demand by increasing taxes or reducing expenditure, thus pre-

Table 6.1 Unemployment insurance: members as a percentage of the labour force in selected European countries		
Country	1925 %	1975 %
Austria	34	65
Belgium	18	67
Denmark	18	41
France	0	65
Ireland	20	71
Italy	19	52
Norway	4	82
Switzerland	8	29
United Kingdom	57	73

Source: Painter, 2002, p.163.

Table 6.2 The Keynesian Welfare State

Primary scale	Economic policies	Social policies	Spatial policies
National, centralized regulation and management of national economy. Some local delivery of welfare services.	Full employment, demand management. Provision of infrastructure to support mass production and consumption.	Collective bargaining and state help to expand mass consumption. Expansion of welfare and redistributionist policies.	Goal of spatial integration and regional balance. Redistribute resources between rich and poor regions.
National	Keynesian	Welfare	Spatial Keynesianism

Source: Adapted from Jessop, 2002, p.59, and Martin and Sunley, 1997, pp.279–80.

Box 6.2

The Keynesian multiplier

In the 1920s and 1930s, Keynes, based at the University of Cambridge in England, began to criticize key elements of the conventional economic wisdom. The cherished assumptions of neoclassical theory were clearly at odds with economic reality during the depression of the 1930s, and Keynes sought to reconstruct economic theory and practice accordingly. The depression was being intensified, Keynes argued, by orthodox economic policy, focused upon balancing the budget and limiting the role of the state, which merely depressed the level of demand still further by restricting expenditure. His aim was to save the capitalist system from collapse through the development of better policies to manage its fluctuations, staving off the threat of prolonged slumps in output and employment (Skidelsky, 2003). Keynes's economic theories were set out in *The General Theory of Employment, Interest and Money*, published in 1936.

The concept of the economic multiplier is a crucial part of Keynesian theory. It refers to the snowball or knock-on effects of additional expenditure or investment within the economy. A new factory funded by state expenditure or private investment will buy inputs and services from other firms, creating employment and income, which is subsequently spent on other goods and services. At the same time, the workers employed in the factory will spend part of their incomes on goods and services, supporting other sectors of the economy such as retail and entertainment. The multiplier effect works in both directions, of course, meaning that relatively small downturns in investment and expenditure can also have significant effects on the overall level of activity in the economy. This makes it all the more important for governments to intervene to counter the fluctuations of the economic cycle.

Aware that his ideas would be opposed by many other economists and officials wedded to traditional doctrines of limited intervention and the balanced budget, the concept of the multiplier allowed Keynes and his disciples to demonstrate that additional expenditure would be self-financing, generating increased income and receipts. This helped to overcome some initial resistance and Keynes's theories gained broad acceptance in the late 1930s and 1940s (Skidelsky, 2003, pp.546–51). An emerging generation of economists were won over as rearmament and programmes like the New Deal in the US led to a process of economic recovery. In this way, the Keynesian revolution redefined economics and provided the intellectual basis for a new form of state intervention in the economy. By the early 1970s, the influence of Keynesianism was so pervasive that the American President Richard Nixon remarked that 'we are all Keynesians now'. Ironically, however, it was just about this time that belief in Keynesian policies faltered as the post-war boom turned to recession, precipating the monetarist revolution of the 1970s and 1980s (section 6.5).

venting inflation. In the context of Fordism, Keynesian policies were relatively successful, delivering sustained growth between the late 1940s and early 1970s (Table 6.3). Long-term growth in output, exports and the stock of fixed capital in 1950–73 was much greater than in earlier periods (Table 6.3). Increases in output out-

Table 6.3 Long-term growth rates, 1820–1970

	Output	Output per head of population	Stock of fixed capital	Exports
1820–1870	2.2	1.0	N/A	4.0
1870–1913	2.5	1.4	2.9	3.9
1913–1950	1.9	1.2	1.7	1.0
1950–1973	4.9	3.8	5.5	8.6

Source: Armstrong *et al.*, 1991, p.118. Refers to the average of the 'advanced capitalist countries' of the US, the UK, Germany, Italy, France, Japan and Canada.

stripped the growth in employment, reflecting rises in productivity associated with the introduction of new machinery and technology (Armstrong *et al.*, 1991, pp.118–19).

The Keynesian Welfare State was associated with national collective bargaining processes where representatives of employers, trade unions and government got together to agree pay rates and awards, often on an annual basis. The basic deal, under Fordism, was that labour would gain higher wages while business gained higher productivity. The expansion of welfare supported mass consumption by establishing a minimum income level, helping to underpin demand. Rather than being regarded as merely a cost of production, workers were now seen as sources of demand, vital to the economic health of the nation. Welfare was designed to be broadly redistributionist, supported by a system of progressive taxation (where the wealthy pay more than the poor), although research has shown that most of the welfare benefits in the UK actually went to the middle classes, having only limited success in alleviating poverty among the working class (Goodin and Le Grand, 1987).

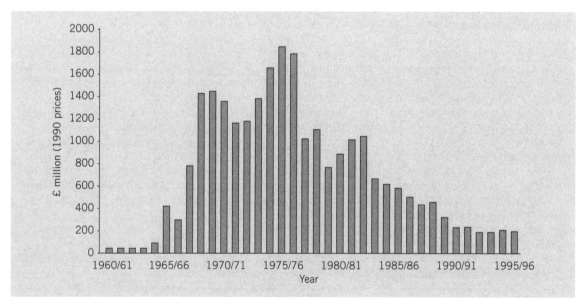

Figure 6.2 Expenditure on UK regional assistance, 1960–96.
Source: 'UK regional policy: an evaluation' in *Regional Studies*, 31, Taylor & Francis Ltd (Taylor, J. and Wren, C., 1997). http://www.tandf.co.uk/journals..

6.3.3 Regional policy

The Keynesian Welfare State also sought to foster spatial integration, creating a regionally balanced economy. The aim of such 'spatial Keynesianism' (Martin, 1989) was to close the gap in income and wealth between rich and poor regions. The establishment of national systems of progressive taxation and welfare expenditure contributed to this aim, transferring resources from core to peripheral areas. This was accompanied by **regional policy**, through which government induced companies to locate factories and offices in depressed regions. It was a highly top-down approach, offering grants and financial incentives to companies to locate factories or offices in depressed regions (Box 6.3). At the same time, development in core regions such as south-east England and Paris was restricted. Firms seeking to expand in such regions had to gain official approval from the government, which was often not forthcoming if the new factory or office could be located in the depressed regions. Classical regional policy reached its peak in the 1960s and 1970s (Figure 6.2), helping to reduce the income gap between rich and poor regions in Europe (Dunford and Perrons, 1994).

6.3.4 The stagflation crisis of the 1970s

The Keynesian Welfare State experienced growing problems from the late 1960s, reflecting the fact that the post-war Fordist growth dynamic based upon the link between increased productivity and rising wages was losing momentum. Mass markets for consumer durables such as cars and washing machines were becoming increasingly saturated in Europe and North America. At the same time, full employment and the increased strength of trade unions meant that wages rose rapidly, outstripping productivity growth by the late 1960s. International trade and investment grew marketedly in the 1960s as capital became increasingly internationalized. Increased international flows of goods and capital undermined Keynesian economic policies, based on the assumption of relatively closed national economies. The position of the US in world manufacturing and trade weakened in the 1960s, in the face of domestic problems and rising competition from Europe and Japan particularly. This meant that the Bretton Woods system of fixed exchange rates (created in 1945) – whereby other currencies were pegged to the US dollar at fixed rates – came under increasing pressure. In 1971, the US government abandoned the fixed exchange rate system, reflecting the fact that it was no longer strong enough to act as the anchor of the system.

The post-war economic boom was only sustained in the years 1969-73 by very loose monetary policy in the US and UK as governments printed money and kept interest rates low in order to sustain growth (Harvey, 1989a, p.145). This meant that inflation rose rapidly, leading governments to tighten policy again in 1973, resulting in reduced growth and rising unemployment. Oil prices quadrupled in the winter of 1973-4, following the Middle Eastern oil producers' decision to reduce exports, raising firms' input costs massively and further fuelling inflation. This helped to trigger a severe economic downturn with industrial production falling by 10 per cent between July 1974 and April 1975 in the 'advanced capitalist countries' (the US, Canada, France, Germany, Italy, the UK and Japan) (Armstrong *et al.*, 1991, p.225). Unemployment began to rise as investment fell and output stagnated, while inflation continued to rise (Figure 6.4). The combination of economic stagnation and rising inflation became known as 'stagflation', a term that has come to symbolize the economic crisis of the 1970s.

One of the consequences of stagflation was to plunge governments into financial difficulties as tax receipts fell in the recession of the mid-1970s, and welfare expenditure continued to grow, fuelled by rising unemployment. This so-called '**fiscal crisis of the state**' (O'Connor, 1973) saw New York City become technically bankrupt in 1975, for example, while the UK was forced to seek an emergency loan from the IMF in 1976 in response to its chronic budgetary problems. The conditions attached to the loan involved the cutting back of government expenditure to reduce the budget deficit and control inflation, prompting the Labour prime minister, James Callaghan, to announce that 'the party is over', referring to the post-war experience of rising wages and increased welfare expenditure.

Box 6.3

French regional policy

French regional policy in the post-war era has been organized in a highly top-down fashion, reflecting the post-revolution tradition of a strongly centralized state. Regional policy, known as '*aménagement du territoire*', was established in the 1950s, addressing the problem of uneven development, particularly the gap between Paris and the provinces (Faludi, 2004). It emphasized national unity during the *Trente Glorieuses* period of sustained economic growth (roughly 1945–75) and the need to redistribute the fruits of growth equally between regions (Ancien, 2005). Regional planning was the responsibility of a specialist state agency, DATAR (*Délégation à l'Aménagement du Territoire et à l'Action Régionale*), charged with formulating and implanting development programmes for the 22 administrative regions. DATAR offered grants of up to 25 per cent for projects that created jobs in the '*zones critiques*' or assisted areas that covered much of the south and west as well as selected parts of the northeast affected by deindustrialization (Figure 6.3). Key investment projects located in the periphery included the Citroën automobile factory in Rennes

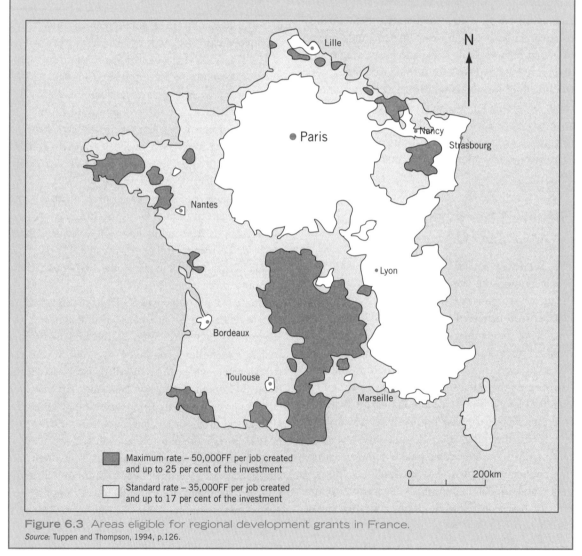

Maximum rate – 50,000FF per job created and up to 25 per cent of the investment

Standard rate – 35,000FF per job created and up to 17 per cent of the investment

Figure 6.3 Areas eligible for regional development grants in France.
Source: Tuppen and Thompson, 1994, p.126.

Box 6.3 (continued)

in western France, employing 12,000 workers.

At the same time, the decentralization of government facilities and departments such as new research centres to the provinces was another important strand of regional policy, along with the designation of eight counterweight 'metropoles' – Lille, Nancy, Strasbourg, Lyon, Marseille, Toulouse, Bordeaux and Nantes – to provide a focus for growth outside Paris (Figure 6.3). Toulouse, for example, became the capital of the aircraft industry, also attracting electronics investment and research laboratories. Restrictions on growth in the Paris region were also introduced, through the '*agrément*', a form of official certification similar to the industrial development certificate used in Britain (Tuppen and Thompson, 1994). Factories and offices were required to look for alternative locations outside the capital, although the majority of the decentralization occurred within a zone of some 200 kilometres around Paris, which contained several new towns designated for this purpose.

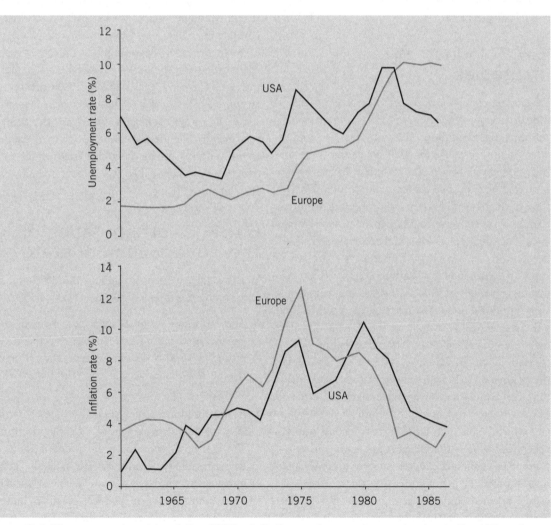

Figure 6.4 The economic crisis of the 1970s: inflation and unemployment rates in the US and Europe, 1960–87.
Source: Harvey, 2005, p.14.

Keynesian economic policies were seen as implicated in the crisis, prompting the development of an alternative neoliberal agenda (section 6.5.1).

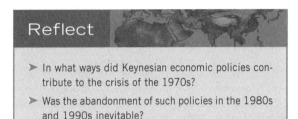

6.4 The developmental state

6.4.1 Industrialization strategies

Developing countries faced a choice of two types of industrialization strategy in the 1950s. **Import-substitution industrialization** (ISI) involves a country attempting to produce for itself goods that were formerly imported. Newly created '**infant industries**' are protected from outside competition through the erection of high tariff barriers, allowing the country's economy to be diversified and dependence on foreign technology and capital reduced (Dicken, 2003a, p.176). By contrast, the second type of strategy is known as **export-oriented industrialization** (EOI), based on producing goods for external markets. EOI is compatible with traditional notions of free trade and comparative advantage (Box 4.1) in contrast to ISI, which involves high levels of protection and state intervention.

Many developing countries followed ISI strategies in the 1950s and 1960s, with only limited success. Outside the larger Latin American countries, such as Brazil and Mexico, domestic markets proved too small to stimulate growth while high tariff barriers against imports generated inefficiency and higher prices for consumers. In response, a number of developing countries began to turn towards EOI from the late 1960s. Aggressive export-oriented strategies have subsequently become particularly associated with the Asian **newly industrialized countries** (NICs) of South Korea, Taiwan, Hong Kong and Singapore.

Crucially, export orientation should not be associated with a reduction of government intervention in the economy; instead, the activities of the state have been crucial in fostering export-led expansion. This has given rise to the notion of a distinctively Asian **developmental state** (Douglass, 1994). The first step was often the devaluation of currencies to make exports cheaper in world markets, followed by a raft of export promotion measures, such as the setting of targets and incentives for firms, the creation of export promotion agencies and the establishment of export processing zones (EPZs). Restrictions on imports remained in place as the state sought to protect strategic infant industries from outside competition until they were strong enough to compete in global markets (Brohmann, 1996, pp.115–16). South Korea, for example, protected strategic infant industries such as chemicals, steel, shipbuilding and electronics during the 1970s and early 1980s (Dicken, 2003a, p.182). Such an approach, influenced by the experience of Japan and some European countries such as Germany, is informed by the interventionist doctrines of the nineteenth-century German economist Friedrich List rather than the classic liberal theory of comparative advantage (Brohmann, 1996, pp.116–17).

6.4.2 Characteristics of the developmental state

Weiss's (2000, p.23) definition of the developmental state emphasizes three key criteria:

➤ The priorities of the state are focused upon enhancing the productive powers of the nation, raising a surplus for investment and ultimately closing the economic gap with the industrialized countries.

➤ Its organizational arrangements are focused particularly on the establishment of a strong government department to coordinate and promote industrial development. Such ministries are staffed by an elite bureaucracy and typically are relatively insulated from both short-term political pressures and the demands of particular social groups. The prototype of the powerful coordinating agency is the Japanese Ministry of International Trade and Industry

Figure 6.5 Hong Kong skyline.
Source: D. Mackinnon.

(MITI), created to promote economic growth and to protect infant industries (Brohmann, 1996, p.117). South Korea's Economic Planning Board and Singapore's Economic Development Board have operated in a similar fashion.

➤ Its links with business stress the value of close cooperative ties rather than arm's-length relations with firms and sectors, allowing the development of new technologies and methods through joint projects. In South Korea, for instance, the state facilitated the development of a small number of large and highly diversified firms – the *chaebol* (large family-based corporations such as Samsung, Hyundai, Daewoo and Lucky-Goldstar) – that have dominated its economy (Dicken, 2003a, p.181).

The first two criteria above reflect the autonomy of developmental states from key interests and groups in society (e.g. landlords, local merchants, the military), helping them to mobilize society towards the goal of economic development. This partly reflects the absence of a strong indigenous capitalist class and the elimination of landed elites in South Korea and Taiwan by land reform programmes in the 1950s, creating an agricultural sector dominated by large numbers of smallholders. This new pattern of landowning led to increased productivity, helping to generate a surplus for the state to invest in industry and stimulating domestic demand for manufactured goods (Gwynne, 1990, pp.185, 187). Cultural factors also seem to have underpinned the success of developmental states in East Asia through common 'Confucian' values emphasizing duty, loyalty and responsibility, providing acceptance for the role of an authoritarian state bureaucracy.

In geographical terms, the NICs' proximity to Japan allowed them to benefit from its powerful growth dynamic in the 1960s and 1970s. Complementarities between Japan, the NICs and a third group of developing countries, including China, Indonesia, Malaysia and Thailand, have fostered the development of an elaborate **regional division of labour** (Brohmann, 1996, p.120). Japan and, increasingly, the NICs (Hong Kong and Singapore in particular – see Box 6.4) have provided capital and focused on high-level managerial and professional activities, promoting themselves as centres for foreign MNCs to locate their regional headquarters while production is carried out in neighbouring lower-wage countries (Figure 6.5).

While the role of the developmental state has clearly been central to the Asian miracle, lifting millions of

Box 6.4

Development states in action: the case of Singapore

By far the smallest of the East Asian NICs, Singapore has been described as the world's most successful economy (Lim, quoted in Huff, 1995, p.1421). In common with the other NICs, the role of the state has been central in driving and shaping development. Although it is a parliamentary democracy, Singapore has been governed by one party, the People's Action Party (PAP) since the early 1960s. Indeed, one powerful politician, Lee Kuan Yew, was in charge until his retirement in 1990,

becoming the longest-serving prime minister in the world. Singapore occupied an important position with the British empire, becoming a major trading centre and port, reflecting its strategic geographical position astride the major East–West global trade route and its fine natural harbour.

The break from Malaysia in 1965 greatly reduced the size of the domestic market, militating against the pursuit of an ISI-based strategy. In response, the PAP regime pursued an

aggressive policy of export-oriented, labour-intensive manufacturing development based on attracting FDI. This was seen as a question of national survival for a newly independent state without any natural resources. The policy was highly successful, with manufacturing employment increasing fourfold between 1967 and 1979 and manufacturing exports growing from 13.6 to 47.1 per cent of GDP over the same period (Huff, 1995, p.1423). Investment was focused in sectors such as electronics, petroleum and shipbuilding. Control of labour was central to the export-based strategy, with the state achieving a rare degree of acquiescence and cooperation from the labour movement. This was based on the incorporation of labour into corporatist bodies such as the National Wages Council and the redistribution of some of the proceeds of growth in the form of employment, housing, education and healthcare programmes (Coe and Kelly, 2002, p.353).

In response to fears about the erosion of its labour costs advantage by low-wage competitors, the government changed policy direction in the 1980s, focusing on the expansion of the financial and business services sector. This approach included an Operational Headquarters Scheme to encourage foreign MNCs to locate high-level management, administrative and research and development functions there (Dicken, 2003a, p.187). This is closely linked to a strategy of 'regionalizing' the economy by establishing Singapore as the control centre of a regional division of labour, hosting high-level functions such as corporate headquarters, business services and research and development while

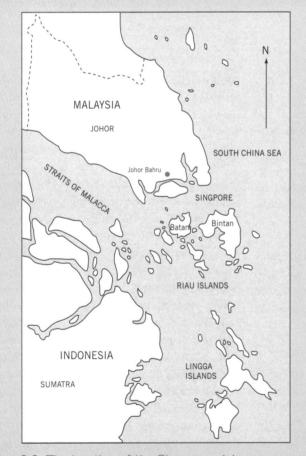

Figure 6.6 The location of the Singapore–Johor Bahru–Batam/Bintan growth triangle in South-east Asia.
Source: Sparke *et al.*, 2004, p.487.

Box 6.4 (continued)

labour-intensive manufacturing and assembly is carried in lower-wage neighbours. The growth triangle formed by Singapore, Johor Bahru in Malaysia, and the islands of Batam and Bintan (Indonesia) is the best-known manifestation of this policy (Figure 6.6).

people out of poverty, there are also negative aspects to this process. These include the perpetuation of authoritarian regimes, the repression of labour and pro-democracy groups, the exploitation of young female workers in export-oriented industries and widespread environmental degradation. The **repression** of large sections of **the labour force** has been one of the main ways in which the NICs have responded to the challenge of maintaining competitiveness through low labour costs while increasing the productivity of labour. South Korea, for example, was governed by a succession of authoritarian, military-backed and strongly nationalist governments between 1948 and 1988, when some liberalization occurred (Dicken, 2003a, p.181). Under the leadership of Park Chung Hee (1961–79), economic policy was directed by the state through a series of five-year plans. In the 1970s, as the popularity of the regime declined, elections were dispensed with, the press silenced, non-state organizations tightly controlled and labour movements subject to direct repression (Douglass, 1994, p.561).

6.4.3 Differences between NIC strategies

At the same time, it is important not to gloss over the considerable differences that exist between the specific strategies adopted by individual NICs. In Taiwan, industrialization involved the direct state ownership of key sectors while the Korean state implemented policy through its close relations with the *chaebol*. In Singapore, by contrast, the attraction of foreign direct investment (FDI) was the central plank of industrialization from the late 1960s (Box 6.4). Hong Kong pursued a less state-directed approach, reflecting its higher level of development in the 1950s with colonialism having bequeathed a legacy of entrepre-

neurialism and a skilled labour force. In both South Korea and Singapore, the state has directly repressed labour and trade unions, while Hong Kong and Taiwan have relied upon more traditional norms of paternalism and loyalty, stressing the mutual obligations between worker and employers within locally owned factories.

Reflect

➤ To what extent does the NIC experience of economic development provide a distinctive model for other developing countries to follow?

6.5 Reinventing the state: neoliberalism, globalization and state restructuring since 1980

Established state structures have undergone considerable restructuring and change since the late 1970s, driven by **neoliberal** reform programmes and the globalization of the economy. In Europe and North America, these structures were based on strong interventionist states managing their economies according to Keynesian theories and dispensing universal welfare services. Most attention has focused on the abandonment of Keynesianism and the reform of welfare states in the developed world, but states in developing countries have faced similar pressures, not least through the activities of international organizations like the IMF and World Bank. Neoliberalism can be seen as providing the basis of the new mode of regulation that has emerged since the early 1980s, informing the introduction of a range of institutional experiments and reforms. Whether it offers the stability and order required for the consolidation of a coherent 'post-Fordist' **regime of accumulation** is, however, highly questionable (Peck and Tickell, 2002).

In considering the impact of neoliberalism and globalization in this section, it is important to recall our earlier discussion of the qualitative state, emphasizing the nature and purpose of its interactions with the economy rather than their magnitude. We are not witnessing the demise of the state, but a set of ongoing and multifaceted changes to its organizational forms and policy functions. While states may have lost powers in some areas, such as trade and the regulation of financial markets, they have maintained or even gained them in other spheres, for example foreign policy, crime, immigration (Peck, 2001). In this section, we examine four key dimensions of contemporary state restructuring: the rise and spread of neoliberalism as a set of ideas shaping state policy; globalization and increased transnational links between states; moves towards a 'competition state' and the introduction of workfare schemes; and the increased prominence of the local and regional scales in terms of economic development policy.

6.5.1 The evolution of neoliberalism

The late 1970s and early 1980s represent a key turning point in the recent history of capitalism (Harvey, 2005, p.1). Following the 1973 military coup, Chile introduced neoliberal economic policies, while the Chinese communist leader, Deng Xiaoping, launched a far-reaching economic reform programme in 1978. A year later, Margaret Thatcher came to power in the UK, espousing a radical new brand of free market liberalism. This was followed by the election of a former Hollywood actor, Ronald Reagan, as US president in 1980. 'From these several epicentres ... revolutionary impulses spread out and reverberated to remake the world around us in a totally different image' as the doctrine of neoliberalism was 'plucked from the shadows of relative obscurity ... and transformed ... into the central guiding principle of economic thought and management' (ibid., pp.1–2).

As a political and economic ideology, neoliberalism

is based on a belief in the virtues of individual liberty, markets and private enterprise. neoliberals are hostile towards the state, believing that its role in the economy should be minimized to that of enforcing private property rights, free markets and free trade (ibid., p.2). Furthermore, in areas where markets do not exist, because of excessive state intervention and regulation, the state should create them through policies of **privatization** (transferring state-owned enterprises into private ownership), **liberalization** (opening up protected sectors to competition) and **deregulation** (relaxing the rules and laws under which business operates). Beyond this, however, the state should not venture, since it cannot possess enough information to second-guess market signals (prices) based on the preferences of millions of individuals (ibid., p.2).

Since the 1970s, neoliberalism has evolved considerably. Three distinct phases can be identified (Peck and Tickell, 2002, pp.387–92). The first, proto-liberalism, refers to its early development in the 1970s when ideas that were deeply unfashionable for most of the twentieth century were developed and promoted by a New Right group of important thinkers and politicians in the UK and US (including the economists Milton Friedman and Friedrich von Hayek) in think tanks, universities and the media. Their views became increasingly influential, appearing to offer radical solutions to the economic crisis of the 1970s in terms of reducing inflation, cutting welfare spending and restricting trade union power while restoring individual liberties through the promotion of free markets. The New Right essentially sought to reassert traditional nineteenth-century liberal principles in the circumstances of the 1970s (hence the term neoliberalism).

After the election victories of Thatcher and Reagan, a second phase of 'rollback' neoliberalism ensued. Neoliberal ideas involving the reduction of state intervention in the economy and the curbing of trade union rights were put into practice. Inflation was tackled by applying the monetarist theory of Friedman, based on reducing the supply of money in the economy. This form of 'shock therapy' succeeded in lowering inflation, in the short term, but at the expense of deepening the recession of the early 1980s and increasing unemployment. State intervention in the economy was reduced through policies of privatization, liberalization and deregulation. Many conservative politicians and commentators criticized the welfare state in the US and UK during the 1980s and 1990s, attacking it for encouraging individuals to become dependent on the state, undermining work incentives and imposing a high tax burden. While successive 'reform' programmes have been launched amid considerable fanfare, it has proved more difficult to achieve significant reductions in welfare expenditure (section 6.5.3).

Since the early 1990s, a new form of 'roll-out' neoliberalism has emerged. By this stage, neoliberalism had become normal, regarded as simple economic 'common sense'. As such, it could be implemented in a more technocratic and low-key fashion by governments and agencies such as the World Bank and IMF. Following the conversion of key figures in the early 1980s, the latter two agencies in particular played a crucial role in spreading neoliberal doctrines across the globe, acting as 'the new missionary institutions through which these ideas were pushed on the reluctant poor countries that often badly needed their grants and loans' (Stiglitz, 2002, p.13). In the early 1990s, neoliberal 'shock therapy' in the form of privatization and liberalization was rapidly implemented in the former communist countries of Central and Eastern Europe.

In this third phase, neoliberalism has become softer and more mainstream, informing the 'third way' policies of centre-left leaders like Blair, Clinton and Schroder, who aimed to find a new path between the conflicting extremes of free market capitalism and state socialism. Marrying the efficiencies of markets to a revived social democratic notion of social justice is the key idea. At the same time, neoliberal principles have been incorporated into the design and operation of particular institutions and agencies. Examples would include the establishment of the WTO to implement further trade liberalization, the creation of the euro and European Central Bank (governed by anti-inflationary goals) and the decision to grant the Bank of England independence (insulating it from alternative political agendas). The implementation of neoliberal policies has been a highly uneven process, however, as individual states have tended to adopt particular aspects of the neoliberal package while ignoring others. Elements of neoliberalism have, moreover, interacted with pre-existing institutional arrangements and practices in complex ways, generating a wide range of distinctive local outcomes (Box 6.5) (Peck and Tickell, 2002).

As a political and economic practice, rather than a 'pure' theory, neoliberalism has generated a number of contradictions and tensions (Harvey, 2005, pp.79–81). In terms of gaining and retaining power, neoliberals

Box 6.5

Neoliberalism 'with Chinese characteristics' (Harvey, 2005, p.120)

Following the momentous decision of the communist leadership to open up to foreign trade and investment in 1979, China has experienced a rate of economic growth almost unsurpassed in recent history, averaging 9.5 per cent a year between 1980 and 2003 (Wolf, 2005, p.4). Reform was initially couched in terms of the 'four modernizations' (referring to agriculture, industry, education and science, and defence), before a period of retrenchment after the Tiananmen Square massacre of 1989. The pace of reform accelerated again in the early 1990s, after an ageing Deng

Xiaoping declared, in a tour of the southern regions in 1992, that 'to get rich is glorious' and 'it does not matter if it is a ginger cat or a black cat as long as it catches mice' (quoted in Harvey, 2005, p.125). The implementation of neoliberal reforms in China has created a curious hybrid of communism and capitalism, creating real tensions between economic liberalization and political authoritarianism, in the form of the continuing control of the Communist Party.

The party leadership initially sanctioned the establishment of four economic zones to attract foreign

investment as local experiments that would have little effect on the rest of the economy. Three of these zones were located in the southern Guangdong province, adjacent to Hong Kong, and the other in Fujian province, across the straits from Taiwan (Figure 6.7). Such experiments proved hugely successful, with the Guangdong province in particular acting as a magnet for foreign capital. Two-thirds of foreign investment in China was being channelled through Hong Kong in the mid-1990s (Harvey, 2005, p.136). Other externally oriented areas such as coastal

Box 6.5 (continued)

cities and export processing zones were created in the 1990s with the growth of Shanghai, China's largest city, proving particularly explosive (Dicken, 2003a, p.190). Meanwhile China's main comparative advantage lay in labour-intensive goods such as textiles, footwear and toys, with hourly wages in textiles in the late 1990s standing at 30 cents compared with $2.75 in Mexico and South Korea, $5 in Hong Kong and Taiwan and over $10 in the US (ibid., p.138).

Rapid economic growth has created a number of social and political contradictions and tensions in China. One of the major difficulties that China faces is how to absorb its huge labour surplus, fuelled by massive migration from the rural areas to the coastal cities, officially estimated at 114 million workers in the reform period (Harvey, 2005, p.127). The principal response in recent years has been the development of massive infrastructure projects, financed by borrowing, such as the Three Gorges Project to divert water from the Yangtze to the Yellow River. Social inequality has increased hugely since the 1980s, with the income divide between the urban rich and rural poor comparable to some of the poorest nations in Africa. Regional inequalities have deepened as the southern and eastern coastal zones have surged ahead of the interior and the north-eastern 'rust-belt' region.

The rapid pace of growth in recent years has made China heavily dependent on imported raw material and energy: China consumes 30 per cent of the world's coal production, 36 per cent of the world's steel and 55 per cent of the world's cement (ibid., p.139). Environmental degradation and pollution have become huge problems with China containing some of the world's most polluted cities. In recent months, the government has recognized the problems caused by rapid growth, expressing concerns about the pace of development and quality of growth, although the extent to which it can now control the process is questionable (Olesen, 2006). While China has maintained controls on capital and exchange rates, protecting it against the dangers of financial speculation, the conditions of WTO membership (achieved in 2001) require these to be phased out over the next few years (Harvey, 2005, p.141). Rapid economic growth and the spread of a modern consumer culture have helped to contain demand for political liberalization, but an upsurge of riots and protests in recent years suggests that the Communist Party may face problems in retaining control.

Figure 6.7 The geography of China's 'open door' trading policy.
Source: Dicken, 2003a, p.190.

have had to foster the loyalty and support of citizens, often prompting an appeal to nationalism that has no place in the theory. At the same time, the political authoritarianism that has often been associated with the implementation and enforcement of neoliberal reforms, sometimes against popular opposition, sits uneasily with the emphasis on individual freedoms. Furthermore, the focus on market freedoms and the commodification of a wide range of cultural activities, social services and environmental resources threaten to lead to social fragmentation and anarchy as traditional social bonds and ties are dissolved in the face of an aggressive individualism. Since 2000, in the US particularly, the rise of **neoconservativism** can be seen as a response to these problems. Neoconservatism emphasizes the need for social solidarity, order and pre-emptive state action in the face of an unpredictable range of external threats and dangers such as global terrorism and 'rogue states' (ibid., pp.81–6). As the problems encountered by the so-called 'war on terror' in Iraq and Afghanistan illustrate only too readily, however, the active pursuit of neoconservative objectives threatens to create further instability and conflict.

6.5.2 Globalization and the state

The dominant view of the relationship between **globalization** and the state has been that of the hyperglobalists. Their theory holds that globalization has eroded the capacity of nation states to regulate their economies, leaving them unable to intervene meaningfully in markets to protect jobs or social conditions. As several commentators have argued, however, this simplistic 'endist' account must be rejected (Weiss, 2000). Rather than offering an objective analysis of state restructuring, it is informed by neoliberal prescriptions, representing a form of wishful thinking that views the reduction of state powers as a 'good thing'. Instead, the 'qualitative state' is experiencing a multifaceted and ongoing process of reorganization, which can partly be viewed as a response to globalization pressures.

The growth of MNCs and global financial markets are certainly associated with changes in state policy since the 1970s, with states exercising political choice in responding to globalization in certain ways. In particular, the huge rise in the volume of mobile capital and the increased openness of national economies led to the widespread abandonment of Keynesian polices in the 1980s. The goals of full employment and modest income redistribution through progressive taxation and high levels of welfare expenditure were dropped. Instead, governments across the world have focused on providing a 'good business climate' for firms, defined in terms of low inflation, reductions in taxation and flexible labour markets. Gaining the confidence of financial markets and investors requires states to ensure low inflation, which is crucial in maintaining the value of financial assets and wealth over the long term (Leyshon, 2000, p.439).

Interest rates were raised by governments such as the US and UK in the early 1980s in order to reduce inflation and attract mobile capital, deepening the recession and leading to increased unemployment. Taxes on higher earnings in particular were reduced in order to increase work incentives and stimulate entrepreneurship. In the US, President Reagan's tax cuts, coupled with increased military expenditure associated with the intensification of the cold war, resulted in a record budget deficit. More **flexible labour markets** were created by curbing trade union powers, relaxing controls on wages and conditions and encouraging local company-level bargaining instead of the integrated national bargaining that occurred under Keynesianism.

As we outlined in section 6.5.1, a more technocratic phase of 'roll-out' neoliberalism has ensued since the early 1990s. Neoliberal policy prescriptions became consolidated into the so-called **Washington Consensus**, reflecting how this agenda has been embraced and enforced by the US Treasury, the World Bank and the IMF, all headquartered in Washington DC. The Washington Consensus consists of the following key elements (Peet and Hartwick, 1999, p.52).

➤ Fiscal discipline: minimizing government budget deficits.

➤ Public expenditure priorities: promoting economic competitiveness, not the provision of welfare or redistribution of income.

➤ Tax reform: lowering tax rates and strengthening incentives.

➤ Financial liberalization: the market determinating interest rates and capital flows.

➤ Trade liberalization: eliminating restrictions on imports.

➤ Foreign direct investment: removing barriers to the entry of foreign firms.

➤ Privatization: selling off state enterprises

➤ Deregulation: abolishing rules that restrict competition.

In adopting the policies outlined above, states have effectively acted as key agents of globalization. Discipline is exercised by the prospect of capital flight (investors withdrawing their money) if alternative policies stressing full employment or the redistribution of wealth are chosen, threatening higher inflation, and by the power of the IMF and World Bank to refuse debt rescheduling for developing countries and to declare them uncreditworthy.

At the same time, states also shape processes of globalization through their relations with a range of organizations such as MNCs, financial institutions, supranational authorities and international agencies. The emergence of a more prominent **supranational tier of government** involving bodies such as the WTO and EU is largely the result of state action in that states have come together to create such bodies. Membership provides a forum for states to assert and extend their power, requiring regular contact and discussion with other states. It is state representatives who negotiate over trade in the WTO while states have sought to exert some supervision over international financial transactions through bodies such as the Bank for International Settlements (O'Neill, 1997, p.297). Regional economic integration offers states access to larger markets and protection against competition from outside the regional bloc. The best example is the EU, which has evolved from a customs union of six states into a monetary union containing 25 states (Box 6.6). Other regional trade blocs include the North American Free Trade

Table 6.4 Major regional economic blocs

Regional group	Membership	Date	Type
EU (European Union)	Austria, Belgium, Czech Republic, Cyprus, Denmark, Estonia, France, Finland, Germany, Greece, Hungary, Ireland, Italy, Latvia, Lithuania, Luxembourg, Malta, Netherlands, Poland, Portugal, Slovakia, Slovenia, Spain, Sweden, United Kingdom	1957 (European Economic Community) 1967 (European Economic Community) 1992 (European Union)	Economic union
NAFTA (North American Free Trade Agreement)	Canada, Mexico, United States	1994	Free trade area
EFTA (European Free Trade Association)	Iceland, Norway, Liechtenstein, Switzerland	1960	Free trade area
MERCOSUR (Southern Cone Common Market)	Bolivia, Colombia, Ecuador, Peru, Venezuela	1973	Common market
AFTA (ASEAN Free Trade Agreement)	Brunei Darussalam, Cambodia, Indonesia, Laos, Malaysia, Myanmar, Philippines, Singapore, Thailand, Vietnam	1967 (ASEAN), 1992 (AFTA)	Free trade area

Source: Based on Dicken, 2003a, p.149.

Box 6.6

The EU and state reorganization

The institutional structure of the EU is unique. It is neither a federation nor a simple intergovernmental organization. Decision-making is centred upon the 'institutional triangle' formed by the European Commission, the Council and the European Parliament. The Council of Ministers is the main decision-making body, representing the member states. The heads of government and foreign ministers meet at least twice a year in the European Council, high-profile summits that attract widespread media coverage. The main roles of the Commission (25 members appointed by the Council) are to initiate legislation and proposals, to implement European legislation, budgets and programmes and to represent the EU on the international stage. The European Parliament, consisting of directly elected MEPs, has powers over legislation and the budget that are shared with the Council. It also exercises democratic supervision over other EU institutions, particularly the Commission.

The EU operates on the basis of agreement and negotiation between independent states, although the member states have delegated some of their functions to the central institutions. Member states continue to shape how the powers of the EU are exercised and implemented, often seeking to further their national interests within the supranational space it provides. This is evident even in areas like trade where the EU is the decision-making body. While the EU is represented as one body within the current Doha round of trade negotiations, different member states have adopted different stances. Britain's liberal free trade views, for example, are countered by the more cautious position of countries like France, making the job of the Commission officials particularly difficult.

Association (NAFTA), incorporating Canada, Mexico and the US, and AFTA in South-east Asia (see Figure 5.5 and Table 6.4). States have certainly lost some powers to the EU, particularly in the economic sphere, such as those over trade, competition and monetary policy (for those within the Eurozone), as well as in sectors such as agriculture and fisheries. Even in these areas, though, state representatives develop policy and national governments continue to exercise some authority. In other spheres there is increased cooperation between national states in the face of threats such as terrorism, drug trafficking and organized crime. The operation of such organizations, then, seems to reinforce and reconstruct state power, rather than usurping it (ibid., p.297).

States in the developing world have also been affected by neoliberalism and globalization. In return for providing grants and loans to poor countries that are short of capital, the World Bank and IMF require governments to meet strict conditions, referred to as 'conditionality', a highly controversial issue in development circles. In response to the debt crisis facing many developing countries in the 1980s and 1990s, the agencies have made the provision of further support dependent on governments undertaking a set of economic reforms, generally known as structural adjustment programmes (SAPs) (section 11.3.3). These are based on the Washington Consensus specified above, requiring countries to devalue their currencies, reduce public spending and inflation, liberalize trade and privatize state enterprises (section 11.3.3). The effects of these policies are often damaging, reflecting how they have been imposed from above, irrespective of economic conditions in developing countries. According to Joseph Stiglitz, the former Chief Economist of the World Bank:

> Forcing a developing country to open itself up to imported products that would compete with certain of its industries, industries that were dangerously vulnerable to competition from much stronger counterpart industries in other countries, can have disastrous consequences. Jobs have systematically been destroyed ... even worse, the IMF's insistence on developing countries retaining tight monetary policies has led to interest rates that would make job creation impossible even in the best of circumstances. And because trade liberalization occurred before safety nets were put in place, those who lost their jobs were forced into poverty.

(Stiglitz, 2002, p.17)

In this way, the burden of adjustment has been placed on the poor countries themselves rather than the Northern institutions and banks that lent the money in the first place. This runs contrary to neoliberal theory, which holds that lenders should also the bear the responsibility of providing bad loans, indicating how the US government has sought to protect the interests of its own banks and financial institutions (Harvey, 2005, p.73). Debt repayments have resulted in the transfer of resources from impoverished developing countries to powerful institutions in developed countries, prompting Stiglitz to comment about the peculiarity of a world in which 'poor countries are in effect subsidising the richest' (quoted in Harvey, 2005, p.74).

6.5.3 Workfare and the 'competition state'

Since the early 1980s, the overriding purpose of the state has been redefined by neoliberals as that of promoting national economic development rather than the provision of welfare services to its citizens. National economic prospects have come to be viewed in terms of ensuring and promoting **competitiveness**. This refers to the underlying strength of the economy in terms of its capacity to compete with other countries, generating increased wealth and improving the standard of living (Bristow, 2005). It is based on the assumption that nations and regions compete for global market share in a similar fashion to firms. Key aspects of competitiveness include the levels of innovation, enterprise and workforce skills and the task of the '**competition state**' is to foster these capacities (Jessop, 1994). Labour market flexibility is an important aspect of competitiveness, requiring workers to be prepared to expand their skills and modify their wage claims and working practices in response to prevailing economic conditions. The emphasis on the promotion of innovation and competitiveness and the subordination of social policy to these primary objectives are central features of the competition state.

Concerns about the welfare state making benefit claimants dependent on the state and eroding their incentives to participate in paid work have prompted the introduction of a number of **workfare** initiatives, particularly in the US and UK. Workfare can be defined as a system that requires people to work in exchange for welfare benefits and payments. Its introduction over the last two decades has involved a number of schemes

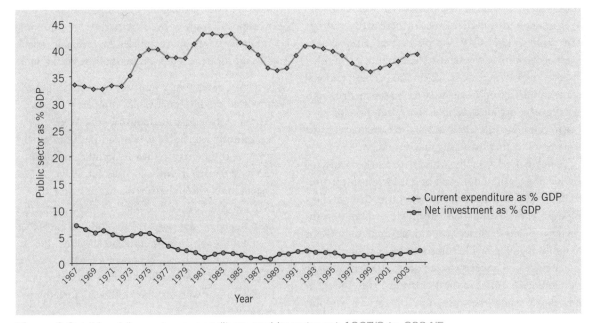

Figure 6.8 UK public sector expenditure and investment 1967/8 to 2004/5.
Source: HM Treasury, 2005, *Public Expenditure Statistical Analyses 2005*. Reproduced under the terms of the Click-Use Licence.

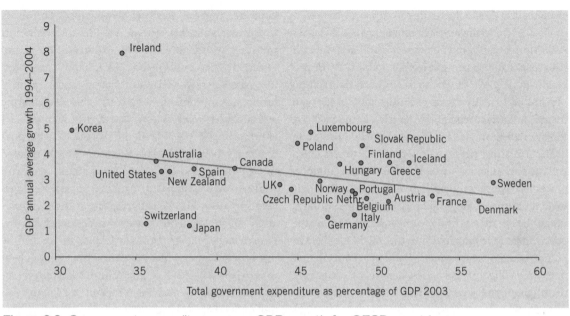

Figure 6.9 Government expenditure versus GDP growth for OECD countries.
Source: Cumbers and Bird, 2005.

that encourage people to come off benefits and into work, education or training, relying on incentives where possible but resorting to compulsion if necessary, exercised by cutting the benefits of those who fail to comply.

US states have pioneered the introduction of workfare schemes, beginning under the Reagan administration in the 1980s. Experiments in states such as Massachusetts, California and Wisconsin generated wide international interest from politicians and officials looking for ways to address the problem of welfare dependency. Under 'roll-out' neoliberalism, workfare has become part of the mainstream centre-left consensus, with Clinton and Blair stressing its central role in the politics of the 'third way' (section 6.5.1) (Peck, 2001). Elements of **international policy transfer** are evident in terms of the introduction of workfare to Britain through national programmes such as the Jobseekers' Allowance introduced by the Conservatives in the mid-1990s and Labour's New Deal programme for unemployed young people.

Workfare programmes reflect the neoliberal conception of unemployment as stemming from the inadequacies of the individual in terms of a lack of suitable skills or an unwillingness to work. By contrast, in

the 1960s and 1970s unemployment was seen as the product of wider economic and social forces such as a lack of sufficient demand in the economy. The introduction of workfare schemes signals a crucial move away from the welfarist notion of universal entitlement to benefits and voluntary programme participation in favour of compulsion, selectivity and active inclusion in the labour market (ibid., p.424). For critics of workfare, it rarely leads to permanent employment and governments would be better advised to create real jobs by expanding the economy. According to Jamie Peck – an economic geographer who has conducted detailed research on the subject – 'workfare is not about creating jobs for people that don't have them; it is about creating workers for jobs that nobody wants' (quoted in Painter, 2002, p.169).

The notion of a workfare-oriented competition state captures some of the underlying shifts that seem to be redefining the role of the state, particularly in North America and Europe. Its importance should not be exaggerated, however. In spite of the rhetoric of reform and the many experiments that have been introduced, the welfare state remains surprisingly resilient (Martin and Sunley, 1997, p.284). Levels of public expenditure in the UK, for instance, have not dropped significantly since the late 1960s, fluctuating in line with the economic cycle (reflecting levels of unemployment) (Figure 6.8). More broadly, the public sector still accounts for between 35 and 55 per cent of GDP in most developed countries (Figure 6.9). The social-democratic tradition remains particularly strong in continental Europe with many countries remaining largely untouched by welfare reform in the face of strong public pressure to protect existing rights and services (Painter, 2002, p.170). For example, in early 2006, French students protested against proposals to introduce a new law making it easier for employers to dismiss workers under the age of 25 without compensation, forcing the government to rethink its claims that this would help to reduce youth unemployment (Hutton, 2006). Rather than signalling the inevitable demise of welfarism, then, the emphasis on workfare and the competition state focuses attention on emerging policy directions and trends.

6.5.4 New forms of local and regional development

One of the contradictory effects of globalization is to have increased rather than diminished the significance of geographical differentiation and place (Harvey, 1989a). A key element of this has been the **resurgence of local and regional levels of government**, particularly as a result of the devolution of powers from the central state. This is closely bound up with the growing emphasis on economic competitiveness and growth that has seen local and regional organizations become increasingly active in seeking to attract investment and to support innovation and entrepreneurship within their regions. As the Keynesian **mode of regulation** associated with **Fordism** has been dismantled in the face of globalization and neoliberalism, regions have become increasingly exposed to the effects of global competition. In particular, the deindustrialization of many traditional industrial regions has forced local authorities and development agencies to focus on the problem of economic regeneration, requiring the attraction of new investment to generate growth and employment.

As part of the move way from Keynesian policies towards a 'competition' state, traditional regional policy (section 6.3) has been downgraded. The objective of spatial redistribution and the interventionist measures associated with it were relaxed in the face of uncertain economic conditions and neoliberal ideology. Regional policy was viewed as an expensive luxury as governments sought to reduce expenditure and adopt less interventionist, market-friendly policies (Knox *et al.*, 2003, p.384). In the UK, the Thatcher government's diagnosis of the 'regional problem' pointed to rigidities in the operation of labour markets with workers and unions demanding excessive wages (Mohan, 1999, p.185), deterring investment and creating unemployment, a view encapsulated by the Employment Minister Norman Tebbit's advice that the unemployed should 'get on their bikes' in search of work (Jenkins, 1987, p.326). The Conservatives cut regional assistance from £842 million in 1979 to £560 million in 1985 (Mohan, 1999, p.185) and reduced the areas eligible for assistance in both 1983 and 1993 (Figure 6.10). In France, restrictions on expansion in the Paris region were dropped in the 1980s as the government realized the need to encourage its development as a world city capable of attracting corporate headquarters and advanced financial and business services.

A range of agencies and organizations have become involved in local and regional economic development since the early 1980s. As well as traditional local authorities, special-purpose agencies were created by central government to address specific problems. For example, Urban Development Corporations were created in certain parts of England and Wales in the 1980s in order to regenerate derelict inner-city areas (Imrie and Thomas, 1999), while Regional Development Agencies were established in 1999 to cover each of the standard regions of England. Other organizations involved in local economic development include business organizations such as chambers of commerce and more specialist trade associations, educational institutions and training agencies (see Figure 6.11). In attempting to address local economic problems, local agencies often work together in 'partnership' where they develop common objectives and share resources. This organizational complexity has prompted the widespread use of the term local **governance** in place of the traditional notion of local

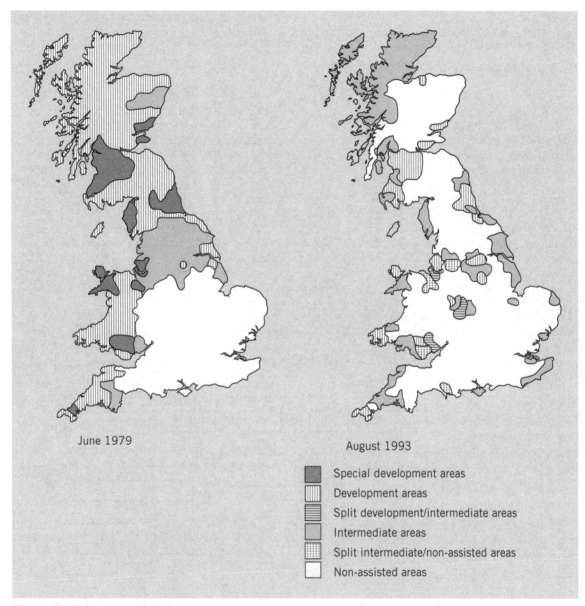

June 1979

August 1993

■ Special development areas
▦ Development areas
▤ Split development/intermediate areas
■ Intermediate areas
▨ Split intermediate/non-assisted areas
□ Non-assisted areas

Figure 6.10 Areas eligible for regional policy assistance, 1979 and 1993.
Source: Mohan, 1999, p.182.

government, incorporating the role of special-purpose agencies, business interests and voluntary organizations alongside local authorities (Stoker, 1999).

In contrast to traditional regional policy, which aimed to direct investment into depressed regions through a range of incentives and controls (Box 6.3), recent approaches have sought to facilitate growth and enhance the **competitiveness** of the regional economy

(Amin, 1999). This **new model of local and regional development** is more 'bottom up' in nature, focusing on the need to develop local skills and stimulate enterprise, giving the regions 'a hand up rather than a hand out' (Amin *et al.*, 2003, p.22). The new approach focuses on internal factors and conditions within regions, viewing these as the key to attracting investment and generating growth. In this sense, 'locally

Box 6.7

Regional development policies in Wales

As an old industrial region that has experienced severe deindustrialization since the 1970s, prompting regional agencies to introduce a range of strategies and initiatives, Wales can be regarded as a suitable laboratory for examining the evolution of regional development policy. Along with Scotland, which also enjoys national status within the UK, Wales has been granted greater institutional autonomy and resources in comparison to the standard regions of England. A territorial department of government, the Welsh Office, was established in 1964 and a special-purpose development body, the Welsh

Development Agency (WDA) created in 1975 in response to deindustrialization and growing support for the Welsh nationalist party, Plaid Cymru. In 1999, Wales was granted devolution within the UK, along with Scotland and Northern Ireland, through the establishment of the directly elected National Assembly for Wales (NAfW), giving it control of unelected agencies (quangos) such as the WDA. The main organizations involved in economic development in Wales following devolution are shown in Figure 6.11, including Education and Learning Wales (ELWa) and Wales Trade International as well as

22 local authorities. The Assembly and quangos have sought to 'regionalize' their operation through the establishment of four official regions. The Assembly decided to abolish the WDA – along with some other prominent quangos – as an independent agency from 1 April 2006, incorporating its functions within the Welsh Assembly Government and giving it more direct control over economic development.

In the early 1980s, Wales lost over 100,000 jobs in the traditional heavy industries of coalmining, steel making and metals manufacturing (Cooke, 1995, p.45), creating huge

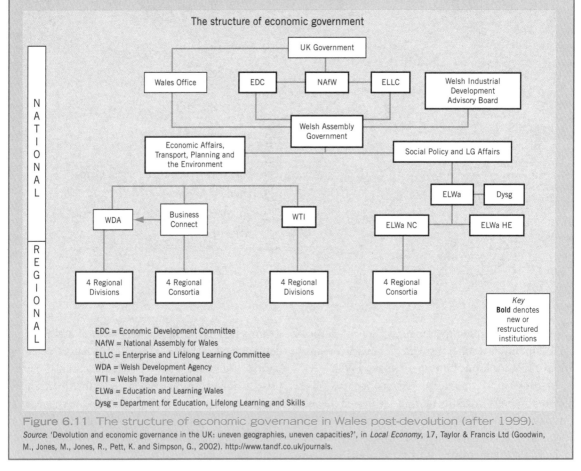

The structure of economic government

EDC = Economic Development Committee
NAfW = National Assembly for Wales
ELLC = Enterprise and Lifelong Learning Committee
WDA = Welsh Development Agency
WTI = Welsh Trade International
ELWa = Education and Learning Wales
Dysg = Department for Education, Lifelong Learning and Skills

Key
Bold denotes new or restructured institutions

Figure 6.11 The structure of economic governance in Wales post-devolution (after 1999).

Source: 'Devolution and economic governance in the UK: uneven geographies, uneven capacities?', in *Local Economy*, 17, Taylor & Francis Ltd (Goodwin, M., Jones, M., Jones, R., Pett, K. and Simpson, G., 2002). http://www.tandf.co.uk/journals.

Box 6.7 (continued)

problems of unemployment, poverty and industrial dereliction in the South Wales valleys particularly, where these industries had been concentrated. The regional state responded by trying to attract new inward investment from the 1960s, following the direction of traditional regional policy. From its inception in 1975 until the early 1990s, the WDA focused on attracting FDI, building new factories and reclaiming contaminated land, becoming Europe's largest industrial property owner and gaining a reputation as a builder of commercial 'sheds'. The FDI-based strategy had considerable success, with the availability of labour, infrastructure and a range of incentives attracting a large number of foreign MNCs to Wales. In the early 1990s, Wales, with just 5 per cent of the UK population, was attracting 20 per cent of new FDI into the UK (Cooke and Morgan, 1998, p.147). Real concerns remained, however, about the creation of a 'branch plant' economy based on routine manufacturing and assembly operations that lacked higher-status and skilled jobs (see section 7.5.2).

Partly in response to the branch plant criticism, the strategy adopted by the WDA changed in the early 1990s, reflecting the broader shift in regional development policy outlined above. A number of supply-side initiatives were developed that aimed to develop local capacities and skills, focusing on encouraging innovation, enterprise and skills. While these initiatives aimed to enhance the competitiveness of the regional economy, the key role of state organizations such as the WDA and NAfW points to an important continuity with old-style regional policy. One high-profile initiative is Source Wales, a programme designed to bring together foreign investors in Wales and indigenous firms with the potential to supply components and services to them (ibid., pp.154–7). The assistance available from the EU has played an important role, with Wales chosen as one of 8 pilot regions for the introduction of regional innovation strategies in the mid-1990s (Morgan, 2004, pp.880–81). Despite such initiatives, however, Wales remains one the three poorest regions in the UK, along with Northern Ireland and north-east England, with its relative position having got worse since the 1970s, although its performance is likely to have been even poorer without the efforts of the WDA and other agencies.

orchestrated regional development has replaced nationally orchestrated regional policy' (ibid., p.22). Key dimensions of policy include initiatives to stimulate innovation and learning within firms, measures to try to increase entrepreneurship in terms of the number of new firms that are being created, and efforts to develop and upgrade the skills of the workforce through a range of training and education programmes (Box 6.7).

These measures are focused on the **supply side** of the local or regional economy, defined in terms of the quality of the main factors of production such as labour (training, skills), capital (emphasizing enterprise and innovation) and land (sites and infrastructure for investors). Improving these supply-side factors is seen as vital to the competitiveness of the regional economy in relation to the other regions and localities against which it is competing for investment and markets. This can be contrasted with the demand-side emphasis of Keynesian policies, stressing the injection of additional purchasing power into the economy through increased investment or government expenditure. At the same time, however, the central state often remains the key funder and coordinator of local and regional development programmes (Box 6.7). The continuities between the 'old' and 'new' models of regional development are evident in initiatives such as France's *poles de competitivité* policy, introduced in 2004/5, where regionally designated growth centres, based largely on high-tech industry, have to compete for central government funding (Ancien, 2005, p.227). In Europe, the EU has become a key funder of regional development programmes, partly compensating for the reduced levels of assistance available from the central state and providing an important focus for the activities of local and regional agencies in lobbying for additional resources (MacLeod, 1999).

The consolidation of neoliberal doctrines of free trade and capital mobility at the national and international scales by national states and bodies such as the EU, WTO and World Bank has established a framework of interregional competition within which local and

Table 6.5 Regional inequalities: variance of the log of regional GDP per capita

Country	Percentage Change		
	1980–90 %	1990–2000 %	1980–2000 %
Developing countries			
China	−16.31	20.21	0.61
India	7.11	16.96	25.27
Mexico	−1.29	13.57	12.11
Brazil	−17.16	1.33	−16.06
Developed countries			
USA	11.75	−2.69	8.74
Germany	2.18	−0.96	1.2
Italy	1.55	3.01	4.6
Spain	−3.92	10.47	6.14
France	8.63	−0.31	8.3
Greece	1.22	0.13	1.35
Portugal		1.82	
European Union		11.25	

Source: Rodríguez-Pose and Gill, 2004, p.2098.

regional development organizations operate. Regions have effectively become 'hostile brothers' directly competing for investment, markets and resources (Peck and Tickell, 1994). The unequal relationship between mobile capital on the one hand and communities requiring employment and income on the other has allowed MNCs to play regions off against one another, on the basis of labour costs and the financial assistance offered by government agencies, sparking fears of a 'race to the bottom' as wage rates and living standards are progressively undermined. Despite this, the prevailing supply-side approach presents regional development as a 'race to the top' where all regions can be winners if they follow the right policies.

A key issue for economic geographers is the relationship between local development strategies and **regional inequalities**. Following a period of regional convergence under 'spatial Keynesianism' in the post-war era, regional disparities have widened since 1975 (Dunford and Perrons, 1994; Rodríguez-Pose and Gill, 2004).

Table 6.5 shows trends in regional inequalities for selected countries, with 10 of the 11 experiencing increased disparities between 1980s and 2000. This outcome is hardly surprising since current policy treats 'unequal regions equally' (Morgan, 2006, p.189), assuming that the same broad strategy should be adopted in all regions, in contrast to the spatial Keynesianism of the 1960s and 1970s, which sought to direct investment to depressed regions and restrict growth in core areas. The magnitude of regional inequalities is generally modest for developed countries but greater for developing countries, with the exception of Brazil where regional inequalities fell.

Reflect

➤ Do neoliberal polices provide the basis for a distinctive post-Keynesian mode of regulation, ensuring prolonged growth and stability (Table 2.2)? Justify your answer.

6.6 Summary

The state refers to a set of institutions that holds sovereignty over a designated territory, exercising a monopoly of legitimate force and law-making ability. The key concept of the qualitative state has been used to frame and inform this chapter, emphasizing that we should focus on the 'nature, purpose and consequences' of state intervention in the economy rather than its extent or magnitude (O'Neill, 1997, p.290). The **Keynesian Welfare State** was the key expression of the state in developed countries over the middle decades of the twentieth century. Shackled to the Fordist system of economic organization, it was associated with over two decades of sustained economic growth between 1945 and the late 1960s. The long post-war boom ended in the 1970s when rising inflation and high unemployment led to a crisis of the state. At the same time, the NICs of East Asia were experiencing rapid industrialization, shaped by the actions of '**developmental states**' that coordinated the process of economic development through powerful government agencies, focusing on the protection of strategic 'infant industries' until they were strong enough to compete internationally, the channelling of investment into these industries and the promotion of exports.

The main political response to the crisis of the Keynesian Welfare State was that of the New Right. Their neoliberal agenda has spread across the globe, underpinning the **Washington Consensus**, implemented through bodies like the WTO, World Bank and IMF. Neoliberalism has sought not only to reduce the role of the state in the economy, but also to reshape the 'internal' structures of the state. Key trends include a move towards a 'competition state', the introduction of workfare initiatives and a focus on enhancing the economic competitiveness of regions through a range of supply-side measures. In geographical terms, the supranational and regional scales of governance have become increasingly important (Jessop, 1994). It would be a mistake to conclude from this, however, that the role of the national state has been reduced. As suggested by the notion of the qualitative state, the increased prominence of certain scales of state action should not be regarded as resulting in the erosion or decline of others. The overriding conclusion of this chapter is that the national state remains a crucial actor in the regulation of the economy and its geography, not least through its role in coordinating activities carried out across different scales. While the nature, purposes and consequences of state action have changed since the 1970s, the overall economic significance of the state has not been reduced.

Exercises

Select a particular region, referring to basic economic statistics (GDP, income, growth, employment and unemployment) available from the appropriate government publications or website to get a basic sense of economic conditions within it. Examine and review the current economic strategy for that region, identifying its strengths and weaknesses.

1. What are the key agencies and organizations involved in the formulation and implementation of the strategy (e.g. in England the RDAs are responsible for regional economic strategies, which have to be approved by the national government)?

2. What are the key elements of the strategy?

3. How realistic or appropriate is it in relation to regional economic conditions and needs?

4. On what assumptions is the strategy based?

5. To what extent is it framed by the new model of local and regional economic development identified in section 6.5.4?

6. Are there any major omissions from the strategy?

7. Are there any potential tensions or conflicts between different objectives?

8. What alternative objectives or priorities would you like to see included?

9. Based on your analysis, sketch your own economic strategy for the region. How would you characterize this strategy (e.g. Keynesian, neoliberal, alternative)?

Key reading

Douglass, M. (1994) 'The "developmental state" and the newly industrialized economies of Asia', *Environment and Planning* A 26: 453-66.

An account of the notion of the 'developmental state', focused on the four NICs of East Asia. The paper offers a good introduction to the concept, examining the key relationships between the state, capital and labour. It highlights important differences between the strategies adopted by individual countries.

Harvey, D. (2005) *A Brief History of Neoliberalism*, **Oxford: Oxford University Press.**

A stimulating account of the impact of neoliberalism across the world from the leading Marxist geographer. Harvey assesses the growth of neoliberal theories, their impact on state polices and the effects on economic growth and development. He views neoliberalism as a project to restore upper-class power and wealth that was developed in response to the economic crisis of the 1970s.

Martin, R. and Sunley, P. (1997) 'The post-Keynesian state and the space economy', in Lee, R. and Wills, J. (eds) *Geographies of Economies*, **London: Arnold, pp.278-89.**

An incisive review of recent debates on state restructuring in Europe and North America. In particular, Martin and Sunley provide a clear definition of the Keynesian Welfare State and review the main claims that have been made about the impact of globalization and the shift to workfare. Their position is sceptical, stressing the exaggerated nature of such claims and the continuities between the Keynesian and post-Keynesian states.

O'Neill, P. (1997) 'Bringing the qualitative state into economic geography', in Lee, R. and Wills, J. (eds) *Geographies of Economies*, **London: Arnold, pp.290-301.**

The key article in which the concept of the qualitative state is developed. O'Neill criticizes the focus on the extent of state intervention, focusing attention on the 'nature, purpose and consequences' of state action (p.290). Instead of heralding a decline in the importance of the state, the period since the 1970s has seen the nature and purpose of its intervention in the economy change significantly.

Painter, J. (2002) 'The rise of the workfare state', in Johnston, R.J., Taylor, P. and Watts, M. (eds) *Geographies of Global Change: Remapping the World*, **2nd edn, Oxford: Blackwell, pp.158-73.**

A concise and accessible introduction to recent debates on the changing nature of the state. Reviews the rise and fall of the Keynesian Welfare State and explores the alleged shift towards a workfare state since the 1990s. Less sceptical about the significance of these changes than Martin and Sunley.

Stiglitz, J. (2002) *Globalization and its Discontents*, **London: Penguin.**

An 'insider' critique of the management of globalization by the former chief executive of the World Bank. The book focuses on the role of international economic organizations, particularly the IMF and WTO. Stiglitz is heavily critical of these organizations for forcing inappropriate neoliberal policies on developing countries, reflecting how the rules of globalization are rigged in the interests of Western countries.

Useful websites

http://www.commerce.ca.gov/state/ttca_homepage.jsp
The official site of the Commerce and Economic Development Program of the State of California. Provides a useful insight into current regional development policy, containing details of a range of programmes and initiatives.

http://europa/eu/index_en.htm
The official site of the European Union. Contains a wealth of information on the EU's activities, divided into specific topic areas. See especially the sections on 'economic and monetary affairs', 'enterprise', 'external trade' and 'regional policy'.

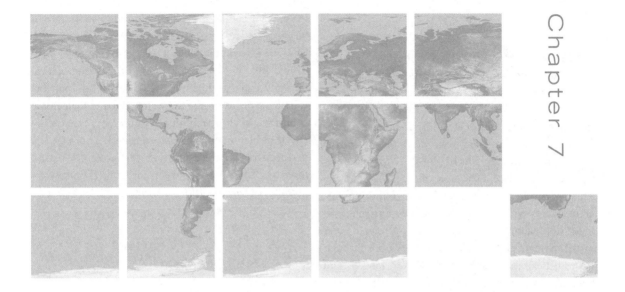

Chapter 7

The changing geographies of the multinational corporation

Topics covered in this chapter

➤ The changing geography of foreign direct investment (FDI).

➤ The emergence of the multinational corporation and the 'new international division of labour' (NIDL).

➤ The 'globalness' of multinational corporations.

➤ Understanding the geographical embeddedness of MNCs and how this influences corporate strategy.

➤ The implications of MNCs for host regions.

Chapter map

We begin the chapter with a definition of the multi-national corporation (MNC), followed by an analysis of their emergence and growing importance in the global economy. Section 7.2 looks at the changing geography of foreign direct investment by corporations since the 1960s while section 7.3 outlines the reasons for the growth of MNCs. Section 7.4 explores the extent to which MNCs have become global entities. The chapter then examines (section 7.5) recent developments in the evolution of MNCs and in particular the increased use of outsourcing and more complicated patterns of corporate organization. In the last two sections of the chapter we consider the relationships between MNCs and the regions they are located in, assessing both the influence that their home regions have on them (section 7.6) as well as the impact they have on the regions they locate in (section 7.7). This leads us to highlight the continued importance of geography in understanding their operation.

7.1 Introduction

Multinational corporations (MNCs) have become one of the dominant actors of globalization and advanced capitalism due to their size and the geographical reach of their operations. Indeed, MNCs are fundamental to the dynamics of the global economy and their organizational geographies are one of the principal mechanisms through which global economic integration takes place. Through their geographically dispersed production networks, MNCs connect up different places in increasingly complex international divisions of labour. In contrast to the hyperreality and virtual spaces of global financial networks, it is through MNCs' organizational spaces that the more tangible and material economic geographies of the global economy are organized. MNCs are increasingly at the heart of 'webs of enterprise' (Dicken, 2003a) that connect headquarters, assembly plants, research and development facilities and increasingly complex supply chains to produce individual products and commodities.

Despite having this critical role in the functioning of the global economy, it is important to dispel the image of all-powerful spaceless organizations that characterizes some of the business and media literatures (e.g. Ohmae, 1990). Instead, as we will demonstrate in this chapter, to understand the workings of MNCs we need to appreciate their diverse organizational geographies. While they may operate transnationally, as organizations whose everyday practices link up spatially dispersed networks of operations, MNCs themselves continue to be embedded in particular organizational geographies, influenced both by the national and regional business cultures from which they originate and by the characteristics of the regions in which they locate. For this reason, we prefer to use the term multinational here rather than transnational. This is to signify corporations that operate in more than one country but are geographically embedded or rooted in particular places, rather than the popular notion that corporations are now footloose and not tied to anywhere, hence 'transnational' (section 7.3).

The chapter begins by tracing the emergence of MNCs as part of the growing importance of **foreign direct investment** (FDI) to world economic development in the period since 1945. We then explore some of the reasons behind the growing internationalization of business activity before considering the extent to which MNCs are becoming fully globalized in the current era. This is followed by consideration of the geographical diversity that characterizes MNCs, focusing upon the ways that they continue to be embedded in their home countries. In the penultimate section we explore the effects of FDI on host regions, considering both critical perspectives and more positive recent accounts. We use three types of case study to consider how FDI plays out in different local contexts.

7.2 The changing geography of FDI

Although international trading connections between firms have been around for a long time (see Chapter 5), the significance of the multinational corporation as a key actor in the global economy can be traced to the 1960s when researchers first began to comment on the 'multinational corporation' as a phenomenon. After the end of the Second World War, the increase in international trade that accompanied peace and political stability in Europe provided a boost for the internationalization of business activities with a dramatic increase in FDI. By the 1960s, rates of growth in FDI – which essentially is a measure of international flows of investment within firms – were growing 40 per cent faster than exports (Dicken, 2003a, p.52). After a brief slump in the 1970s, FDI levels experienced another dramatic growth from the early 1980s onwards (see section 5.4.2).

The changing geography of FDI over the past century paints a particularly revealing picture of both the continuities in the pattern of global economic development and the shifts in recent years. If we consider, first, the stock of outward FDI (i.e. from the country of origin) the period from 1914 up to the late 1970s marks the decline of the UK as the dominant source and the growth of the United States (Table 7.1a). The expansion overseas of US MNCs was particularly dramatic after the end of the Second World War, when the development of a stable trade regime under the Bretton Woods agreement, the recovery of the Western European economy and the dominance of

Table 7.1a Foreign direct investment outward stock by leading investing countries as a percentage of world total, 1914–2001

	1914 %	1960 %	1978 %	2001 %
Developed countries	100.0	99.0	96.8	92.2
US	18.5	49.2	41.4	21.1
Western Europe	–	–	–	56.8
UK	45.5	16.2	12.9	14.3
Germany	10.5	0.8	7.3	7.9
France	12.2	4.1	3.8	7.8
Japan	0.1	0.7	6.8	4.6
Developing countries	neg.	1.0	3.2	11.8

Source: derived from Knox and Agnew, 1989, Tables 8.4, 8.5, pp.255–6, and UNCTAD, 2002, Annex Tables B.3, B.4, pp.310–17.

Table 7.1b Foreign direct investment inward stock by world region and selected countries as a percentage of world total, 1914–2001

	1914 %	1960 %	1978 %	2001 %
Developed countries	37.2	67.3	69.6	65.8
of which:				
US	10.3	13.9	11.7	19.3
Western Europe	7.8	22.9	37.7	40.6
Japan	0.2	0.2	1.7	0.7
Developing countries	62.8	32.3	27.8	31.9
of which:				
Latin America	32.7	15.6	14.5	10.1
Africa	6.4	5.5	3.1	2.3
Asia	20.9	7.5	7.0	19.4
Central and Eastern Europe	N/A	N/A	N/A	2.3

Source: derived from Knox and Agnew, 1989, Tables 8.4, 8.5, pp.255–6, and UNCTAD, 2002, Annex Tables B.3, B.4, pp.310–17.

the US dollar as a global exchange currency, presented new market opportunities (Wright, 2002, p.71). The period since the 1960s charts the growing importance of overseas investments by Japanese firms in particular, but also the growth of European countries such as France and Germany. Despite these developments, the US and UK are still the leading source countries for FDI.

In terms of inward FDI (see Table 7.1b) – desti-

nation countries for investment by MNCs – in the early part of the twentieth century, developing countries were the dominant destination, largely linked to resource and raw material extraction to supply the industries of the developed world, a set of relations that we can term the 'old international division of labour'. But after the First World War, and especially since the end of the Second World War in 1945, there has been a shift towards greater investment flows between

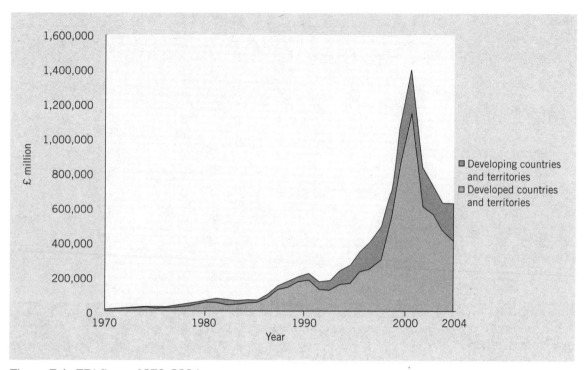

Figure 7.1 FDI flows, 1970–2004.
Source: Adapted from UNCTAD database www.unctad.org.

different regions of the developed world, with FDI linked more to the setting up of manufacturing plants either for reasons of market access or to take advantage of low-wage labour. Western Europe in particular became the major destination for FDI from the US in the period from the late 1940s up to the mid-1970s, with over 40 per cent of the world share by the latter period (Dicken, 2003a, p.59). Between the mid-1970s and 2001 there was a marked growth in FDI to Asia, particularly East Asia, reflecting the rapid economic development of the NICs (Box 5.3) and China, while other parts of the world, such as Africa and Latin America, saw their share of foreign investment fall or remain static. Overall, Western Europe continues to dominate as the destination for inward investment, but there has been a marked drop in FDI globally since 2000 (Figure 7.1).

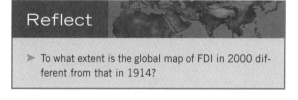

Reflect

➤ To what extent is the global map of FDI in 2000 different from that in 1914?

7.3 Understanding the emergence of the MNC

There are two main ways in which firms internationalize: through the setting up of new plant or facilities; and through the acquisition of existing factories or firms (Dicken, 2003a). The choice of strategy in turn will be dependent on the motivations behind the initial decision to internationalize. For example, if a firm is attracted to a country because labour is cheap, it might set up a new production plant there that is integrated into its existing operations at home, supplying the domestic market. Alternatively, a firm might prefer to transfer its existing management methods and production strategies by setting up a completely new venture. Conversely, a firm may have identified a new growth market in the destination country that can provide a better return than some of its existing activities at home but, for reasons of cost and uncertainty, decided that the acquisition of an existing plant is a less risky option than developing a new plant from scratch.

Acquisition might also be driven by a desire to reduce the competition in its market by taking over a foreign rival.

7.3.1 Why do firms internationalize?

In addressing the issue of why firms internationalize, we can return to some of the insights about the functioning of the economy in Chapter 3. Firms operating in a capitalist system have the ultimate aim of realizing surplus value or profit. The rate of profit must increase over time for a firm to reinvest and stay ahead of the competition, as we saw through the operation of the **circuit of capital**. In successful firms this is usually accompanied by expansion both at the level of the organization itself, with economies of scale providing increasing returns on capital, and growth in geographical terms as firms seek to overcome the initial restrictions of place.

Initial increases in surplus value are achieved through increasing the productivity of existing operations, by increasing the exploitation of labour or by developing new technology or new forms of organization (as we noted in Chapter 3). This can continue successfully for some time, but at some point the returns on capital from remaining in its original location begin to fall, through the growing organization and resistance of labour, increased competition from other firms or the achievement of productivity limits in existing technologies and methods of production. Addressing these falling profit rates requires firms to seek out cheaper locations than existing arrangements. The need to resolve this dilemma and develop a new geography of production has been termed the **'spatial fix'** (Harvey, 1982), where capitalists seek to set up operations elsewhere. Typically, this will usually involve relocation within the existing national economy in the first instance, creating a new spatial division of labour (Massey, 1984), but over time, as new limits to profitability are reached at the national level, firms will start to internationalize their operations.

Investment overseas is usually driven by the desire for either market access or tapping cheaper supplies of labour. In this respect, it is likely that market access accounts for the largest share of FDI, given that around 67 per cent of foreign investment flows are between the world's developed economies (Dicken, 2003a, p.58). Certainly, the flow of US investment into Europe in the 1950s and 1960s was linked to taking advantage of the opportunities in the rapidly growing European economy, which was experiencing an average growth rate over the period above that of the US (Williams, 1995). Equally, the wave of inward investment from Japan and other East Asian MNCs from the 1980s onwards into North America and Western Europe was driven by market access, although protectionism of home-based companies also played a part. In order to overcome barriers to market entry, principally involving local sourcing and production rules, Asian firms gradually shifted from export-based strategies to setting up overseas plants. For the first time, the United States became a major destination of new inward investment.

7.3.2 The new international division of labour

While market access has been the dominant factor, illustrated by the FDI trends above, there has been a growing level of FDI activity linked to the search for cheaper labour costs. This has led to a growing level of investment in the developing world, associated with a phenomenon know as the **'new international division of labour'** (NIDL) (e.g. Froebel et al., 1980; Hymer, 1972), replacing or more accurately running alongside the OIDL. The simple model envisaged by economist Stephen Hymer, and dating from the 1960s, was one in which MNCs developed a new geography of production by reorganizing the division of labour within the firm. Higher-level decision-making and research and development activities would remain concentrated in the major metropolitan regions of the advanced world (e.g. London, New York, Paris) while the more routine production activities would be dispersed, depending upon skill and levels of technology, to more peripheral locations at home (see Table 7.2). Over time, the more basic production functions would be relocated to cheaper locations overseas with a 'race to the bottom' encouraging the search for low-cost locations in the Third World or global South. The developed countries, however, would retain the higher value-added activities within production systems.

147

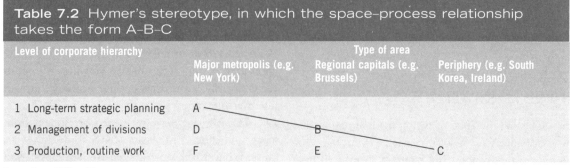

Table 7.2 Hymer's stereotype, in which the space–process relationship takes the form A–B–C

Level of corporate hierarchy	Type of area		
	Major metropolis (e.g. New York)	Regional capitals (e.g. Brussels)	Periphery (e.g. South Korea, Ireland)
1 Long-term strategic planning	A		
2 Management of divisions	D	B	
3 Production, routine work	F	E	C

Source: Sayer, 1985, p.37.

The archetypal example of production organized through the NIDL is in the related industries of clothing and textiles. Global wage rates vary enormously, from an average of $17 per hour for the European Union to less that $0.50 per hour in Bangladesh (ILO, 2000). The sector is still dominated by US and European manufacturers and retailers, who control the marketing, design and technologically more sophisticated production processes, but an increasing amount of the more routine production is located overseas. For European MNCs, a first wave of relocation in the 1960s and 1970s to low-cost production sites in geographically proximate areas (e.g. Portugal, Greece and North Africa) has been followed since 1980 by a further wave to even cheaper locations in South and East Asia. As a consequence, employment in Europe in the clothing sector fell by 50 per cent between 1980 and 1998, while growing in Asia by 35 per cent. At a global level, China has been the largest employer since 1980 and by 1998, with a clothing workforce of over 3.6 million, had almost four times that of the next largest country, the US (ibid., p.19).

For developed countries, these gains in employment are often associated with poor working conditions and low pay. Thus, a new form of exploitation replaces the old while, at the same time, the working classes in the core economies are undermined.

Three factors have shaped this trend, according to Wright (2002, p.73):

➤ the fragmentation of work tasks in labour processes into simple routine functions that need minimal training (see section 3.3.2);

➤ the development of advanced transportation tech-

nologies that allow the production of final assembly or finished/semi-finished goods to be undertaken virtually anywhere;

➤ the release of agricultural labour as a low-wage resource on to urban labour markets in less-developed economies through the modernization and intensification of agriculture under the 'Green Revolution'.

The NIDL now applies to routine work in some service sectors, as well as manufacturing activities, particularly in areas such as banking and telecommunications, where new information technology allows 'back-office functions' to be relocated overseas to lower-cost regions (see section 8.7.2).

7.3.3 Variations in internationalization

While market access and labour supply explain why some firms choose to internationalize production, this is clearly only part of the story and in practice many firms choose not to internationalize production at all. An alternative approach is the 'eclectic' paradigm of international business theorists such as John Dunning (1980), which emphasizes that there are no simple models of MNC development, but rather that a range of motives, related to 'accidents of history' and the particular path dependencies of individual firms, explain why and when internationalization occurs. It is important to remember, however, that the vast majority of firms do not internationalize – even in the contemporary economy. Overseas investment, particularly in a new and unfamiliar environment, involves many risks for the individual firm, such as concerns about the quality of the labour force and the abilities of local management. The lack of such business knowledge about foreign economies remains an important deterrent.

The extent of transnationalization also varies between different sectors of economic activity, reflecting variations in the nature of the production process and markets being served. According to Dicken, MNCs are particularly evident in three types of industry: high-technology industries such as pharmaceuticals and electronics, large-volume, medium-technology industries such as motor vehicles, and mass-consumer products such as jeans and T-shirts. These are industries that require both high levels of technology and resources (both monetary and labour) but for which demand is highly variable, therefore suiting firms that can organize at the transnational level and take advantage of variations in local conditions (Dicken, 2003a).

Within individual industrial sectors, firms can choose completely different geographical strategies when faced with the same market conditions, suggesting the importance of particular corporate cultures and even individual management decisions in shaping firm action (Schoenberger, 1997) (see Box 7.2 for an example).

7.3.4 A 'newer' international division of labour and changing organizational forms

By the early 1980s it was evident that a significant shift had occurred in the organization of production within MNCs. One pair of geographers at the time, Taylor and Thrift, talked of a 'truly global system of production', remarking that,

> With a global network of subsidiaries set up, the question for most large multinationals had changed from what would be the most profitable area in which to expand next to which of the existing areas could be relied upon to produce the highest returns.
>
> (Taylor and Thrift, 1982, p.1)

The implication of this remark was that MNCs were becoming increasingly flexible and footloose in their organizational geographies. Underlying such developments, the rather simplistic set of bilateral relations underpinning the NIDL was beginning to give way to a much more complex set of organizational arrangements for MNCs.

Coffey (1996) has suggested that a new phase of MNC development has been under way since 1980 as part of what he terms the **newer international division of labour.** MNCs are increasingly shifting strategy from setting up their own production facilities in developing countries to subcontracting work out to locally based firms. Such a change provides MNCs with greater locational flexibility to switch suppliers and avoid the **sunk costs** that are involved in committing investment to fixed assets. Sunk costs refer to the costs of investment that are not directly recoverable if a firm were to pull out of a particular location (Clark, 1994). The clothing firm Nike is the oft-quoted example of an MNC that maintains spatial flexibility through a vast web of 'independent' suppliers. Nike has only 8,000 direct employees, mostly in marketing, advertising and sales, and no in-house production, most of which now takes place in subcontractors' factories in Asia (Coffey, 1996). The geography of its supply chain has changed over time with continual relocation between countries in search of lower costs: for example, suppliers in Japan and Hong Kong were abandoned in the 1980s and new producers in Thailand, China and Indonesia were brought in (Donaghu and Barff, 1990). The Italian clothing firm Benetton is another much-cited example of a producer that outsources its manufacturing, but the trend has also spread to other industries, even automotive production (see Box 7.2).

As Dicken (2003a) has noted, an increasing amount of MNC activity takes place through external relationships with other firms, leading to a growing literature on **global commodity chains** (GCC) (Gereffi, 1994) (see also section 1.2.1), which focuses upon the product or commodity and explores the connections that link different localities and firms within the production chain. The GCC literature is concerned with identifying where power resides within commodity chains. It focuses upon the 'lead firms' that organize production networks, the implication being that 'the type of lead firms that drive a commodity chain, and therefore the type of governance structure that characterizes it, will shape local development outcomes in those areas where the chain touches down' (Bair and Gereffi, 2001, p.1888).

As well as the vertical relations within a commodity chain, Dicken notes an increased amount of collaboration and partnership among MNCs themselves with a growing number of **international strategic alliances** between firms who are competitors in the same markets. One study recorded a six-fold increase in strategic alliances from 1,000 at the start of the 1990s to over 7,000 by the end of the decade, with North American firms dominating the practice (Dicken, 2003a, p.258). Such alliances tend to be more pronounced in sectors that require high levels of research and product development and therefore have high start-up costs for new products, such as computers, biotechnology, automotive production, aircraft manufacture. But they also reflect the vulnerability of MNCs to changes in global markets and the desire to minimize uncertainty (see Box 7.1).

Reflect

➤ What are the main factors encouraging firms to internationalize their activities?

➤ What have been the main trends in the reorganization of the geography of MNCs since 1980?

7.4 How global are they?

An important debate has been waged regarding the 'globalness' of MNCs. From the late 1980s business school theorists began to talk about the shift in corporate organization from the 'multinational' defined as 'an enterprise that engages in foreign direct investment and owns or controls value-adding activities in more than one country' (Dunning, 1993) to a 'transnational' form, defined as an organization with the ability 'to strut across the global stage in a footloose manner [with] an attachment to any one particular country cast off' (Allen and Thompson, 1997, p.214). Bartlett and Ghoshal's classic 1989 work, *Managing Across Borders*, predicted the emergence of firms increasingly transnational in character, reflected in the dispersed and decentralized geography of corporate operations, the international nature of ownership and the diverse range of nationalities involved in the senior management of firms. The image portrayed was one of 'placeless' corporations, no longer dependent upon the activities of any one state or territory, but capable of global roaming, playing off one location against another in search of competitive advantage.

In assessing these claims it is useful to draw upon the United Nations **Transnationality Index (TNI)**, which compiles data on the international extent of MNC operations by calculating the average of three ratios for individual corporations: foreign assets to total assets; foreign sales to total sales; and foreign employees to total employees. From this data, the economic geographer Peter Dicken (2003b) has undertaken an analysis of the extent to which firms are becoming more global in their operation. His findings reveal that between 1993 and 1999 there was no significant increase in the 'transnationality' of firms. He also noted that the majority of the world's leading MNCs continued to have over 50 per cent of their activities located in their home countries. He did find considerable variation by country. Comparing the major advanced industrial nations, Japan and the US had relatively low levels of internationalization among their leading firms, while the UK recorded one of the highest. Such variations reflect differences in the size of the domestic economy, the way activity is organized and the openness of the economy to foreign trade. Japan has traditionally operated a protectionist industrial policy that has encouraged the investment in large-scale industrial complexes at home, while the US, as one of the world's largest economies, records a low score because of its size rather than a lack of internationally oriented MNCs. The UK has both a relatively open economy and few restrictions on outward or inward investment, which reflect its long imperial and trading history.

While the Transnationality Index is a useful measure, it arguably does not necessarily reveal the most

Table 7.3a The world's most 'transnational' MNCs and selected other MNCs, rankings, 2003

Transnationality Index (TNI) Rank	Index	Corporation	Home economy	Industry	Foreign share of Assets %	Employees %
1	95.4	Thomson Corp.	Canada	Publishing/printing	98.6	92.5
2	95.2	Nestlé	Switzerland	Food/beverages	88.9	97.2
3	94.1	ABB	Switzerland	Food/beverages	88.2	96.3
4	93.2	Electrolux	Sweden	Electrical	92.9	90.4
5	91.8	Holcim	Switzerland	Construction materials	91.9	93.4
6	91.5	Roche Group	Switzerland	Pharmaceuticals	90.4	85.6
7	90.7	BAT	UK	Food/tobacco	84.0	96.8
8	89.3	Unilever	UK/Netherlands	Food/beverages	90.4	90.5
9	88.6	Seagram Co.	Canada	Beverages/media	73.1	
10	82.6	Akzo Nobel	Netherlands	Chemicals	85.0	81.0
75	36.7	General Electric	US	Electrical	34.8	46.1
77	36.1	Ford	US	Motor vehicles		52.5
83	30.7	General Motors	US	Motor vehicles	24.9	40.8
18	73.7	BP	UK	Petroleum	74.7	77.3
22	68.0	Exxonmobil	US	Petroleum	68.8	63.6

Source: UNCTAD, 2005, Table 3.

Table 7.3b The world's top 15 non-financial MNCs, ranked by foreign assets, 2003

Corporation	Home economy	Industry	Assets Foreign US$m	Total US$m
General Electric	US	Electrical	258900	647 483
Vodafone	UK	Telecomms	243839	262 581
Ford	US	Motor vehicles	173882	304 594
General Motors	US	Motor vehicles	154466	448 507
BP	UK	Petroleum	141551	177 572
Exxonmobil	US	Petroleum	116853	174 278
Royal Dutch Shell	Netherlands	Petroleum	112587	168 091
Toyota	Japan	Motor vehicles	94161	189 503
TotalFinaElf	France	Petroleum	87840	100 989
France Telecom	France	Telecomms	81370	126 083
Suez	France	Electricity	74147	88343
Electricité de France	France	Electricity	67069	185 527
E.On	Germany	Electricity	64033	141 260
Deutsche Telekom	Germany	Telecomms	62624	146 601
RWE Group	Germany	Telecomms	60345	98592

Source: UNCTAD, 2005, Table 3.

powerful and influential international actors in the world economy. Looking at the most recent data, based upon the 2003 UNCTAD survey, the top 10 rankings include a number of firms that would not immediately spring to mind as global 'movers and shapers' (Table 7.3a) (Dicken, 2003a, pp.221–4). In contrast, the largest MNCs in terms of actual size are in most cases far more embedded geographically in their own domestic economies. The largest automobile producers in particular, while containing extensive foreign operations, are still heavily rooted in their countries of origin.

A problem with the TNI is that it is based on the proportion of foreign activities relative to total firm operations rather than absolute size. A distortion that arises as a result is that firms based in smaller national economies are likely to be over-represented because of the necessity to internationalize at an earlier stage in their development, as they quickly reach the capacity limits of the domestic market. Thus firms from Switzerland and Canada appear in the top 10 (see Tables 7.3a and b). A more useful indicator of the most

powerful MNCs is asset values (Table 7.3b), which reveals a more familiar list of names, pertinent in demonstrating where power lies in the global economy and who the more internationally oriented firms are. The United States dominates the top 10 firms, with motor vehicles, petroleum and telecommunication the leading sectors.

The UNCTAD figures are a useful reminder of both the extent and limits of the global integration of business activities. But they remain a rather crude measure of the geographical networks within which MNCs are embedded, particularly given the increased use of foreign suppliers over direct investment that we noted above. For example, a firm can be predominantly nationally oriented in its own operations, but have enormous global influence through its purchasing and sourcing strategies. The classic example used to illustrate this point is the US retailer Wal-Mart (see Dicken, 2003b), which employs around one and a half million workers, three-quarters of whom are in the US (UNCTAD, 2004). Taken overall, however, the data

Box 7.1

The limits to global corporate power: MNC restructuring in the global oil industry

The oil industry is perhaps the most globally integrated of industries and indeed has been organized on a transnational basis through the activities of dominant multinationals for a long period. As early as 1870, the US oil industry was selling two-thirds of its output to foreign markets (Vernon, 1971, p.28) and by the mid-1950s the leading seven multinationals – known as the 'Seven Sisters' because of their perceived market collusion – controlled over 90 per cent of the world's oil reserves and 70 per cent of refining capacity (Jacoby, 1974). In today's global economy, it comes as no surprise that the leading oil MNCs

record high scores on the TNI measure with 11 oil companies in the top 100, ranked by size of foreign assets, and the top five MNCs (British Petroleum (BP), Exxon Mobil, Royal Dutch Shell, Total and Chevron Texaco) all recording TNI figures of over 50.

From 1986 onwards the industry underwent an intensive period of restructuring, driven initially by low oil prices and the long-run decline in reserves. Typically, firms have attempted to shed their less profitable non-core assets in established regions such as North America and Western Europe, particularly in the

'downstream' activities of marketing, refining and associated petroleum products, to free up resources for 'upstream' exploration in new and growing oil producing regions, such as the former Soviet Union and West Africa. This resulted, on the one hand, in a streamlining of operations with the 10 leading firms shedding over 140,000 jobs between 1993 and 1998 (Table 7.4) and, on the other, a wave of merger and acquisition activity (Table 7.5).

Rather than being a sign of the industry's strength and increased global dexterity, the merger wave represented recognition of the

Box 7.1 (continued)

Table 7.4 Employment change in 10 largest integrated oil companies, 1991–1998

Company	1993	1998	Change %
Shell	117,000	102,000	−12.8
Exxon	91,000	79,000	−13.2
Elf	94,300	58,300	−38.2
Mobil	61,900	41,500	−33.0
BP	84,500	96,650	+14.1[1]
ENI	93,076	78,906	−15.2
Chevron	47,576	39,191	−17.6
Texaco	32,514	24,628	−24.3
Amoco	46,317	–	–
Total Oil	49,772	48,831	−1.9
Total	717,955	569,006	−20.1

[1] Figure for merged BP Amoco group.
Source: Cumbers and Martin, 2001, p.38.

Table 7.5 Key oil and gas mergers and acquisitions in period since 1998

Date of deal	Companies involved		Comments
August 2002	Phillips	Conoco	Merged firm worth $53.5 billion
September 2001	Chevron	Texaco	$45 billion takeover by Chevron
February 2000	Total	Elf	'Friendly merger' (company annual report, 2000, p.3)
February 1999	BP Amoco	Atlantic Richfield	$26.8 billion takeover by BP
December 1998	Total	PetroFina	$7.0 billion takeover of PetroFina
December 1998	Exxon	Mobil	$75.3 billion takeover
August 1998	BP	Amoco	$48.2 billion takeover

Source: BP Amoco website http://www.bpamoco.com/; *Financial Times*, 21 December 1998, p.22; various company reports.

geographical embeddedness and the vulnerability of many MNCs. Indeed, despite their global aspirations, the reality confronting most MNCs was a continuing dependence upon their home countries or particular world regions for oil supplies and markets. Apart from Exxon, other US MNCs in particular were still heavily dependent upon producing oil from ageing and increasingly costly US fields. The sudden spate of mergers in the late 1990s reflected both the imperative to reduce costs and the desire for a greater global reach. The BP–Amoco merger is a case in point, fusing BP's Eurasian orientation with Amoco's predominantly US interests (the latter was still dependent on US-based resources for 60 per cent of its total revenue prior to the merger).

While the newly merged BP can be considered to be one of the most globalized of firms, with a TNI figure of 73.7 and activities on every continent, it remains embedded in important ways in its home base. Over 60 per cent of company shares are still held by British institutions and individual shareholders. Its ability to successfully 'globalize' was also underwritten by its own government, which, after intensive lobbying from BP and the other British oil MNC, Shell, reduced tax rates on oil extracted from the North Sea in March 1993, allowing the corporation to finance new exploration and investment overseas (Woolfson *et al.*, 1997, p.310). The attempt to develop a greater global reach has on the whole been achieved through joint venture and partnership activities. It operates in partnership with Norwegian company Statoil in Azerbaijan, Kazakhstan, Vietnam,

Box 7.1 (continued)

Angola and Nigeria, and in a joint venture with Russian company TNK in Siberia (company annual reports). The desire to work increasingly in partnership or in alliance, both with competitors and suppliers, reflects the continued vulnerability of MNCs to geopolitical uncertainties, given that many of the world's major oil producing regions are characterized by political instability.

illustrates the continued importance of geography in shaping corporate strategy. While the global economy is becoming more integrated – or perhaps interconnected is a better term – MNCs are still heavily embedded in and reliant upon their domestic economies and states (see Box 7.1).

Reflect

➤ How useful is the transnationality index for measuring the globalization of MNCs?

➤ What other indicators might be used to assess the global reach of MNCs?

7.5 The embedded geographies of MNCs: the continuing influence of home countries on MNC strategies

As the the oil industry example shows, even the most transnational of firms continue to be geographically embedded in particular places and regions. While there has been an increasing level of integration of the global economy, as we noted in Chapter 5, MNCs continue to be economically, politically and culturally embedded in their countries of origin, in a range of different ways, from the nationality of the directors and their families, to the legal frameworks binding company action, to the financial system governing firm decision-making, to particular national cultures of 'doing business' (Dicken,

2003a, pp.225–35). For example, while there is an emerging elite of multinational corporate executives (see Chapter 8), most firms are still dominated by managers from their home countries at the highest levels. One recent international survey of the directors of MNCs, for example, found that only 10 per cent of board members were foreign nationals (Knox *et al.*, 2003, p.84). Among Japanese firms, foreign board members were almost non-existent, being described as 'as rare as British sumo wrestlers' (ibid., p.84).

At a very basic level, all MNCs develop from particular local contexts and, while the local scale itself has an influence upon the practices and strategies of the firm, for MNCs it is the national scale that continues to be critical in shaping firm actions. Despite increased global economic integration, researchers continue to highlight diversity in forms of capitalism between places and between MNCs (Albert, 1993; Hutton 1995; Weiss 1997; Whitley, 1999). For example, a distinction is often made between US, German and Japanese MNCs (see Table 7.6). Whereas US MNCs are largely financed by stock markets that discipline firms to deliver increasing returns and dividends to shareholders, thereby instilling a 'short-termist' approach, German and Japanese MNCs have greater freedom due to the role of banks in providing finance on a longer term and more stable basis. A range of other motivations, such as increasing market share, safeguarding employment and a commitment to training and modernizing production often dominate over narrow profit concerns. In locational terms, US MNCs have shown a greater propensity to invest overseas, both in basic production activities but also some R & D. In contrast, Japanese MNCs still do very little R & D overseas and up until the 1980s were essentially domestic producers for export markets (see Box 7.2).

Table 7.6 Differences between US, German and Japanese TNCs

	US TNCs	German TNCs	Japanese TNCs
Corporate governance and corporate financing	Constrained by volatile capital markets; short-termist perspectives. Finance-centred strategies	Relatively high degree of operational autonomy except during crises. Long-term perspectives. Conservative strategies	Bound by complex but reliable networks of domestic relationships. Long-term perspectives. Market-share-centred strategies.
	High risk of takeover. 90% of firm shares held mainly by individuals, pension funds, mutual funds. Less than 1% held by banks	Low risk of takeover. Firm shares held mainly by non-financial institutions (40%). Significant role of regional bodies	Very low risk of takeover – mainly confined to within network
	Banks provide mainly secondary financing, cash management, selective advisory role	Banks play a lead role. Supervisory boards of companies are strongly bank-influenced	High degree of cross-shareholdings within group. Lead bank performs a steering function
	Ratio of bank loan/ corporate financial liabilities = 25–35%	Ratio of bank loan/corporate liabilities = 60–70%	Ratio of bank loan/corporate liabilities = 60–70%
Research and development	Corporate R&D expenditure peaked in 1985 at 2.1% of GDP. Declining	Corporate R&D expenditure declined steeply in late 1980s/early 1990s. At 1.7% of GDP lower than US and Japan	Corporate R&D grew very rapidly in 1980s. Overtook US in 1989. Peaked at 2.2% of GDP. Real cuts made only as last resort
	Diversified pattern; innovation-oriented	Narrow focus	High-tech and process orientation
	Some propensity to perform R&D abroad	Some propensity to perform R&D abroad	Very limited propensity to perform R&D abroad
Direct investment and intrafirm trade	Extensive outward investment. Substantial competition from inward investment	Selective outward investment. Moderate competition from inward investment	Extensive outward investment. Very limited competition from inward investment
	Moderate intrafirm trade; high propensity to outsource	High level of intrafirm trade	Very high level of intrafirm and intragroup trade

Source: Based primarily on material in Pauly and Reich, 1997.

Box 7.2

Contrasting geographies of the global automobile manufacturers

While the automobile industry is undoubtedly one of the most global of industries with some of the largest MNCs, it continues to be influenced by considerable variations in spatial strategies. These in turn reflect recognizably different strategies employed by firms at the national level. The greatest contrast until recently was in the strategies of US firms compared with Japanese firms. In the 1980s, while the two largest US car producers, GM and Ford, were already producing a high proportion of their cars overseas (41.9 per cent and 58.7 per cent in 1989) (Dicken, 2003a, p.374), Toyota, the largest Japanese firm, was still primarily a domestic producer, with only around 8 per cent of its operations overseas. Like other Japanese firms, it had been highly successful at competing overseas without having to relocate production from its home base. As Dicken notes:

> Before the early 1980s, Japanese automobile producers had shown

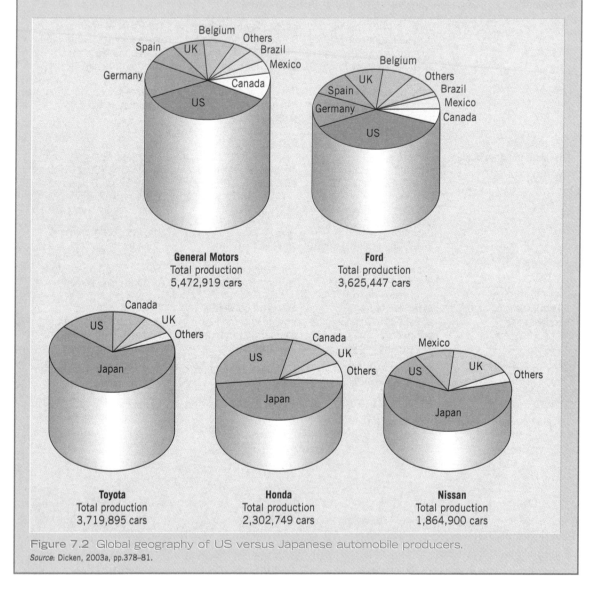

General Motors
Total production
5,472,919 cars

Ford
Total production
3,625,447 cars

Toyota
Total production
3,719,895 cars

Honda
Total production
2,302,749 cars

Nissan
Total production
1,864,900 cars

Figure 7.2 Global geography of US versus Japanese automobile producers.
Source: Dicken, 2003a, pp.378–81.

Box 7.2 (continued)

themselves perfectly capable of serving the North American and European markets by exports from the Japan. The price competitiveness of Japanese ... exports was based upon extremely large-scale flexibly organized and highly automated production plants in Japan [aided by a] very high degree of vertical integration with Japanese component suppliers.

(Dicken, 2003a, p.380) Japanese auto producers also sub-contracted out more of their production than US firms, 80 per cent of inputs in 1989. The result was that Toyota's sales turnover was half of GM's in 1989 but its work-force was 11.9 per cent of the US

MNC's (Hayter, 1997, p.356). At a more local level, the need for closer interaction with suppliers meant that Japanese firms tended to concentrate their production systems in localized clusters while US firms tended towards dispersal, still basing their locational decisions largely on cost. Japanese firms began to expand their production overseas in the 1980s, largely due to growing political opposition to imports in North America and Western Europe, so that they are now more globalized in their operations, but they continue to have a majority of production in Japan, in contrast to the two US MNCs (see Figure 7.2)

Changes in the organization of the industry in the 1990s have diluted

the national character of many MNCs. There have been a series of mergers and acquisitions, the most important being the takeover of Chrysler by German firm Daimler-Benz, the purchase of Swedish firm Saab by GM, and the buying of a 36.8 per cent stake in Nissan by the French firm Renault. Additionally, the increase in cooperation between firms in recent years means that the industry at a global level has become what Dicken describes as 'a veritable spider's web of strategic alliances ... that stretches across the globe' (Dicken, 2003a, p.376). Nevertheless, national variation and embeddedness continue to play an important role in shaping MNC strategies.

Researchers have also highlighted continuing diversity in the approach to labour relations by MNCs of different countries. While MNCs in continental Europe generally still accept trade unions as legitimate social partners, US and UK MNCs tend to display a strong anti-trade union bias, in many cases failing to recognize the rights of workers to collective action, as well as demonstrating a greater preference to use strategies of employment flexibility (see Chapter 9).

Particular local or national cultures of production (section 2.5.3) are also 'exported' as corporations internationalize their operations, although this does not subsume existing cultures of production in the host countries. Indeed, Gertler's research into German machine manufacturers supplying markets in North America found that they were unable to reproduce their home production practices in their own American plants, let alone those customers they were supplying. Flowing in the other direction, researchers have noted the frustration of US firms setting up in Germany, eventually having to accept German norms on employment relations and working practices such as the recognition of trade unions and more restrictive working hours than in the US (Gertler, 2004).

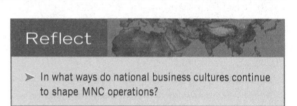

Reflect

➤ In what ways do national business cultures continue to shape MNC operations?

7.6 The impact of MNCs on host regions

There have been continuing debates in economic geography going back to the 1970s about the benefits of inward investment by MNCs for **host regions**. For some, inward investment helps to transform regional economies, bringing new jobs, skills and knowledge as well as helping connect local economies up to wider global networks (Table 7.7). For others, however, the effects are more pernicious, leading to a loss of local control over economic development in return for jobs that are often low skilled and always vulnerable to plant closure if MNCs decide to invest elsewhere (e.g. Firn, 1975; Massey, 1984). Here, we consider the effects in different types of region: those in more advanced economies, those in developed country contexts and those in the transition states of Eastern Europe.

Table 7.7 Advantages and disadvantages of MNC inward investment for host regions

Advantages	Disadvantages
Local Economy	*Local economy*
Injection of investment and income opportunities	Development increasingly dependent on external control
Transfer of global knowledge and management techniques – enhancing competitive advantage in context of globalization	Greater vulnerability to closure and job loss
	Deskilling/downgrading of local economy
Improves skills base through training, employment policies	Production enclaves linked to broader global production systems but with few local benefits
Contributes to region's tax base	Marginalization/displacement of other sectors – economy linked to narrow development trajectory
Local firms	*Local firms*
New opportunities to supply inward investors	Increased competition in local markets
'Piggy-backing' on MNCs into export markets	Increased competition for labour, land, capital drives up factor prices
Learning through imitation of best practice in FDIs	Become tied in to dominant client firms as suppliers
Local Community	*Local community*
Upgrading of local infrastructure that may not otherwise have occurred (e.g. transport and communication links)	Disruption/destruction of local culture/society
	Resource displacement – investment on FDI-specific infrastructure rather than community projects/initiatives
FDI 'social investment' (e.g. in local schools, community services)	
Employment	*Employment*
New jobs for local people	Introduction of more coercive management strategies
Wages often in excess of average local wage	Jobs often low skilled, routine
Training and more progressive approach to human resource management	Training limited to company-specific skills and not transferable
Political Implications	*Political implications*
Influx of powerful non-local actors enhances power of region in context of inter-regional competition for state resources	Regional development agenda becomes hijacked by the Inward Investing Service Class (IISC) – Phelps *et al.* (2003)
	Alternative political discourses marginalized – democratic deficit

Source: Adapted from Pavlinek, 2004, p.48.

7.6.1 Branch plant regions in developed economies

Early studies of FDI in developed economies, influenced by a Marxist perspective, suggested that inward investment could lead to relations of dependency as hitherto locally controlled economies became increasingly subservient to the needs of large foreign corporations. In the 1970s, when the focus of research was upon the impact of foreign inward investment (alongside domestic firms from outside the region)

into some of the older industrial regions of Western Europe, particularly the UK, the term '**branch plant economy**' was coined to reflect the way particular regions were being repositioned within the wider space economy. Large corporations, both foreign and domestic, were instrumental in the emergence of a **spatial division of labour** (see Chapter 3). With the decline of traditional industries such as steel production, shipbuilding, textiles and engineering, and the closure of local firms, regional industrial specialization and local ownership were being replaced by a situation

Box 7.3

From boom to bust: the changing fortunes of East Asian investment for the UK's old industrial regions

The UK was the dominant location in Europe for inward investment from East Asian MNCs setting up to gain access to the emerging European Single Market, receiving over 40 per cent of the European total during the 1980s and early 1990s (Mason, 1994, p.32). Critics questioned the motivations behind the approval pursued by the Conservative Government of the time. The inward investment strategy was viewed as an attempt to undermine the trade unions, which were strongly represented in traditional industries, by encouraging new foreign management with more hostile attitudes (Garrahan, 1986). There was also concern about the longer-term implications of the types of jobs being created and the degree of embeddedness of such firms in the local economy. Nevertheless, there were considerable short-term benefits in terms of the number of jobs created in areas that had been suffering industrial decline. The two major

regions receiving investment, South Wales and the north-east of England, had received 13,000 and 11,000 jobs from Japanese investment alone by the mid-1990s (Cumbers, 1999).

Opinions are mixed as to the reasons behind inward investment but a combination of low labour costs – by EU standards – and government grants appear to have been important (Stone, 1998). What is clear, however, is that few of the investments could be considered to be linked to high-technology or high-performance production. Recent research has also questioned the embeddedness of inward investment, finding little evidence that firms were setting down longer-term linkages in the regions with a lack of additional investment after the initial start-up (Phelps et al., 2003).

A reminder of the dangers of over-reliance upon FDI has been brought home in the period since 1998 with the closure or relocation of many of

the Asian branch plants, linked to the Asian financial crisis (Box 5.3) on the one hand, but also to the availability of new sources of cheap labour in Eastern Europe on the other, which has also led to the closure of earlier branch plants. The collapse of the market for certain electronics components was felt heavily in the north-east of England, where both German firm Siemens (1,100 jobs) and Japanese firm Fujitsu (600 jobs) closed semiconductor plants in 1998. In Scotland, the crisis led to the collapse of plans by Korean firms Chungwa and Hyundai to set up new plants and, in 2002, the closure of a Mitsubishi plant outside Edinburgh (500 jobs). Companies that relocated from the UK's regions to the Czech Republic included Japanese firm Matsushita (from Wales, costing 1,400 jobs) and US firms Compaq (Scotland, 700 jobs) and Black and Decker (northern England, 600 jobs) (Pavlinek, 2004, p.53).

in which regions were being recast in terms of their different functions within individual corporate production systems. In the UK context, the recession of the early 1980s seem to bear out many of the critics' concerns, with various studies indicating that rates of firm closure and job loss were highest in externally owned plants, with foreign-owned plants experiencing the greatest levels of decline (e.g. Healey and Clark, 1986; Lloyd and Shutt, 1985).

Controversy over the merits of foreign inward investment, as a strategy for regional development in the US and UK, were reignited during the late 1980s and early 1990s, accompanying a wave of new inward investment into North America and Western Europe by Japanese companies and later by other East Asian cor-

porations. Aside from issues of external dependence (see Box 7.3), the amount of money national and regional governments are prepared to spend to attract FDI has been queried by some commentators. When the Nissan car company set up a manufacturing plant in the north-east of England in 1986, it received £112 million from the UK government for its first phase of development (Garrahan and Stewart, 1992). Although it has subsequently expanded to employ over 5,000 employees with an estimated value of £1.2 billion (Conte-Helm, 1999), it is a moot point as to whether the investment would not have been better spent helping to modernize existing firms and protect the jobs of local workers in traditional industries. A later project, set up by the Korean conglomerate LG in 1996

Figure 7.3 Nissan factory, Sunderland, north-east England.
Source: D. Mackinnon.

to develop a television and components plant in Newport in South Wales, received over £247 million of public money and made enormous demands on the local infrastructure (Phelps *et al.*, 1998, p.125). The promised 6,000 jobs never materialized, however, as LG was severely hit by the Asian financial crisis of 1997–8. Less than 1,000 jobs were eventually established at the plant, which was closed in May 2003 (http://news.bbc.co.uk/1/hi/wales/3049261.stm).

The negative branch plant stereotype was questioned by some researchers in the 1980s and 1990s (e.g. Morgan, 1997; Munday *et al.*, 1995). Compared with the low-skill assembly line activities of previous US investments, the new wave of inward investment into the UK and US from Japanese and German companies was geared towards the creation of locally integrated industrial complexes that were more firmly tied to their host regions because of the large investments made by MNCs. These investments were seen as more strategic and long term in nature, not being purely concerned with the cost advantages of a particular location, but instead were geared towards drawing upon key 'social resources' deriving from local skills and 'know-how'.

This more optimistic scenario hinges around a number of organizational changes taking place in MNCs themselves in response to globalization, characterized as a shift away from centralized bureaucratic hierarchies to flatter and more decentralized structures in which more decision-making powers and higher-level operations will be devolved to local branch plants. Such developments reflect the prerogative of MNCs to be 'globally efficient, multinationally flexible, and capable of capturing the benefits of worldwide learning at the same time' (Dicken *et al.*, 1994, p.30). In these circumstances, it is argued, host regions become home to key forms of knowledge that are valuable to the firm (Schoenberger, 1994). Learning-by-doing and learning-by-using activities are in effect territorialized, through 'the everyday experiences of workers, production engineers and sales representatives' (Lundvall, 1992, p.9), and locationally fixed (at least in the short term) in the sense that they 'remain tacit and cannot be removed from [their] human and social context' (Lundvall and Johnson, 1994, cited in Morgan, 1997, p.493).

As a result of such developments, regions can use the global knowledge networks of MNCs to upgrade and improve their competitiveness. For example, Japanese branch plant investments in the Great Lakes Industrial Belt of the US during the 1980s were seen as instrumental in rejuvenating the regional economy through the transfer of 'best practice' management and production techniques to local firms (Florida, 1995). In particular, the **high-performance manufacturing** model employed by Japanese firms, based upon **just-in-time** (**JIT**) production, continuous improvement and team working, was viewed as superior to the assembly-line techniques associated with Fordism (see Chapter 3) (Florida and Kenney, 1993).

The evidence to support such claims is mixed and in the longer term (as Box 7.3 demonstrates) the question of the vulnerability of plants to closure as corporate strategies and priorities change remains. In-depth studies of Japanese branch plants, for example, have found Japanese MNCs transferring high-performance manufacturing to their foreign subsidiaries to be the exception rather than the rule, with activities more often than not characterized by the kind of routine assembly work associated with earlier forms of FDI (Danford, 1999).

7.6.2 The impact of FDI in developing economies

Concerns about the vulnerability of MNC branch plants in advanced economies are amplified considerably in less-developed economies where host regions often lack the political and social resources (e.g. independent labour unions, strong forms of local democracy and participation, effective employment and health and safety legislation) for local communities to negotiate the terms of inward investment. The extension of production networks by MNCs into the developing world as part of the NIDL has received much critical scrutiny from researchers concerned at the new forms of exploitation that have emerged. The term '**export platform**' (Smith, 2000) has been used to define production enclaves that are set up by western MNCs (or their suppliers) in parts of the less-developed world to supply markets in advanced economies. Not only do these 'platforms' tend to be totally disconnected from the wider local economy but they are often strictly policed and controlled by their national state governments who discourage the formation of labour unions. Such work is often associated with the exploitation of

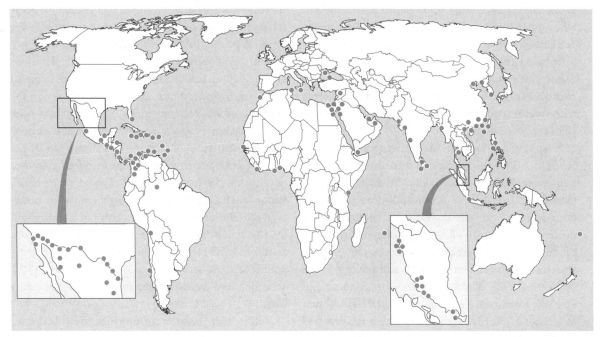

Figure 7.4 Map of export processing zones in developing countries
Source: Dicken, 2003a, p.180.

women workers in low-skilled and poorly paid work, sometimes referred to as 'labour control regimes' (Kelly, 2001).

Many 'Third World' countries have set up 'export processing zones' (see Figure 7.4), which effectively offer foreign firms tax and investment incentives to locate there. Typically these will include up to 100 per cent rebates on local taxation, the provision of all infrastructure and the relaxation of the usual rules governing foreign ownership (Dicken, 2003a, pp.179–80). Some of the most notorious of these zones have been set up in Mexico along the border with the United States. Under the Mexican government's Border Industrialization programme, established in the 1960s, branch plants known as 'maquiladoras' (literally translated as 'assemblers') have been set up in the region to supply the US consumer market. The plants are known as 'in-bond' plants, which are allowed to import components and materials from the US for the production of finished products, provided the product is then exported back to the US. Spectacular growth has ensued, with US MNCs attracted by the lower wages and lax environmental and safety conditions over the border. A further stimulus to growth was the setting up of the North American Free Trade Agreement in 1994 when the removal of all trade barriers with the US, allied to a fall in the value of the Mexican peso, led to further spectacular growth, with plants being established away from the Border region in parts of central Mexico. Bair and Gereffi (2001, p.1890) estimate that over 400,000 jobs were created between 1994 and 1998, with an estimated 600,000 now employed in the plants (Dicken, 2003a, p.80).

Despite the growth in jobs and exports following NAFTA – with Mexican exports increasing from $51.8 billion in 1994 to $166.4 billion by 2000 (Bair and Gereffi, 2001, p.1885) – the maquiladora policy has been highly controversial, not least because of the documented evidence of poor and often abusive conditions of employment, especially among women workers who are subject to sexual and physical abuse (Wright, 2001). While there is some evidence of upgrading occurring (e.g. Bair and Gereffi found that as a result of NAFTA, some US clothing retailers had relocated the entire production process to Mexico) the plants are still vertically organized and controlled by US

MNCs, with little evidence of employment conditions improving. Indeed, the vulnerability of Mexico's global connections has been exposed in the period since 2000 with the country's economy being hit hard by competition from China. An article in the *International Herald Tribune* (3 September 2003) estimated that around 500 maquiladora plants had closed with the loss of around 218,000 jobs between 2001 and 2003.

7.6.3 FDI as regional development panacea in Eastern Europe

Nowhere have the merits of FDI as a strategy for the modernization and increased competitiveness of regional economies been made so forcefully than in Eastern Europe. After the fall of communism and the entry of the states of Central and Eastern Europe into the world capitalist economy in 1989, the region experienced an upsurge in FDI from Western and predominantly European MNCs. Because of the perceived benefits from FDI in terms of job creation and industrial modernization, there was fierce competition between the different countries to attract foreign companies. In some cases, this led to substantial subsidies and tax breaks, with as much as $5,000 for every job created in the case of some overseas investors in the Czech Republic (Pavlinek, 2004, p.59).

Initially, investment was driven by both the availability of cheap labour and market access, but in recent years cost considerations have increasingly driven the strategies of Western companies. The effects have been highly uneven between both countries and regions within countries. Generally, the lion's share of FDI has been concentrated in a few regions with four countries dominating investment flows: Poland, the Czech Republic, Hungary and Russia, although when measured in terms of per head of population, Hungary and the Czech Republic dominate. A general pattern within countries is for FDI to be concentrated in large metropolitan regions, developed industrial regions and those areas that have borders with Western European countries.

Within areas benefiting from FDI, the effects have been mixed. The positive stimulus of jobs, new investment, Western management expertise and

Table 7.8 Footloose MNCs: recent examples of relocation/closure of FDI plants in Central Europe

Firm	Location	Nationality	Product	Destination of relocation	Number of job losses	Additional details
Mannesmann	Hungary	German	Steel tubes	China	1,100	
Shinwa	Hungary	Japanese	Electronic equipment	China	500	
Solectron	Hungary	US	Electronics	Romania	Unknown	
Flextronics	Hungary and Czech Republic	Singapore	Electronics	China	1,200 in Hungary, 2,500 in Czech Republic	Czech plant opened in 2000, closed in 2002. Labour costs cited as reason.
Varta Arku	Czech Republic	German	Telecoms	Unspecified (Asia)	344	Labour costs cited.
Massive Production	Czech Republic	Belgian	Light production	China	Unknown	
Takta Petri	Czech Republic	Japanese–German	Steering wheels	Unknown	Unknown	

Source: adapted from Pavlinek, 2004, pp.55–6.

contributions to the local tax base have been offset by the displacement of local firms, who find themselves in a position where they are losing their skilled labour' to incoming firms paying higher wage rates. Additionally, with the exception of the few local firms that are acquired by foreign investors, the majority find themselves lacking the capital to modernize so that they face a continuing struggle to survive against better-resourced competitors in receipt of government grants. There is also relatively little evidence that incoming firms are becoming firmly embedded within regional economies, with many new start-ups serving as final-assembly 'turnkey' plants for components shipped in from Western Europe.

A growing concern among local policy-makers relates to the longer-term sustainability of the FDI wave in the region in the face of cheaper foreign competition, with a number of recent examples of firms relocating production elsewhere (with China being the main beneficiary) as costs have risen (Table 7.8). One strategy of Western companies has been to retain ownership of machinery in supplier companies in order to avoid **sunk costs** (Clark, 1994) and stay 'footloose'. In this case, this would include the training of local

labour or the investment in buildings or infrastructure. By owning the machinery but not the plant, and by not directly employing the workforce, firms can avoid such sunk costs. As one director of a local firm supplying a major auto producer noted: 'We could be replaced at any time as we replaced the previous suppliers. It is possible to take away the machinery but impossible to move the building' (Pavlinek, 2004, p.58).

7.6.4 MNCs and global production networks

While our examples here have tended to reinforce the negative perception on branch plant economies, it is important nevertheless to stress the continuing diversity that characterizes MNC investment overseas. Not all MNC investments are driven by narrow cost considerations and, at the same time, host regions are differentially placed in their abilities to control or mediate MNC investment. There is a tendency in some of the literature to give agency to MNCs and to see regions as relatively passive receptors of development processes.

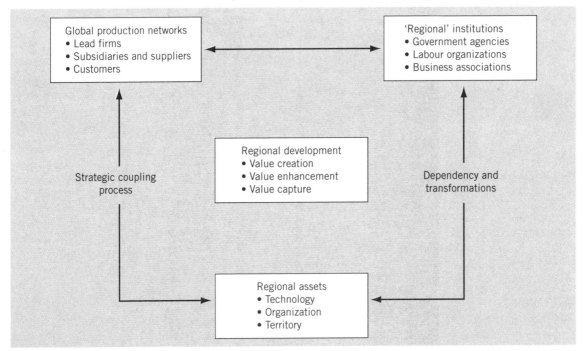

Figure 7.5 The strategic coupling of regions and global production networks.
Source: Coe *et al.*, 2004, p.470.

In this respect, an important insight into the relationship between MNCs and host regions has been provided by recent work on **global production networks** (GPNs) (Coe *et al.*, 2004). A GPN perspective takes this analysis a stage further by focusing not on the firm as the main agent but instead explores the 'strategic coupling' between firms and regions. This is important in giving a sense of agency back to regional actors (see Figure 7.5). The analysis then explores how the different actors within the firm and within the region interact in the processes of value creation and capture. A key point is that regional actors need to match up their assets, in terms of knowledge, labour reserves, training, etc., with the strategic needs of firms at any one point in time. Regional development is also influenced by the strategies of other non-local actors such as regional and national government, labour unions and business associations.

The implications of this analysis are that having substantial regional assets (in terms of skilled labour, good infrastructure, active local development agencies and cooperative employment relations) is not enough if they do not match up with the strategic interests of lead firms within GPNs. But the analysis does suggest that regional development outcomes are always contingent upon particular circumstances and the changing balance of power within business networks, rather than being predetermined by MNC strategies.

> ### Reflect
>
> ➤ What are the benefits and disadvantages of FDI for host regions?
> ➤ How might the relationship between MNCs and host regions vary in different parts of the world?

7.7 Summary

In this chapter, we have traced the emergence of the MNC as an important agent within the global economy. In particular, we have sketched out the changing patterns of FDI and explained the motivations behind the internationalization of business activity. Contrary to some of the claims about globalization, much MNC activity is driven by market access, although the desire for cheaper labour has been an important spur to a new global division of labour.

However, contrary to some of the hyperbole in the business studies literature about the placelessness and mobility of MNCs, we have stressed their continued geographical connection to place. While there has been increasing integration of the global economy, there are limits to the 'globalness' of MNCs. Even the most powerful of MNCs continue to be embedded in their own domestic economies in important ways and often lobby their home governments to defend their interests overseas. The geographical imprint of MNCs is also reflected in their operations, which continue to be characterized by cultures and practices that reflect their countries of origin. There remains in this sense continuing diversity in the practice of MNCs. Despite the increasing transnational nature of many firms, national characteristics remain important, to the extent that many MNCs face problems in trying to displace existing local business cultures and institutions in host countries and regions.

Nevertheless, MNCs have a pronounced impact on host regions, creating relations of dependency in the eyes of some observers, although offering the opportunities for upgrading according to others. Recent reorganization and the growing use of supplier firms in wider networks present MNCs with greater spatial flexibility. For host regions, the vulnerability to competition and MNC relocation seem to be common themes in a range of different circumstances, although it is important not to view regions as passive victims of corporate restructuring. Instead, regional development outcomes are the product of the interaction of regional actors with MNCs in global production chains, dependent upon processes of strategic coupling where a region's needs are mapped on to those of the MNC.

Exercises

Using the most recent annual reports of two MNCs from different countries provided in Table 7.3b (most reports are now available on the Internet), construct detailed corporate geographies. Start by sketching out a map detailing the firms' locations and the type of activities (by sector and function) in different places. Then examine other material (share-holdings, number and type of employees, etc.). Having developed as thorough a picture of the geography of each MNC as possible (you will notice that the level and quality of information provided will vary), consider the following questions.

1 How is each MNC's internal spatial division of labour organized?

2. What does this tell us about the economic and political relationships between employees in different places?

3. To what extent are the two MNCs tied to particular places?

4. Compare the operation of the two MNCs and reflect upon the influence of the home country in shaping the strategy of each.

Key reading

Coffey, W. (1996) 'The newer international division of labour', in Daniels, P. and Lever, W. (eds) *The Global Economy in Transition*, Harlow: Longman, pp.40–61.
Good, though a little dated, reference for the changing international division of labour and the processes behind it.

Dicken, P. (2003a) *Global Shift: Reshaping the Global Economic Map in the Twenty-first Century*, 4th edn, London: Sage, Chapter 8, Webs of enterprise.
An excellent chapter on the changing organizational forms of MNCs and the networks within which they are located.

Dicken, P. (2003b) 'Placing firms: grounding the debate on the global corporation', in Peck, J. and Yeung, H.W. (eds) *Remaking the Global Economy*, London: Sage, pp.27–44.

Contains a useful analysis of the global orientation of MNCs, highlighting the variations between country and sector.

Morgan, K. (1997) 'The learning region: institutions, innovation and regional renewal', *Regional Studies* 31: 491–504.
This a seminal text on the progressive potential for regions in bedding down MNCs through smarter knowledge-based development strategies. It highlights the way recent changes in MNC organization may provide opportunities for traditionally less-favoured regions.

Pavlinek, P. (2004) 'Regional development implications of foreign direct investment in Central Europe', *European Urban and Regional Studies* 11: 47-70.
An up-to-date and valuable analysis of the impact of FDI restructuring on host regions containing useful data on transition economies in Central Europe. Provides some good case studies of individual MNC strategies towards foreign investment. Is a counterweight to Morgan's paper above.

Wright, R. (2003) 'Transnational corporations and global divisions of labour', in Johnston, R.J., Taylor, P. and Watts, M. (eds) *Geographies of Global Change: Remapping the World*, Oxford: Blackwell.
A good overview of the emergence of MNCs and their role in the global economy.

Useful websites

http://www.unctad.org
United Nations site containing annual World Investment Reports that provide data on FDI and the transnationality index of MNCs.

http://www.ft.com
Financial Times site containing detailed data on MNCs and international business reports.

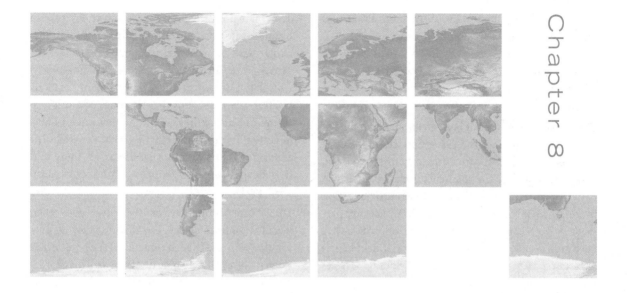

Geographies of the new service economy

Chapter map

We begin the chapter by defining services as a group of economic activities and indicating how geographers have approached their study. This is followed by a classification of the main groups of service activity and an assessment of the links between services and manufacturing. We then identify the main reasons for the growth of service industries since the 1960s. Section 8.4 assesses aggregate patterns of trade and foreign direct investment (FDI) in services. Given the diversity of services, our coverage in the second part of the chapter is necessarily selective, highlighting key developments and locational trends. We examine the concentration of high-level financial and business services in large urban areas in section 8.5, focusing on world cities in particular. This is followed by an exploration of the emergence of new ways of producing and consuming services in terms of the growth of the Internet, e-commerce and the digital economy. Section 8.7 is concerned with the growth of IT-enabled services such as call centres and their spatial dispersal to lower-cost locations, including developing countries.

8.1 Introduction

The growth of services has been an important theme of economic change since the 1960s. Services now dominate the employment structures of most developed countries, eclipsing the manufacturing sector, which has contracted in size as a result of **deindustrialization** processes since the 1960s (Figure 8.1). Services are particularly dominant in the UK and the US, where they have grown most rapidly and the effects of **deindustrialization** have been most pronounced. Specific services that play an important role within the economy include banking, insurance, accountancy, advertising, transport and travel, communications services, retail, hotels, bars and restaurants, education and health (section 8.2).

While specific services are easy to recognize, they remain notoriously difficult to define as a category (Illeris, 1989, p.8). At a general level, three key features of services can be identified:

➤ The intangible or **immaterial nature** of **service outputs**, often based on the performance of a particular task or operation. Examples include the provision of advice to companies by management consultants or lawyers, the arrangement of vacations by travel agents and the cleaning of an office complex. This contrasts with the physical and tangible nature of manufactured goods, recalling *The Economist's* definition of a service as 'anything sold in trade that cannot be dropped on your foot' (Allen, 1988, p.94).

➤ Direct interaction between the buyer and seller, usually of a brief or ephemeral nature, lasting only for the period of the transaction. The interaction often involves an exchange of expertise. This characteristic means that the provision of services is embodied in the individual worker, making their personal characteristics part of the transaction. Think of notions of good service in a shop, restaurant or hotel, for instance, where the friendliness and helpfulness of the individual provider is regarded as important in addition to the general atmosphere of the establishment.

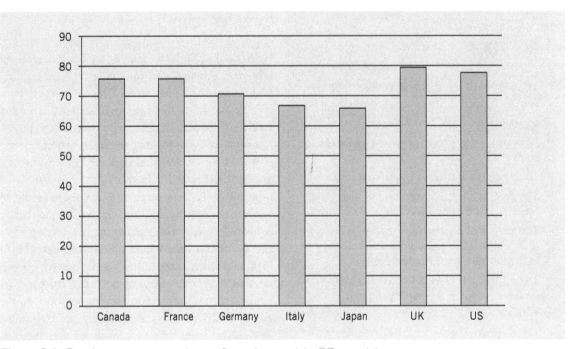

Figure 8.1 Services as a percentage of employment in G7 countries.
Source: OECD in Figures, OECD, 2005.

➤ Service outputs cannot be stored, owned or exchanged, reflecting their intangible nature. For example, financial or management advice cannot be stored (although it is often accompanied by documentation that can), nor can the experience gained through a vacation or attendance at an art gallery or concert.

These characteristics need to be linked together because exceptions to any one of them can often be found. Some service products do generate a tangible output, for example a meal or a concert, but these are provided through a direct personal encounter between the buyer and seller where the quality of the accompanying service and experience is crucial. The same is true of industries like retail, involving the direct sale of goods to consumers.

Traditionally, geographers assumed that the geographical distribution of services simply reflected the distribution of manufacturing industry and the population. From the 1980s, however, geographers became increasingly interested in services as a dynamic growth sector in its own right, focusing on advanced 'producer' services in particular (Marshall *et al.*, 1988). Processes of geographical concentration and dispersal are of particular interest to economic geographers. The impact of information and communication technologies (ICT), which offer new ways of delivering services remotely by telephone or over the Internet, is a key issue. For several prominent economists and business writers (Cairncross, 1997; O'Brien, 1992) digital technologies and the Internet allow work to be broken down into specific components that can be moved to lower-cost countries (Friedmann, 2005). As we show, however, the notion that geography no longer matters is mistaken, ignoring the 'social life of information' (Brown and Duguid, 2000), which means that spatial proximity and face-to-face contact remain important, particularly in advanced services such as finance and business services.

8.2 The nature and scope of the service sector

8.2.1 Classifying services

Traditionally, services were regarded as something of a residual category, encompassing a range of tertiary activities outside the primary (agriculture, forestry and fishing) and secondary (mining, manufacturing and construction) sectors of the economy. As services have expanded, however, the limitations of this traditional sectoral classification have become increasingly apparent. Services are highly diverse as an economic sector, making the tasks of classification and analysis particularly challenging. One influential classification draws a distinction between **producer** and **consumer** **services**. These are defined according to the sources of demand for their products, stemming from other businesses in the case of the former and individual consumers in the case of the latter. Although useful, this distinction is not wholly satisfactory since some service industries, for example finance (often regarded as a producer service) and transport and communications, serve both producers and consumers.

While it is important to acknowledge these definitional problems, the key task is to identify and assess the development of the main service activities. The so-called service sector is made up of the following industries (Bryson *et al.*, 2004, p.7):

➤ Finance, insurance and real estate, including commercial and investment banking, insurance of all types and the commercial and residential real estate industry.

➤ Business services that are provided, primarily, to other businesses (sometimes called producer services). These include accountancy, legal services, research and development, management consultancy, advertising and marketing. Some of these, particularly legal services, are also provided to private individuals.

➤ Transport and communication services, comprising the transport industry (truck haulage, buses, trains, airlines, taxis), covering freight and passenger traffic, and postal, courier, telecommunications and media services.

➤ Wholesale and retail trades that mediate between producers and consumers, being concerned with the distribution and sale of goods.

➤ Entertainment, hotels and leisure, mainly made up of the main elements of tourism, alongside local leisure pursuits.

➤ Education, health and social services, generally provided through the state. Often these activities have not really been seen as part of the service economy, despite their importance in providing employment and income.

➤ Non-profit agencies such as charities, churches, museums and galleries. This is the so-called 'third' sector outside the private and public sectors.

In this chapter, we are particularly concerned with **financial** and **business services**, which have been important growth sectors in developed economies over the past 25 years. The scope of our coverage is somewhat broader than this, however, since IT-enabled services – referring to administrative, clerical and customer service tasks that are being increasingly delivered remotely – are assessed in section 8.7. This category runs across those listed above, related to the technology through which services are delivered rather than the nature of the service itself. Similarly, the impact of ICT on **cultural industries** such as music and film – which run across the business services, wholesale and retail trades, and entertainment and leisure categories – is considered in section 8.6. Tourism and public services

are not examined, since they are dealt with in Chapters 12 and 6 respectively.

8.2.2 Services and manufacturing

As we indicated in section 8.1, the rise of services in developed countries is accompanied by a process of **deindustrialization** as manufacturing declines in importance. As such, the nature of employment has changed considerably for many people, with manual work in factories, mills and mines giving way to administrative and clerical jobs in offices and shops. As Table 8.1 shows, the process of **deindustrialization** is common to the major industrialized nations, although its magnitude has varied, with the UK experiencing the most far-reaching shift since the 1970s and Japan the least change. Deindustrialization provides the backdrop for discussions about the emergence of a new 'post-industrial' society (Box 8.1).

The shift to services has been explained in terms of different levels of productivity in different sectors of the economy and rising incomes in society. As economies grow, according to the influential **Fisher–Clark thesis**, originally developed in the 1930s, rising levels of productivity through the development of new technologies in one sector mean that workers move into another sector (Allen, 1988, pp.104-5). Thus, increased productivity in manufacturing in the post-war period was releasing labour that became employed in services. A direct parallel was drawn between this and increases in agricultural productivity in the eigh-

Table 8.1 The decline of industrial employment in selected G7 countries: percentage of workforce employed in industry

	1976 %	1986 %	1996 %	2004 %	Fall %
France	37.9	31.4	26.0	23.0	39.3
Germany	44.9	40.8	35.4	31.0	31.0
Japan	35.8	34.5	33.3	28.4	20.7
UK	39.6	34.1	27.4	22.3	43.7
US	30.8	27.7	23.8	20.0	35.1

Source: OECD, Labour Force Statistics, 2000, 2005b.

Box 8.1

The rise of post-industrial society

The theory of post-industrial society was developed and popularized by the American sociologist Daniel Bell in the late 1960s and early 1970s (Bell, 1973). It was based on the observation that services were becoming the dominant sector of the economy in developed countries like the US and the UK, taking over from manufacturing. Bell linked this fundamental change in the structure of the economy to the emergence of a new kind of post-industrial society. Three key claims can be identified. First, the rise of services was associated with a shift from blue-collar to white-collar employment, with professional, scientific and technical occupations playing a key role. The role of human and professional services such as research and development (R & D), advertising, management consultancy and advanced finance was stressed. The basis of the workforce was shifting,

Bell argued, from the traditional industrial worker to white-collar occupational groups.

Second, and in line with this, knowledge and information are the key resources shaping economic and social development. Abstract and theoretical knowledge such as scientific research had become increasingly important to economic development. Third, Bell argued, the move to post-industrial society is liberating, freeing people from the drudgery associated with industrial labour in factories, mines and mills. Instead, an increasingly educated workforce would be able to work far shorter hours in more intrinsically rewarding jobs, giving people a much higher quality of life as they spent more time engaging in leisure and travel pursuits.

This bold sociological vision of a new post-industrial age brought about by the development of technology and

the advancement of knowledge proved highly contentious. In particular, the notion that the nature of work has changed, becoming more knowledge-intensive, skilled and rewarding, is questionable. This may be the case for certain groups of professional and technical workers, but there are many insecure and low-paid forms of service employment such as contract cleaning, catering and security guards (Allen and Henry, 1997). These kinds of service jobs are often lower-paid than traditional jobs in manufacturing industry, questioning Bell's reading of the shift from the latter to the former as a form of progress. Instead of liberation, many people's experience of work over the last 25 years has been one of intensification in terms of longer hours, increased pace and more pressures to acquire new skills (section 8.5.2) (Green, 2001).

teenth and nineteenth centuries, which enabled workers to move into the manufacturing sector. As economies develop, there is a tendency for people to spend a higher proportion of their income on services rather than food or manufactured goods.

In recent years, the emphasis on a structural shift from industry to services has given way to a growing recognition of links between services and the production of goods (Walker, 2000). Increasingly, the production of commodities involves combinations of manufacturing and service functions, accompanied by the continued extension of the division of labour between firms (Daniels and Bryson, 2002, p.977). In general, service functions have become increasingly important as research and development, design, branding, advertising, financial packages and service agreements have become key sources of profitability (ibid., p.978). Manufacturing and services have become increasingly intertwined and complementary, as

demonstrated by a number of well-known manufacturing companies that have moved into services. One prominent example is IBM, the US-based computer company, which has moved into the provision of software packages and support services such as logistics, invoicing and call centres (London, 2004). These activities, rather than its traditional manufacturing operations, now account for most of the company's profits. IBM acquired the management consultancy arm of PriceWaterhouseCoopers, the accountancy and business service group, for $3.5 billion in 2002 (ibid.), while its personal computer division was sold to a Chinese company in December 2004 (Skapinker, 2004).

Reflect

➤ Critically assess the concept of post-industrial society in relation to changing employment trends.

8.3 The growth of services

8.3.1 Key factors

A number of reasons for the growth of services in developed countries since the 1960s can be identified (Bryson *et al.*, 2004, pp.11–14):

➤ Increased consumer demand, reflecting rising per capita incomes. An important point to appreciate here is that the demand for many services is what economists call **income-elastic** (ibid., p.11). This means that demand rises with increasing incomes, unlike many agricultural goods and raw materials, where demand tends to remain static. Rising middle-class affluence in developed countries since the 1950s has therefore increased demand for services. Services with high-income elasticities include entertainment and retail – especially specialist and luxury products – and travel, helping to explain why tourism has become the world's biggest industry.

➤ Increased business demand for specialist services. This is related to specific factors such as the increased volume of information in circulation, the speeding up of communications, more intense competition from abroad and the deregulation of rules and laws. As a result, the economic environment in which firms are operating has become increasingly complex since the early 1980s. They have responded by investing in the resources that enable them to deal with this uncertain market and legal environment. These resources include specialist research, consultancy and advisory services such as market research, legal support, marketing and management consultancy.

➤ The increasing demand for such services is linked to changes in how they are provided. Instead of maintaining such expertise internally, large corporations have sought to buy it in from specialist firms. This is the much-vaunted process of **outsourcing**, which is evident in services as well as manufacturing, with corporations buying in a range of specialist advisory and consultancy services, fuelling the growth of business services in particular.

➤ Information and communications technology has encouraged the growth of most kinds of services. On one level, technology allows activities located in distant locations to be closely linked and integrated. This has been especially important for financial services, facilitating almost instantaneous communication and interaction between the major markets of New York, London and Tokyo. Technology also offers new ways of providing and purchasing services remotely by telephone or the Internet, fuelling international trade and investment in services (section 8.5). Examples of products purchased in this way include airline tickets, hotel rooms and items such as books and music recordings through sites such as Amazon, not to mention the wide array of goods available through eBay.

➤ Demand for health and education services. This is slightly different from demand for other kinds of consumer services since it is more a reflection of demographic changes and social expectations than income per se. Demand for healthcare services has increased steadily as life expectancies have grown in developed countries (ibid., p.12). At the same time, a growing awareness of the importance of qualifications and demands for a more skilled workforce has fuelled demand for education, especially in developed countries. In the UK, for instance, the government has set a target of 50 per cent of school leavers gaining university degrees.

➤ The role of the public sector. A number of important services such as education and health are largely provided by the state, accounting for a significant share of employment in many countries. While governments have sought to reduce social expenditure, the welfare state remains largely in place, particularly in Western Europe (Chapter 6). The rules, laws and regulations established by states also provide work for specialist service providers such as lawyers, accountants and a range of consultants and advisers.

8.3.2 Geographical variations in service growth

Beneath the notion of a universal shift from manufacturing to services we can also find an uneven geography of change. Services have grown in virtually all developed countries, but the magnitude of the shift varies somewhat. As we saw, the dominance of services is particularly pronounced in the UK and the US, for instance, with the manufacturing sector remaining more important in Germany and Japan (Table 8.1). Service sector employment is correlated with the distribution of economic wealth between countries with a correlation coefficient (R) of 0.67 (Figure 8.2), indicating that a strong positive relationship exists (both variables increase together). The most service-oriented economies also tend to be those with the highest GDPs, in Europe, North America, Japan and Australasia, while agriculture is generally the dominant sector in the world's poorest countries, in sub-Saharan Africa and Asia. At the same time, though, it is important to appreciate that official figures tend to underestimate the significance of service employment in developing economies. This is because they generally ignore the

informal sector that exists outside the officially recognized, money-based and state-regulated economy. It is estimated to account for between 25 and 40 per cent of output in developing countries, consisting mainly of low-value services such as street hawking, shoe shining, car washing and many others (Potter *et al.*, 2004, p.376).

Reflect

➤ Why do you think there are variations in the size of the service sector among developed countries?

8.4 Global patterns of trade and investment in services

8.4.1 Trade in services

Services are commonly assumed to serve only local markets, being inherently less mobile and exportable than manufactured products due to their intangible

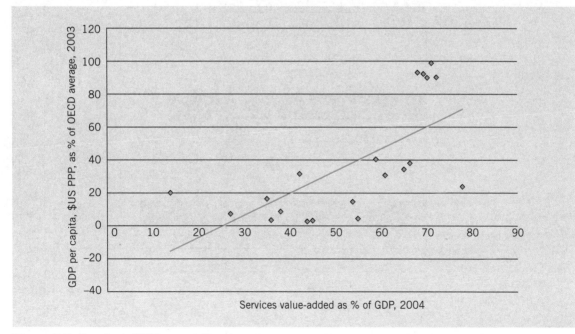

Figure 8.2 Services–GDP correlation.
Sources: Adapted from World Bank, 2006, pp.296–7; UNDP, 2005a, p.220–2.

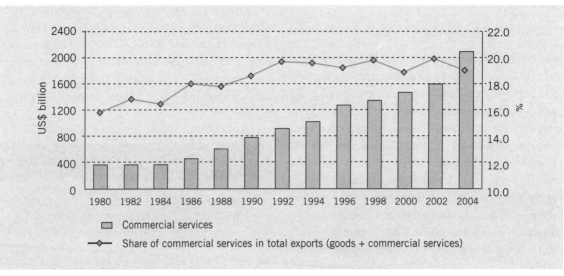

Figure 8.3 World trade in commercial services, total exports, 1980–2004.
Source: WTO, 2006, p.9.

nature and the need for physical proximity between the supplier and the consumer. In recent years, however, it has become apparent that the assumption of local boundedness no longer holds. In reality, **globalization** is about flows of services, information and knowledge as well as material goods, facilitated by the growth of ICT. Tourism, transport and communications and financial services are the most obvious examples of

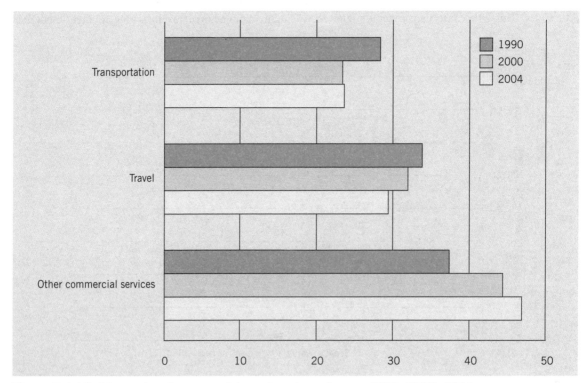

Figure 8.4 World exports of commercial services by category, 1990, 2000, 2004.
Source: WTO, 2005, Chart IV.2.

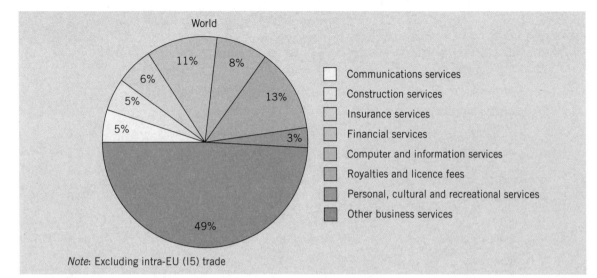

World

- Communications services
- Construction services
- Insurance services
- Financial services
- Computer and information services
- Royalties and licence fees
- Personal, cultural and recreational services
- Other business services

Note: Excluding intra-EU (I5) trade

Figure 8.5 World exports of 'other commercial services' by category, 2003.
Source: WTO, 2005, Chart IV.19.

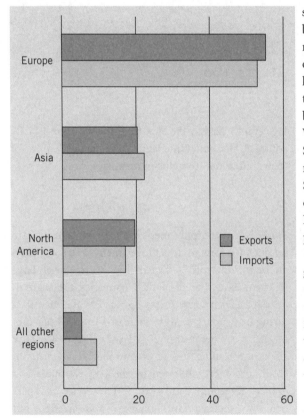

Figure 8.6 Regional shares in world trade in 'other commercial services', 2004.
Source: WTO, 2005, Chart IV.18.

services that are traded internationally. Increasingly, business services such as accountancy, legal services, marketing and retail are becoming internationally organized. The growth of international trade in services has been encouraged by the inclusion of services within the so-called **Uruguay Round** of trade negotiations between 1986 and 1994. With the establishment of the WTO in 1995, the General Agreement of Trade in Services (GATS) was established as a framework for negotiating agreements (Bryson *et al.*, 2004, pp.235–6). Services have been a key focus of discussion during the current **Doha round** of trade negotiations, launched in November 2001, with the EU and the US in particular pushing for further liberalization.

International trade in commercial services has grown on average by 7.6 per cent per year since 1980 (Figure 8.3), outstripping the growth of trade in goods (6.6 per cent on average) over the same period (WTO, 2006, p.9). The majority of services continue to be traded within national borders, however, and commercial services accounted for approximately 19 per cent of total world trade in 2004, compared with 16 per cent in 1980 (ibid., p.38). The share of commercial services exports by WTO category is shown in Figures 8.4 and 8.5.

The global map of services exports shows great geographical unevenness, with Europe (EU 25), North America and Asia accounting for 92 per cent of exports

Table 8.2 Leading exporters and importers in world trade in commercial services, 2004 (percentage and rank)

Rank	Exporters	Share %	Rank	Importers	Share %
1	United States	15.0	1	United States	12.4
2	United Kingdom	8.1	2	Germany	9.2
3	Germany	6.3	3	United Kingdom	6.5
4	France	5.1	4	Japan	6.4
5	Japan	4.5	5	France	4.6
6	Spain	4.0	6	Italy	3.8
7	Italy	3.9	7	Netherlands	3.5
8	Netherlands	3.4	8	China	3.4
9	China	2.9	9	Ireland	2.8
10	Hong Kong, China	2.5	10	Canada	2.7
11	Belgium	2.3	11	Spain	2.6
12	Austria	2.3	12	Korea, Republic of	2.4
13	Ireland	2.2	13	Belgium	2.3
14	Canada	2.2	14	Austria	2.2
15	Korea, Republic of	1.9	15	India	2.0

Source: WTO, 2004, Table 1.7.

of other commercial services (excluding transport and travel) in 2004 (Figure 8.6). Both Western Europe and North America run trade surpluses in services, compared with deficits in manufacturing, while the reverse is true for Asia. At the same time, services exports are less geographically concentrated than goods in terms of individual countries, with 66.6 per cent generated by the top 15 countries in 2004 compared with 69.4 per cent for goods (WTO, 2005, Tables 1.5, 1.7). The US is by far the most important exporter, with 15 per cent of world exports of commercial services, while the UK is also more important as an exporter of services than manufactured goods (Table 8.2). The converse is true for Germany and Japan (OECD, 2005, p.155). This reflects how the likes of the US and the UK have gained a **comparative advantage** in the export of financial and business services in particular since the 1970s, although this owes more to the creation of information resources, knowledge, skills and reputation than to pre-existing factor endowments, corresponding more closely to the dynamic notion of **competitive advan-**

tage emphasized by the 'new trade theory' (Box 4.1). By contrast, Germany and Japan have retained more of their traditional strengths in manufacturing.

8.4.2 FDI in services

Trade is not the only mechanism of globalization in services, for **foreign direct investment** (FDI) also plays an important role, reflected in the growth of large service MNCs. For the OECD countries, the share of outward FDI in services (as opposed to the manufacturing or primary sectors) increased from 50 per cent in 1992 to 67 per cent in 2002 (p.43) (Figure 8.7). The US is the leading source of overall FDI in terms of absolute volume, followed by the UK, Germany and France (Table 8.3). Until 1990, FDI in services was particularly focused on finance and trade, but subsequently other services such as business services, electricity, telecommunications and water supply have experienced substantial growth (UNCTAD, 2004, pp.98–9). In the airlines sector, by contrast, FDI is of

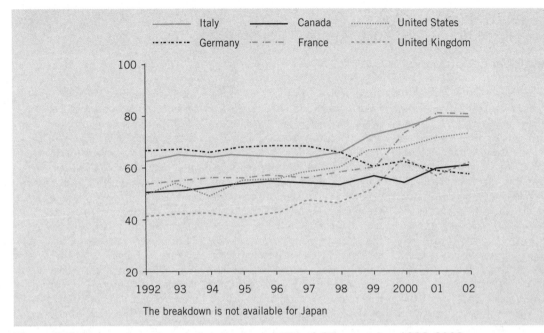

Figure 8.7 Share of services in total outward FDI of G7 countries, 1992–2002.
Source: Figure B.5.4. Share of the service sector in the total inward FDI positions of OECD countries. OECD International Direct Investment Statistics, June 2005, reproduced in *Measuring Globalization: OECD Economic Globalization Indicators*, © OECD, 2005.

little importance, with airlines relying, instead, on the development of alliances (e.g. code sharing, frequent-flyer programmes, cargo agreements) with other operators, the number of which increased from 20 worldwide in the early 1990s to 1,222 by 2001 (ibid.,

Table 8.3 Outward FDI in services by country, 2001

Country	Stock of outward FDI, 2001 $USm
United States	941,738
United Kingdom	496,109
Germany	422,931
France	371,581
Hong Kong	303,838
Netherlands	185,687
Switzerland	176,601
Canada	144,920
Italy	91,410
Denmark	54,700

Source: UNCTAD, 2004, p.306.

p.108). The turnover of countries' MNC affiliates located abroad in services is generally higher, compared with the total services exports of these countries, than the same ratio in the manufacturing sector. This demonstrates that FDI is a more important means of serving overseas markets than exporting in service industries (OECD, 2005, p.90). It reflects the fact that much service provision is embodied in the individuals supplying it, requiring either direct investment in new overseas offices to provide services to clients or indirect expansion through affiliations and agreements with foreign firms (Box 8.2).

8.4.3 The globalization of financial services

The sector in which globalization is most advanced is financial services, reflecting the highly mobile nature of money, a basic property that has been greatly enhanced by modern communications technology. As we indicated in section 5.4, there has been a massive increase in global financial flows since the 1970s. For example, daily turnover on the major foreign exchange markets grew from over $10–20 billion in 1973, about twice the

Box 8.2

The global accountancy industry

Many business service firms have internationalized by following their main MNC clients overseas, building on existing relationships to serve the needs of these clients in new locations. The accountancy industry, for example, has become highly global, mirroring the transnationalization of the various industries that use its services. The industry is dominated by the 'big four' firms (Table 8.4), which have grown and expanded through the establishment of networks and partnerships with local accounting firms under one brand name and through mergers. This process of concentration increased following a series of mergers among the 'big eight' firms in the 1980s and 1990s and the dissolution of one leading firm, Arthur Andersen, in

2002, after it was caught up in the Enron scandal. The big four are much larger than other firms, with their annual revenues reaching into billions of dollars (Table 8.4). They perform both accountancy and management consultancy services, although there is a trend towards the separation of the two. Their combined revenue amounted to 33 per cent of the global market for accountancy services in 2002, and they are among the most global of business service corporations. The 'big four' audit the bulk of publicly listed companies in developed countries: 78 per cent in the US, 80 per cent in Japan, 80 per cent in Italy, 90 per cent in the Netherlands and an estimated 95–98 per cent in the UK.

Historically, firms moved abroad to

service the needs of clients from their home countries in overseas markets. Their specific mode of expansion has been shaped by the nature of the accountancy industry, particularly the legal and regulatory environment in which firms operate. Firms that wish to expand internationally face various barriers in terms of nationally specific regulations about trade and commercial presence, accounting standards and the recruitment of staff. As a result, firms have typically expanded through the formation of partnerships, adding new members to a network of firms that are usually legally separate, locally owned and locally managed. (UNCTAD, 2004, p.110)

Table 8.4 The top ten accounting firms, ranked by total revenue, 2003

Name	Headquarters	Total revenue US$bn	Employees	Number of host countries
PriceWaterhouseCoopers	New York	16.0	122,820	139
Deloitte Touche Tohmatsu	New York	15.1	119,770	144
Ernst & Young	New York	13.1	103,000	140
KPMG	Amsterdam	12.2	98,900	148
BDO	Brussels	2.6	23,230	99
Grant Thornton	Chicago	1.8	21,500	110
RSM	London	1.8	20,000	80
Moores Rowland	High Point, NC	1.8	20,850	92
Howarth	New York	1.5	16,680	86
Baker Tilly	London	1.5	17,000	67

Source: UNCTAD, 2004, p.110.

value of trade, to an average of $1.5 trillion in 1998, about or just over 70 times the value of trade (Pollard, 2005, p.347).

Three key forces behind the process of **financial globalization** can be identified:

1. Deregulation through neoliberal policies. Conventionally, the financial system was tightly controlled by governments through a range of detailed rules and restrictions that separated different kinds of activities (e.g. banking and insurance) and limited entry of firms into the financial sector, particularly foreign ones. But an unprecedented 'crumbling of walls' has taken place since the 1970s, through major deregulation programmes. These involved the abolition of exchange and capital controls – which essentially limited trade in foreign exchange markets and the volume of capital that could be exported – and the removal of barriers between different financial activities and limits on the entry of foreign firms. For example, in 'the United Kingdom the so-called "Big Bang" of October 1986 removed the barriers which previously existed between banks and securities houses and allowed the entry of foreign firms into the Stock Exchange' (Dicken, 2003a, pp.448–9).

2. The development of advanced ICTs. Computers have transformed payment systems, allowing electronic money to be moved around the world at great speed. The introduction of chips (microprocessors) allows customers to pay for their purchases using plastic in the form of credit and debit cards. Technological developments have facilitated 24-hour global trading, exploiting the overlap between the trading hours of the world's major financial centres. Above all, ICT has greatly increased the pace of financial activity, allowing almost instanteous trading between distant centres (ibid., p.14). Finance is a crucial agent of time–space compression, obliterating the friction of distance in terms of the rapid movement of capital and information across space (Harvey, 1989).

3. The number of financial and monetary products has grown massively. Of particular significance is the rise of new financial instruments known as

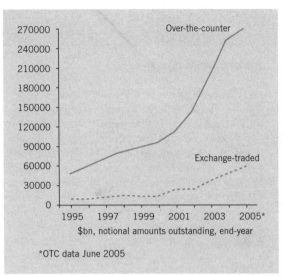

Figure 8.8 Growth of international derivatives markets, 1995–2005.
Source: International Financial Services London (IFSL), March 2006.

derivatives; they are traded, and have made it easier to move money across the globe (Figure 8.8). The term encompasses a highly complex range of instruments, but they are essentially 'contracts that specify rights/obligations based on (and hence derived from) the performance of some other currency, commodity or service used to manage risk and volatility in global markets' (Pollard, 2005, p.347). Such risk and volatility are associated with changes in prices of commodities, currencies and instruments such as interest rates. The basic forms of derivatives include: 'simple futures (where an agreement to buy a commodity at a given date and given price is made, allowing purchasers to purchase a degree of certainty)' (Tickell, 2000, p.88); swaps (where the parties to an agreement exchange particular liabilities, e.g. interest repayments, usually through an intermediary such as a financial institution); and options 'which, in exchange for a premium, allow buyers to trade assets at a predetermined price at a point in future' (ibid.). Derivatives are traded in two ways: on organized and regulated exchanges such as the London International Financial Futures Exchange or the Chicago Mercantile Exchange, and 'over the counter' between traders and clients. Since the early 1980s, derivatives have become symbolic of a highly monetized and

Box 8.3

Financial exclusion

Financial exclusion can be defined as 'those processes that serve to prevent certain social groups and individuals from gaining access to the financial system' (Leyshon and Thrift, 1995, p.314). It is based on income, compounding and accentuating the difficulties facing low-income individuals and groups. An estimated 1.5 million people in the UK, for instance, do not have a bank account, while a further 4.4 million (20 per cent) exist on the margins of the financial system, operating little more than a basic bank account (Kempson *et al.*, 2000, p.7). There is an important geographical dimension to financial exclusion in the sense that people are denied access to financial products and services because of the social and economic characteristics of the area in which they live, attracting the interest of economic geographers (Leyshon and Thrift, 1995). Inner-city neighbourhoods in the US and the UK are typically characterized by financial exclusion.

Two key mechanisms of financial exclusion can be identified:

1. The **closure of bank branches**. Established banks have sought to reduce costs in the face of increasing competition from supermarkets and large retailers offering services such as 'cash back', for example, and through

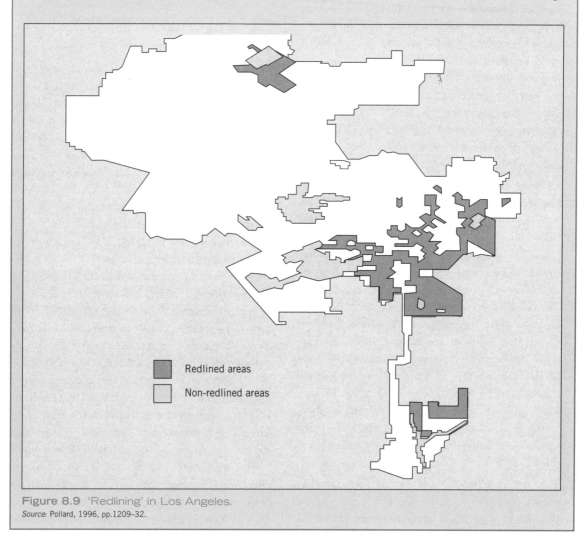

Redlined areas

Non-redlined areas

Figure 8.9 'Redlining' in Los Angeles.
Source: Pollard, 1996, pp.1209–32.

Box 8.3 (continued)

the expansion of telephone and Internet banking particularly. One of the principal ways in which banks have reduced costs is by closing a large number of branches, also fuelled by mergers between banks. A 27 per cent reduction in the number of banks and building societies in Britain occurred between 1988 and 1998 (calculated from Kempson *et al.*, 2000, p.19). The pattern of closure was geographically uneven, focusing particularly on low-income areas in urban environments. Recent research has found that the process of branch closure has continued over the last few years, with banks and building societies closing one in five of their branches since 1995, again affecting low-income areas disproportionately (ESRC, 2006).

2. '**Redlining**', which is defined as cases 'where goods and services are made unavailable, or made available only on less than favourable terms, to people because of where they live' (Squires, quoted in Pollard, 1996, p.1209). 'Redlining' is a product of the risk-assessment procedures employed by financial institutions where computer-based analysis has replaced the personal assessments of bank managers and insurance underwriters. Contemporary methods classify geographical areas (based on postcodes) according to a variety of social and economic factors. Typically, low-income inner-city areas are regarded as high risk and are 'redlined' or marked out on a map. As a result, people living within these areas are denied

access to credit and insurance services or, more commonly, are charged higher premiums. This process is most advanced in the US, where mainstream financial infrastructure has been withdrawn from many deprived inner-city areas (Figure 8.9), although it is also evident in the UK. In such areas, the withdrawal of mainstream financial services has resulted in the growth of unregulated '"second tier" financial sector[s]: pawnbrokers, moneylenders, cheque-cashing firms and hire purchase shops' (Tickell, 2005, p.250). Such enterprises tend to highly exploitative, lending money at cripplingly high rates of interest and using intimidation and violence against defaulters.

financially driven world economy, in which ICT has rendered financial instruments increasingly mobile and separate from flows of material goods (Tickell, 2003).

Yet the **globalization** of finance and the associated process of **time–space compression** have been highly uneven, both socially and geographically. While groups such as financial traders, brokers, investors and executives have benefited hugely, large sections of the world's population have been left behind. Groups such as the poor in developing countries and low-income groups in developed countries are experiencing growing marginalization from the financial system. Increased financial flows have forged closer connections between key financial centres, but other places have moved further apart (Leyshon, 1995), subject to processes of **financial exclusion** where low-income

groups are denied access to financial products (Box 8.3). In developed countries such as the UK, the problem of financial exclusion has attracted growing attention from governments in recent years The debt crisis afflicting many developing countries can be seen as a global problem of financial exclusion and marginalization, with such countries only gaining access to further credit on the conditions imposed by international organizations like the IMF and World Bank (Chapter 11).

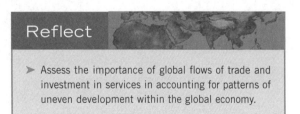

Reflect

➤ Assess the importance of global flows of trade and investment in services in accounting for patterns of uneven development within the global economy.

8.5 Business and financial services and world cities

8.5.1 Defining world cities

The tendency for high-level business and financial services to be geographically concentrated in major metropolitan areas is a key theme of recent research on services (Daniels, 1995). It is in these large urban areas that such 'producer' services have grown most rapidly. For example, two out of every three business and professional services jobs in the US were located in large metropolitan areas in 1986; almost half the information services, research and advertising jobs in Japan are located in Tokyo; and almost one-quarter of the business service jobs in the UK are found in London (Bryson *et al.*, 2004, p.96). Cities such as London, New York and Tokyo contain a very marked concentration of business and financial services. They are commonly described as **world cities** (Friedmann, 1986).

World cities are defined by their function within the world economy, not their size (Hamnett, 1995, p.109). Three main economic functions can be identified. First, they operate as command centres for large MNCs (ibid., pp.110–14). Second, world cities are defined by the concentration of business services found within them. This reflects the fact that MNCs are often the main clients of firms in areas such as accountancy, legal services, management consultancy and research and development. The third key dimension to the growth of world cities is their role as financial markets (Sassen, 1991), containing an intense concentration of a wide variety of international financial businesses and transactions, including banking, insurance, foreign exchange markets, stock exchanges and derivatives trading. Since the 1970s, the power and influence of the leading global financial centres has grown through the globalization of financial markets. At the same time, world cities are important centres of social and cultural production, often controlling the information, news and film industries.

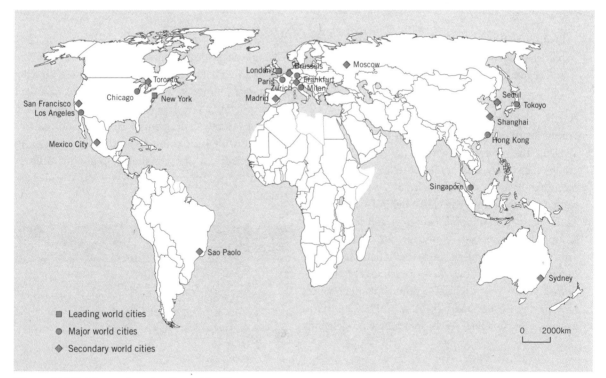

Figure 8.10 The network of world cities.
Source: Adapted from Beaverstock *et al.*, 1999, p.456.

In the early 1990s, New York, London and Tokyo were regarded as the three leading centres (Tickell, 2005, p.246). Subsequently, however, London and New York have moved ahead, regarded as the only two truly global financial centres, in contrast to regional centres like Frankfurt, Paris and Hong Kong (Z/Yen Ltd, 2005). Tokyo has lost ground and respondents to a recent survey of financial executives and managers in London believed that Shanghai was the most likely city to emerge as a third global financial centre, reflecting the rapid growth of the Chinese economy. The contemporary structure of world cities, divided into three distinct tiers, is shown in Figure 8.10.

8.5.2 Explaining the concentration of producer services within world cities

In addition to identifying the key functions of world cities and listing examples, there is a need to explain the tendency for advanced producer services to cluster within them. While globalization has seen the dispersal of production to low-cost locations, this is paralleled by a growing concentration of corporate power in the major cities (Sassen, 1991). Such power is symbolized by the distinctive urban landscape of skyscrapers, as demonstrated by the dramatic skyline of Manhattan (Figure 8.11). As Hamnett (1995, p.120) puts it,

'telecommunications has intensified, not eliminated, the historic role of certain cities as centres of specialist knowledge, information and power'.

In standard economic terms, global financial centres benefit from classic **agglomeration economies** – advantages derived from the concentration of firms in a particular location (section 4.3) – such as the availability of skilled personnel, proximity to a large number of customers, access to specialist suppliers of services and the provision of an advanced infrastructure. Financial managers and executives in London, for instance, emphasized the importance of skilled personnel, the degree and quality of (government) regulation, access to international financial markets, the availability of business infrastructure (advanced telecommunications and ICT as well as transport links) and access to customers as the key factors giving the city a competitive advantage in global financial markets (Z/Yen Ltd, 2005). As growth becomes self-reinforcing (see Box 4.2), the size of the market and the existence of a large pool of highly skilled and highly paid workers attracts new financial institutions, and helps to generate innovations in the form of new products and services and the development of ICT (Tickell, 2005, p.246). Beyond this, less tangible factors are also important and three more 'social' reasons for the concentration of business and financial services in world cities can be identified (Thrift, 1994):

Figure 8.11 Lower Manhattan skyline.
Source: D. Mackinnon.

1. The need for spatial proximity between specialist service firms. Such firms generally work through small teams of expert consultants and advisers, sometimes involving cooperation with people working for other firms. Much of this work is project-based, involving small teams putting together packages and deals for their clients, often large MNCs. There is a need for frequent and rapid communication between these teams and their clients, and face-to-face contact is particularly important, allowing complex and specialist forms of knowledge to be exchanged. Advertising is a good example of a project-based industry structured by the personal relationship between account managers from advertising agencies and marketing managers from client firms (Grabher, 2002). While agencies operate within wider global communication networks, the London industry is highly concentrated in the Soho district, colloquially known as 'adland', incorporating a range of 'supplier' services such as printing, lithography and graphic design.

2. The need for trust between groups of specialist individuals and teams. In the City of London, for example, such trust traditionally flowed from the shared social background of many of the main players, drawn from the public schools and Oxbridge colleges upholding an ethos of 'gentlemanly capitalism'. Following an influx of foreign firms bringing different practices and attitudes since the 1960s, the make-up of the City is far more complex and diverse. Financial deals and packages are made between people from a range of different social and geographical backgrounds. Again, this reinforces the importance of face-to-face contact in terms of the need to size people up through direct relationships. As a result:

> Places such as Wall Street or the City of London are not simply aggregations of people and institutions who carry out specialist tasks: they are nodes of information and experience, alive with messages and social contacts. This dense clustering of people and information in a small area is crucial to the functioning of such places.
>
> (Hamnett, 1995, p.117)

Although neglected by some influential writers who place more emphasis on the economic logic of clustering, an appreciation of this social and cultural dimension is crucial to an understanding of the concentration of business and financial services in world cities.

3. Reinforcing this social and cultural dimension, world cities also operate as **centres of interpretation**, focused on information about the performance of markets and financial products. As the ex-chairman of Citibank, Walter Wriston, famously commented, 'information about money has become almost as important as money itself' (ibid., p.117). The quality of interpretation depends on a body of accumulated knowledge and expertise that has been built up in places like the City of London and Wall Street over a long period of time. Such knowledge is part of the distinctive cultures of these centres, embodied in key people and organizations. As such, it cannot be easily duplicated by other places, indicating that the dominance of financial centres is likely to continue despite the spread of ICT (Box 8.4).

Box 8.4

The concentration of financial services in the City of London

The City of London is the leading financial centre in the world, having pulled slightly ahead of New York between 2003 and 2005 (Z/Yen Ltd, 2005, p.13). It is also more internationally oriented than New York, which sources a larger volume of business from its domestic market (ibid., p.17). The City generates £753 billion of foreign exchange turnover each day (31 per cent of global turnover), accounts for 43 per cent of the global foreign equity market and is the world's leading market for international insurance (City of London Corporation, undated). Its share of cross-border bank lending, foreign exchange turnover and 'over the

Box 8.4 (continued)

counter' derivatives turnover has increased, while the share of foreign equities turnover, income from marine insurance and international bonds has fallen (Table 8.5). London accounted for 29.5 per cent of financial employment in the UK in 2005, compared with 26.8 per cent in 1987 (IFS, 2006). The proportion of managerial and professional jobs in the City increased markedly between 1993 and 2000, while employment in clerical and related occupations fell (I. Gordon et al., 2005, p.51).

Traditionally, financial services have been concentrated within the 'Square Mile' itself, with a range of formal and informal rules requiring firms and banks to locate in the area, extended by the recent development of Canary Wharf in the East End. Such spatial concentration has always coexisted with the international scope of the

City's activities, suggesting that there is little new about the apparent paradox of global localization. The culture of financial centres such as the City of London has often been summarized by the slogan 'information, expertise, contacts', stressing the importance of gaining information, developing the expertise to interpret it and the contacts to build trust and mutual understanding (Thrift, 1994, p.334). Firms and managers must be sociable to survive and compete with contacts providing access to business and information. In this sense, 'who you know' is regarded as part of 'what you know' (ibid., p.335).

Crucially, the growth of electronic communication seems to have enhanced rather than diminished the role of the City. The greatly increased volume of information on financial

markets and products now in circulation demands additional expertise and contacts to interpret it. Electronic communication seems to act as a supplement to face-to-face communication. As one respondent in interviews conducted by Cook et al. commented:

Face-to-face contact is very important, absolutely, you're never going to replace face-to-face contact, you can't pick up body language; you can't build relationships truly over the phone and with video-conferencing. They help because they make things more efficient but it's never going to replace the face-to-face contact and that's face-to-face contact in all aspects, even internally there is a point when you have to go and meet people even if you've seen them on a

Table 8.5 London's share of international financial markets

	Cross-border bank lending %	Foreign equities turnover %	Foreign exchange dealing %	Exchange-traded derivatives turnover %	OTC derivatives turnover %	Marine insurance net premium income %	International bonds – primary market %	International bonds – secondary market %
1989	17	–	25	5	–	–	75	80
1992	16	64	27	12	–	24	60	70
1995	17	61	30	12	27	21	–	70
1998	20	65	33	11	36	14	–	70
1999	19	56	–	8	–	15	60	70
2000	19	48	–	8	–	17	60	70
2001	19	56	31	7	36	18	60	70
2002	19	56	–	6	–	16	60	70
2003	19	45	–	6	–	21	60	70
2004	20	44	31*	7	43*	15	60	70
2005	20**	43***	–	–	–	–	–	–

* April; ** March; *** January–September
Source: International Financial Services London (IFSL), 2006.

Box 8.4 (continued)

video-conferencing basis because you don't get a feel for that relationship aspect.

(Cook *et al.*, 2003, p.8)

For all the reasons stressed above, the role of the City of London is unlikely to be diminished by the growth of electronic communications. A recent study showed that the much-vaunted 'offshoring' of services to developing countries such as India was likely to have only a limited impact on the City since the kinds of routine and standardized services upon which this process is largely focused tend not to be located in the City in any case (I. Gordon *et al.*, 2005). As we have emphasized, its high-value activities depend on frequent face-to-face contact between clients, suppliers, customers and col-laborators, providing embodied knowledge and market intelligence that cannot be replaced by electronic interactions over long distances. Whether the City retains its pre-eminence as a global node depends rather more on how it responds to competition from other financial centres, particularly within Western Europe (Z/Yen Ltd, 2005, p.13).

8.5.3 Social and spatial polarization within world cities

The role of world cities as centres of corporate wealth and power does not mean that all their inhabitants are prosperous. Even within such cities, there is a stark contrast between the gleaming corporate headquarters and the elegant suburbs of the wealthy and the squalid and run-down neighbourhoods of the urban poor. As we emphasized in Chapter 1, **uneven development** occurs at the local scale as well as the global. For some writers, such as Saskia Sassen and the urban sociologist Manual Castells, world cities are defined by increasing social polarization. This **'dual city' hypothesis** claims that a growing divide between rich and poor has emerged as a result of changes in the economic base. Relatively well-paid manufacturing jobs have been lost as factories shut down or moved out of the city in search of lower costs. The new occupational structure is characterized by high-status, high-paid jobs in business and financial services and a range of routine, low-paid jobs, such as cleaners, porters, waiters and bartenders, which exist to serve the lifestyle needs of the rich. World cities attract large numbers of immigrants, many of whom are drawn into low-grade services.

As Hamnett (1995, p.123) notes, this is a persuasive argument, but not a wholly convincing one. There is evidence of the opposite trend occurring in certain world cities such as London, where professionalization has seen the number of professional and managerial jobs increase. This has bolstered the number of middle-class rather than highly wealthy people, and led to a fall in the number of low-grade jobs at the bottom end (ibid., p.125). In other European cities there are few indications of increased social polarization. The 'dual city' thesis seems to rely too much on the experience of New York and other large American cities, extending the argument from these to world cities in general. World cities certainly contain concentrations of extreme wealth and extreme poverty, but increases in inequality seems to be confined to certain cities in the US, the transition economies and developing countries.

Reflect

➤ Explain why globalization seems to be accompanied by an increased centralization of corporate power in world cities.

8.6 Digitization and the Internet economy

As we stressed in earlier chapters (sections 1.2, 4.5.2), the growth of information and communications technologies since the 1970s has reduced the costs of transmitting information across space, serving as a key vehicle of **time–space compression** (Harvey, 1989a). The combination of **digitization** and the Internet is of huge significance, allowing information to be collected, packaged and distributed more rapidly and efficiently than before. One crucial question that has sparked much speculation and debate concerns the impact of the Internet and **digitization** on existing forms of economic organization. Many commentators and business people have assumed that the effects will be far-reaching, conferring huge 'first-mover' benefits on the firms that initiate and pioneer the technology and penalizing established corporations that fail to innovate and adapt (Leyshon, 2001, p.56). In this sense, the Internet economy can be seen as the latest wave of creative destruction within capitalism (*Financial Times*, 2005). It was the prospect of such first-mover benefits that fuelled the soaring share values of high-technology companies in the late 1990s when few if any had yet made profits.

8.6.1 The growth of e-commerce

The spread of advanced ICTs has led to much discussion among economic commentators and policy-makers of a '**new economy**', symbolized by the soaring valuations of 'dot.com' companies by the stock markets in the late 1990s (Daniels, 2003). While the term has become less widely used in the wake of the dot-com crash in share prices since 2000/2001, Internet-based activity has continued to expand, offering new ways of delivering services (Waters, 2005). The big four Internet businesses are Amazon, the online bookstore, the auction site eBay and the search engines Google and Yahoo. A key trend has been the growth of **e-commerce**, which can be defined as 'trade that actually takes place over the Internet through a consumer visiting a seller's website and making a transaction there' (*The Economist*, 2000, quoted in Leyshon, 2001, p.55). E-commerce is divided into business-to-business (b2b) and business-to-consumer (b2c) transactions. The latter are more familiar, particularly through the growth of online retailing, requiring only an Internet connection and a credit card. Online shopping has grown rapidly, with spending in the UK rising by 29 per cent in 2005, compared with a 1.5 per cent increase in general retail spending, although it still only accounts for 3.1 per cent of total retail expenditure in the UK (Palmer, 2006).

One of the main effects of e-commerce is to create increasingly frequent and rapid direct interaction between businesses, other businesses and consumers, leading to an increased emphasis on connectivity, speed and customer service (Bryson *et al.*, 2004, p.151). It seems to reduce the costs of economic coordination, enabling consumers to search markets rapidly and conveniently for information about suppliers and products. In this way, the Internet is associated with the **disintermediation** of production networks, removing the need for intermediaries between the buyers and sellers of goods. At the same time, however, electronic technologies also lead to **reintermediation**, reshaping the structure of markets and creating opportunities for new or small firms, often at the expense of established firms that are unable to react quickly enough (Leyshon, 2001, p.57).

8.6.2 The cultural industries

The impact of new digital technologies on what are termed the cultural or creative industries (Pratt, 1997) has been the subject of particular debate (Leyshon *et al.*, 2005). The **cultural industries** have been defined as:

> performance, fine art, and literature; their reproduction: books, journal magazines, newspapers, film, radio, television, recordings on disk or tape; and activities that link together art forms such as advertising. Also ... the production, distribution and display processes of printing and broadcasting, as well as museums, libraries, theatres, nightclubs, and galleries.
>
> (Pratt, 1997, p.1958)

The cultural industries have grown significantly in terms of both employment and contribution to GDP since the early 1990s. In the UK, software, computer games and electronic publishing experienced particularly rapid growth in gross value added and exports between 1997 and 2004 (Department of Culture, Media and Sport, 2005).

The cultural industries have been characterized as innovative, flexible and creative, operating at the intersection of the local and the global scales, often by packaging and selling local distinctiveness through global distribution networks (Lash and Urry, 1994, pp.113–43). This characterization of the cultural industries emphasizes the tendency for them to become geographically concentrated in particular locations, particularly large metropolitan regions. In the US, for example, over 50 per cent of employment is located in metropolitan areas with a population of 1 million or more, with the majority of this found in just two centres, New York and Los Angeles (Scott, 2001, p.16). Many cultural industries are grouped together in dense clusters of firms close to the central business districts of major cities, giving them access to suppliers and large markets for their products. Notable examples of such clustering in major urban centres include the motion-picture complex of Hollywood (Box 10.4), the media complex of Manhattan and the publishing industry of London (ibid., p.16).

8.6.3 The impact of digital technologies on the cultural industries

The cultural industries are being restructured by the spread of new technologies. As Gibson and Kong (2005, pp.541–61) put it, 'the media through which cultural

products are consumed are increasingly reliant upon a common digital platform'. This trend is particularly evident for products such as music, computer games, film and television, but it also influences publishing and advertising, among other activities. One very significant digital application is MP3, a compression programme that reduces the size of digital audio files, allowing them to be made and distributed more easily (Leyshon, 2001, p.51). The ease with which MP3 files can be downloaded from the Internet and exchanged between users has profound implications for industries such as music and film, threatening established forms of organization (Box 8.5). In this context, the Internet has been described as a '"disruptive technology" ... that destroys old business models by producing products more cheaply and efficiently' (*Financial Times*, 2005).

The control of information is crucial in enabling firms in cultural industries such as music, film and publishing to generate and maintain a competitive advantage over their rivals (Gibson and Kong, 2004, p.545) As such, the protection of **intellectual property rights** (copyright) is vital. In the music industry, for example, copyright law enables those who claim ownership of the music and its sound recording – the

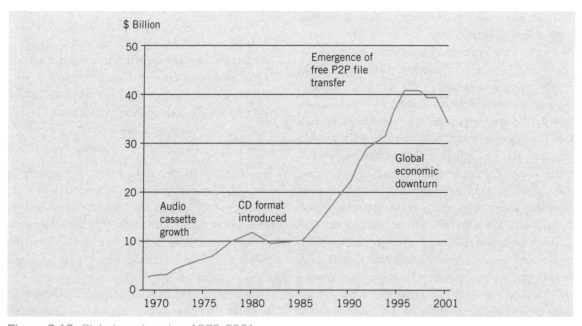

Figure 8.12 Global music sales, 1970–2001.
Source: Leyshon *et al.*, 2005, p.178.

publishing and recording arms of the record companies – to earn royalties on each unit (e.g. CD or DVD) sold (Leyshon *et al.*, 2005, p.186). It is this system that is threatened by the widespread availability of music files across the Internet through digital file-sharing systems such as peer-to-peer networks of the kind pioneered by Napster.

This software has hugely increased the circulation of music files; in 2002 it is estimated that there were almost 1 billion music files available to be downloaded on the Internet, and that 27 per cent of Americans and 13 per cent of Europeans regularly downloaded music in this way (ibid., p.179). Some of these files are downloaded illegally, in breach of copyright law, with this 'piracy' sparking strong condemnation and legal action from the music industry (Box 8.5). Such action has been fuelled by a significant fall in sales since 2001, leading to cuts in output and employment (Figure 8.12). The growth of Internet-based 'piracy' has also been a major problem for related industries like film (Currah, 2003).

The effects of new digital technologies have been to further spatially 'stretch' industries like music and film

'across a force field of global and local relationships', with production increasingly concentrated in clusters of creative activity and output channelled into ever more spatially extended networks of consumption.

(Currah, 2003, p.71).

Research on the new media cluster of 'Silicon Alley' in Manhattan, for instance – originally centred on the Flatiron district, and stretching south to Soho and Tribeca – has demonstrated the importance of spatial clusters in the Internet-based economy, explaining this in terms of the same need for face-to-face communication and direct interaction that is evident in business services (section 8.5) (Pratt, 2000). In the music industry, the emergence of Internet-based companies and labels seems to have fostered the growth of new clusters around San Francisco and San Diego in California, reinforcing the western end of the long-established London–New York–Los Angeles axis (Figure 8.13) (Leyshon, 2001, p.69). Production activities are organized on an increasingly transnational basis

Box 8.5

The Internet and the crisis of the music industry

The global music industry is dominated by four large corporations: AOL-Time Warner, Sony/BMG, Universal and EMI, who were responsible for 80 per cent of global music sales in 2003 and had significant interests across the media, entertainment and technology sectors (Leyshon *et al.*, 2005, p.177). While Internet 'piracy' has attracted much of the blame, there are other factors behind the crisis of the music industry. These include the rise of genres like dance music (which are difficult to control through conventional means because of the facelessness of the artists and the emphasis on clubbing rather than buying recorded music), the increasing links between music and

other media such as advertising and film (with music viewed as less distinctive and less intrinsically valuable) and the fact that popular music is no longer as important to the key 14–24 age group as it once was, with other products such as mobile phones and computer games, capturing more of their expenditure (ibid., pp.183–5).

The industry has reacted in a range of ways to the growth of Internet-based downloading, with record companies taking legal action against file-sharing networks and the Internet service providers that they claim allow it to happen. At the same time, they have emphasized the virtues of traditional arrangements for producing music where the record companies provide the con-

tacts and funds to bring together specialist workers (ibid., p.63). Existing firms have established their own subscription and downloading services, while new companies have emerged to fill particular niches, for example the provision of subscription packages or brokering between artists and record companies. The major record companies have thus far resisted more radical organizational changes that might allow them to better accommodate the impact of new software formats and Internet distribution systems (ibid., p.202). While some have managed to restore profits in the short term, it is questionable whether they have developed more sustainable long-term solutions.

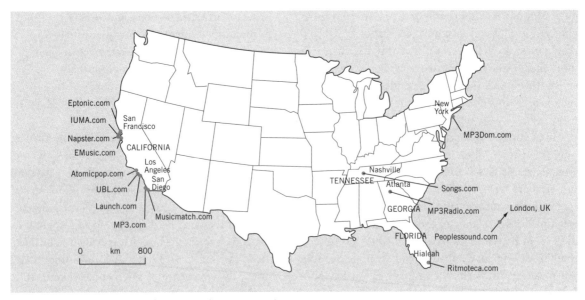

Figure 8.13 Leading MP3 distribution companies.
Source: Leyshon, 2001, p.69.

in industries like music and film, with activities such as film-shooting or CD manufacture dispersed to lower-cost locations (Leyshon, 2001; Scott, 2002). Further geographical dispersal of standardized activities may be encouraged by digitization, which creates more spatially extensive networks of distribution as producers can access consumers in distant locations more easily.

Reflect

➤ Will the effects of digitization and the Internet undermine established media centres such as Los Angeles and New York in the long term?

8.7 Globalization and the geographical dispersal of services

8.7.1 Forms of dispersal

Since 2002–3, the process of '**offshoring**', where firms relocate operations to low-wage economies, has attracted much media attention in the UK and the US. To avoid confusion, the inherently geographical process of '**offshoring**' can be distinguished from the organizational trend of '**outsourcing**', where companies decide to contract out a particular service or function to another firm (Fraser, 2003). Pure offshoring as an internal process based on the relocation of functions from a parent company to its foreign affiliates (sometimes referred to as 'captive' offshoring, involving FDI) can be distinguished from the external reorganization of services that are outsourced internationally to third-party service providers (UNCTAD, 2004, p.147).

In geographical terms, a distinction can also be made between '**offshoring**' involving a shift of operation to low-wage countries, '**near-shoring**' referring to the relocation of services to (usually neighbouring) countries of a broadly similar level of development to one's own (e.g. UK–Ireland, US–Canada) and '**onshoring**' when services are relocated within the same country (Bryson, 2006). The distinction between 'near-shoring' and 'offshoring' is not precise, referring more to wage differentials – moderate and very large respectively – than physical distance per se (I. Gordon *et al.*, 2005, p.9). '**Blended-shore**' strategies refer to situations where companies combine these different forms of geographical shift.

8.7.2 Processes of dispersal

There is nothing particularly new about the spatial dispersal of services, with **'back office' functions** – routine clerical and administrative tasks such as the maintenance of office records, payroll and billing, bank checks and insurance claims – long having been subject to relocation out of city centres to surrounding suburbs where property (rent) and labour costs are typically much lower. What is new is that ICT opens up the possibility of relocation at the regional and, especially, global scales, creating new **spatial divisions of labour** in services. One of the earliest examples of the international relocation of services was New York financial firms relocating administrative work to western Ireland, actively encouraged by the Irish government. These firms sent life insurance claims to a site located near Shannon airport for completion and checking before the processed data was sent back to New York via fibre-optic cable or satellite (Figure 8.14). Similarly, the Caribbean has become a favoured destination for US back-office functions. This reflects its combination of low wages and geographical proximity, giving it a similar relationship to the US in terms of back-office functions as Mexico holds for manufacturing (albeit on

a smaller scale). In the example shown in Figure 8.15, American Airlines sent its tickets for processing to a subsidiary in Barbados, a process that began in 1981. The subsidiary opened a second office in the Dominican Republic, where wages were half as high as in Barbados, in 1987. While relocation from New York to Ireland can be defined as **near-shoring**, the example of relocation to the Caribbean represents **offshoring**, emphasizing wage differentials not distance as the critical factor.

Recent processes of offshoring seem to reflect a combination of three main factors (I. Gordon *et al.*, 2005, p.19):

➤ technical developments in telecommunications and Internet provision, ensuring that 'many clerical tasks have become increasingly footloose and susceptible to spatial variations in production costs' (Warf, 1995, p.372);

➤ increased pressures to reduce costs across a range of service industries, as the sustained growth of the 1990s gave way to a period of slower growth and increased competition from 2000;

➤ the discovery and, to some extent, the creation of new pools of skilled labour at sites within certain low-wage economies.

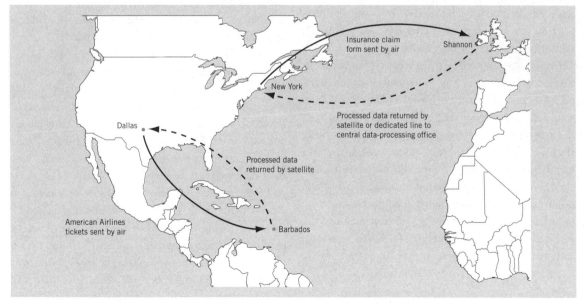

Figure 8.14 Offshore processing in the airline and insurance industries.
Source: 'Telecommunications and the changing geographies of knowledge transmission in the late twentieth century', in *Urban Studies*, 32, Taylor & Francis Ltd (Warf, B. 1995). http://www.tandf.co.uk/journals.

As a consequence, the spatial division of labour between regions and nations can be extended into the delivery of consumer services as firms take advantage of geographical variations in the costs of labour. Three main groups of service activity are now subject to off-shoring (ibid., p.372):

➤ **Call centres**, focusing on marketing, routine customer enquiries and more sophisticated technical support. Recently rebranded as contact centres or customer service facilities, they have grown very rapidly since the early 1990s. Call centres were pioneered in the financial sector and have subsequently grown in areas such as travel, telecommunications, mail order and the deregulated utilities;

➤ IT functions, including data processing, code-checking, software development and modification, operations support, publishing and statistical analysis;

➤ Wider business support functions, including accounting/payroll operations, paralegal work and record maintenance.

The offshoring of service functions accelerated rapidly from the mid-1990s, with the total market for all offshore service exports estimated at $32 billion in 2001. It marks the next stage in the evolution of the **international division of labour**, described by economic geographers as a **second 'global shift'** following the relocation of manufacturing activities in the 1970s and 1980s (Bryson, 2006). The process is likely to accelerate further over the next few years, driven by powerful pressures of cost reduction, quality enhancement, consolidation of services to realize economies of scale and the accessing of certain skills and markets. Cost is not always the overriding motivation: enhancing the quality of services and the realization of economies of scale through the consolidation of operations are also important drivers. While costs tend to be the most important factor determining call centre locations, access to markets, the availability of a skilled workforce and proximity to customers are more important for IT-related services (UNCTAD, 2004, p.166).

8.7.3 The geography of dispersal

Among developed countries, the UK became a particular focus of growth during the 1990s, gaining a reputation as 'the call-centre capital of Europe' with the number reaching 5,320 by the end of 2003 (Snow, 2005, p.526). Call centres were particularly important in old industrial regions such as north-west England, north-east England and Scotland, where labour costs are lower than in the South-East (Bishop *et al.*, 2003, p.2759). On an international scale, the practice of **offshoring** began in the US, with companies from that country still dominant, followed by the UK. In general, companies from other European countries have shown less inclination to offshore services (UNCTAD, 2004, p.168). Ireland, India, Canada and Israel are the most important destinations for relocated services (ibid., p.153). Like trade and FDI in services more generally, most international relocation occurs between developed countries, demonstrating that it is not primarily a North–South issue as often reported in the media. At the same time, however, the share of developing and transition economies is growing rapidly, from 37 per cent in 2002 to 51 per cent in 2003 (ibid., pp.160, 164). Many service companies are now following **blended-shore** strategies that combine offshore, onshore and near-shore operations depending on the knowledge and skill requirements of individual service tasks.

India has emerged as the favoured destination among developing countries for British and American companies in particular, rapidly becoming the 'world's back office' (Bryson *et al.*, 2004, p.127). This reflects its advantages in terms of skilled, English-speaking labour and ICT infrastructure. In the 1990s, British Airways and American Express pioneered this trend by moving their customer service divisions to Mumbai (Bombay) and Delhi respectively before the process accelerated in 2003-4 (Table 8.6). These companies are attracted by the cost savings involved when graduates can be employed in India for starting salaries of around £2,500 compared with around £12,500 in the UK. Labour costs comprise only a third of total costs in Indian call centres, compared with around 50–70 per cent in the UK (P.J. Taylor and Bain, 2004, p.10). Despite such differentials, however, some companies, such as the

Table 8.6 Selected offshoring cases, UK, 2003–2004

Service	Company	Function	Country and number of jobs or value involved
Financial services	HSBC	Back-office processing jobs	4,000 by end of 2003 in India, China and Malaysia. Another 3,500 were announced in June 2004
	Norwich Union/Aviva	Administrative insurance jobs; 350 in call centres, 2000 in back office and administration.	2,350 in India by end of 2004
	Lloyds TSB	Call-centre jobs	1,500 jobs in India by end of 2004
	Barclays	Back-office staff	500 to India
	Axa		700, some to India
	Abbey National	Back- and front-office work	400 jobs to Bangalore
Distribution services	Tesco	Business support centre	350 to India
Telecommunication services	BT	Call centre	2,200 by 2004 to India
Transport services	Network Rail	National Rail enquiries	600 to India
Health services	NHS	Fast-track centres offering surgery to NHS patients. Foreign providers run mobile operating units. Netcare plans to bring over surgical teams from South Africa once every 11 weeks	Non-UK healthcare providers, including Netcare of South Africa, amounting to a total of £2 billion
	NHS	£896 million IT contract to modernize the NHS	Tata Consultancy Services (India) part of a consortium
Other government	Greater London Authority	Software for toll charging	A $10 million contract to Mastek

Source: UNCTAD, 2004, p.168.

Royal Bank of Scotland (in contrast to competitors like Barclay's and HSBC), have consciously decided not to shift services abroad, citing the need to maintain customer relations. Other offshore destinations that have attracted call-centre investment include the developing Asian countries of Malaysia, the Philippines and China, the transition economies of Central and Eastern Europe, such as the Czech Republic, Poland and Hungary, Brazil, Chile, Costa Rica and Mexico in Latin America, and South Africa.

Box 8.6

The growth of the Indian IT services industry

India's service exports have grown rapidly in recent years, based on software development, and back-office and call-centre services, with the value of exports of these services growing from under $0.5 billion in the mid-1990s to £12 billion in 2003/4 (UNCTAD, 2004, p.169). India's share of the global market for offshore IT-related services was estimated at 25 per cent in 2001 – second only to Ireland – and is likely to have increased subsequently (ibid., p.170). The US and the UK accounted for 82 per cent of India's exports of software and IT-enabled services in 2003/4 (ibid., p.173). Particular services relocated by overseas MNCs to India include customer care, finance, human resources, billing and payment services, administration and content development. Total employment in IT-enabled services stood at 245,000 in 2003–2004 (ibid., p.172) The industry is estimated to be growing at a rate of about 70 per cent a year (Bryson *et al.*, 2004). The main concentrations are in the Delhi region and Mumbai (Bombay) with Bangalore and Chennai (Madras) and then Kolkata (Calcutta) and Hyderabad also important (Figure 8.15).

In India, call centres are seen as 'cool' and 'hip' workplaces, staffed by the young, educated and cosmopolitan, whereas such work is widely regarded as routine and of low status in developed countries like the UK and the US, attracting students and temporary workers. The training in British culture and society offered by Indian call centres has attracted much media attention. Trainees applying for work with centres such as Spectramind attend pronunciation classes and are educated in British history, customs, celebrities and slang. Videos of soap operas such as *The Bill*, *Emmerdale*, *Coronation Street* and *EastEnders* are required viewing, along with knowledge of Robbie Williams, the Royal Family and the climate (Harding, 2001). The latest sports results are also part of the crash course curriculum, while the tabloids are downloaded daily so that operatives can keep up with the latest news. Similarly, US-oriented staff are trained in the nuances of baseball and American television.

If an applicant for call-centre employment is successfully trained and accepted (many are rejected for not meeting speech and accent tests), they spend their time fielding calls from customers in the UK or the US. Bryson *et al.* (2004, p.127) cite the example of Anjali Maindiritta, aged 25, speaking to a middle-aged man in North London about his car insurance. She is based in Bangalore and has never visited England, calling herself 'Angela' when dealing with English customers. Over a nine-hour shift Anjali will receive around 150 calls from customers of one of the UK's leading insurance companies, and only one caller a fortnight will realize that she's not based in Britain. Interestingly, while media accounts suggest that Indian graduates are delighted to accept such work – in contrast to the reluctance of low-paid workers in the UK – research indicates that turn-over rates are similar

Figure 8.15 Major service centres in India.
Source: McFarlane, 2004, p.891.

Box 8.6 (continued)

to the UK (P.J. Taylor and Bain, 2004, p.11). Wage rates are lower than the IT/software sector but higher than for other white-collar professionals (ibid., p.12). As with call centres generally, the work is highly intensive and pressurized, creating problems of exhaustion, withdrawal and burnout.

8.7.4 The future of offshoring

While offshoring has largely focused on the more routine, standardized operations, known as 'business process outsourcing', there has been a sharp rise in the relocation of high-value services such as medical research and financial analysis since late 2004, known as 'knowledge process outsourcing' (Merchant and Johnson, 2005). In recent years, US legal firms have started to relocate tasks such as case study research, analysis of documents, contracts and patent applications, with an estimated 12,000 legal jobs moved to developing countries, particularly India, in 2004 (Jain, 2005). Recent media coverage has picked up on this, portraying it as an inevitable consequence of globalization and the rise of the 'emerging giants' of India and China. In reality, offshoring is limited by a number of constraints such as the impossibility of standardizing and digitizing some service activities (often those associated with innovation and creativity that typically require face-to-face communication), the need for proximity to existing customers, the nature of information that is personal, sensitive and confidential, and national standards and regulations (UNCTAD, 2004, p.152).

One analysis suggested that the maximum number of jobs potentially subject to offshoring from the US was around 11 per cent of all jobs (14 million), although the actual number is likely to be much lower (ibid., p.154). A recent report from Deloitte Touche Tohmatsu (the accountancy and management consultancy firm) suggested that another 150,000 largely white-collar service jobs would be relocated from the UK, mainly to India, by 2010 (Pym, 2006). While these are likely to include more skilled work, such as financial analysis and legal research, the number of jobs is relatively small against a UK workforce of 29 million. Developing countries such as India are likely to make further inroads into more complex and high-value tasks, as a result of knowledge process outsourcing, but the offshoring of services is not a zero-sum game and such employment is likely also to continue growing in developed countries as companies pursue sophisticated blended-shore strategies.

Reflect

➤ Compare and contrast the offshoring of services since the late 1990s with the new international division of labour (NIDL) in manufacturing in the 1970s and 1980s (section 7.3.2).

8.8 Summary

Services have come to dominate the employment structures of most developed countries since the 1960s, reflecting an underlying shift away from manufacturing industry. They can be defined by the intangible or **immaterial nature** of their **outputs** and the basis of service provision in direct interaction between buyers and sellers. Traditionally, geographers neglected the importance of services, assuming that their distribution merely reflected that of industry and the population in general. Since the mid to late 1980s, however, the importance of services in shaping the economic landscape has been recognized. The impact of ICTs on existing locational patterns has been a question of major interest for economic geographers. Services have been subject to globalization in the form of increased international trade and FDI, overturning the assumption that they are always locally traded. Financial services are probably the most globalized sector of the economy, acting as key agent of **time–space compression**. At the same time, new digital technologies have had a major impact on creative industries like music and film, enabling consumers to download material directly from the Internet – encouraging illegal 'piracy' – and threatening to undermine established forms of production and distribution.

The geography of services is defined by two opposing processes. First, advanced business and financial services tend to be concentrated in major metropolitan areas, especially the world cities from where the global economy is coordinated and controlled. This can be explained in terms of the continuing reliance on face-to-face contact, allowing complex interaction between specialist firms to occur. This generates the trust that underpins business transactions. Financial markets in places such as Wall Street and the City of London function as centres of interpretation, drawing on the expertise that they have accumulated over the decades (Thrift, 1994). At the same time, ICT has had a dramatic impact on certain administrative and consumer services through the relocation of back offices and **call centres** to low-wage regions, increasingly in developing countries. India, in particular, has emerged as 'the world's back office' (Bryson *et al.*, 2004, p.127). As such, an increasing spatial concentration of 'command and control' functions in world cities is matched by the dispersal of manufacturing and routine administrative and consumer services (Sassen, 1991). While appearing contradictory, these two processes are closely interrelated, creating a new **international division of labour** in services. The trend of **offshoring** routine services to low-wage locations may well continue for some time yet, but there is little real prospect of the major economic centres of the world dissolving in the face of a 'global space of flows' (Castells, 1989).

Exercises

Select a service industry (e.g. finance, advertising, management consultancy, telecommunications, film, retail), and focus on a particular country of your choice. Collect statistics, reports and academic articles on this industry (perhaps using the websites listed below as a starting point). Review the development of this industry over recent decades, addressing the questions listed below.

1. How important is it in terms of employment, value added and exports?

2. What is its geography: is it concentrated in a small number of centres or more widely dispersed?

3. In what ways has the industry been restructured in recent years?

4. What are the main forces driving this restructuring?

5. What has been the impact of new technologies?

6. Have exports grown?

7. Is offshoring important?

8. What is the future geography of this industry likely to be?

Key reading

Bryson, J., Daniels, P. and Warf, B. (2004) *Service Worlds: People, Organizations, Technologies*, London: Routledge.
The most recent economic geography textbook on services by a group of active researchers in the area. Reviews theories and classifications of services in addition to assessing the production and consumption of different service industries. The impact of globalization and ICT is a key theme

Hamnett, C. (1995) 'Controlling space: global cities', in Allen, J. and Hamnett, C. (eds) *A Shrinking World?: Global Unevenness and Inequality*, Oxford: Oxford University Press, pp.103–42.
An engaging, if by now somewhat dated, assessment of the role and significance of world cities in the global economy. Examines the concentration of business and financial services in world cities and reviews the argument that they are characterized by increased social and spatial divisions.

Pollard, J. (2005) 'The global financial system: worlds of monies', in Daniels, P., Bradshaw, M., Shaw, D. and Sidaway, J. (eds) (2005) *Human Geography: Issues for the Twenty-first Century*, 2nd edn, Harlow: Pearson, pp.337–58.
An accessible and up-to-date introduction to the geographies of finance, emphasizing the social, cultural and political significance of money and outlining processes of global integration since the 1970s.

Thrift, N. (1994) 'On the social and cultural determinants of international financial centres: the case of the City of London', in Corbridge, S., Thrift, N. and Martin, R. (eds) *Money, Power and Space*, Oxford: Blackwell, pp.327–55.

An excellent example of economic geographers' work on the concentration of high-level financial functions in world cities. Stresses the importance of social and cultural factors in shaping both the general operation of financial markets and the specific tendency for them to focus on a small number of centres.

UNCTAD (2004) *World Investment Report 2004: The Shift Towards Services*, **New York and Geneva: United Nations Conference on Trade and Development.**
Part of an annual series, this World Investment Report focuses on services. Provides a detailed examination of investment trends, packed with figures and case studies. Chapter IV on offshoring processes is particularly relevant. Available at http://www.unctad.org/Templates/Page.asp?intItemID=1485 &lang=1.

Warf, B. (1995) **'Telecommunications and the changing geographies of knowledge transmission in the late twentieth century',** *Urban Studies* **32: 529–55.**
A key article assessing the implications of telecommunications, digitization and the emerging Internet for the location of economic activity. One of the earliest analyses by an economic geographer that has informed subsequent research. Includes the example of New York-based insurance companies dispersing routine processing tasks to western Ireland.

Useful websites

http://www.cityoflondon.gov.uk/Corporation/business_city/research_statistics/
The website of the City of London Corporation. The research section contains a wide range of useful publications and statistics on financial services in the City, containing very topical and up-to-date information.

http://www.census.gov/econ/www/servmenu.html
The US Census Bureau provides up-to-date statistics on services on an annual and quarterly basis.

http://www.sitrends.org/
Service Industry Trends, providing a range of up-to-date statistics on service industries and trade.

http://www.wto.org/english/res_e/statis_e/its2005_e/its05_bysector_e.htm
http://www.wto.org/english/tratop_e/serv_e/serv_e.htm
The WTO website. The former part covers 'International Trade Statistics' (Section V on commercial services) while the latter is the 'Services Gateway', which contains material on trade agreements, negotiations and regulations in services.

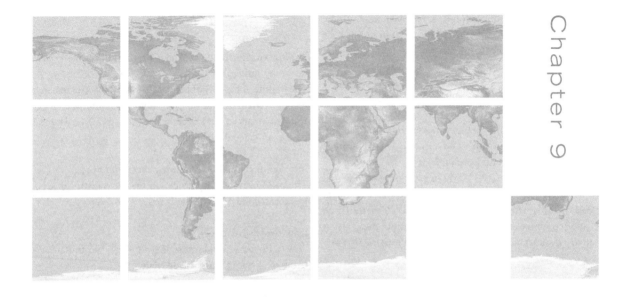

The transformation of work and employment

Chapter map

The chapter begins with some basic definitions of work and employment and a brief examination of how the relationship between them varies in time and space. It then proceeds to explore some of the key changes to the world of work that have characterized the contemporary era (stemming from the late 1970s), critically assessing different theoretical perspectives on these changes and their geographies. In particular, it considers the merits of the concepts of post-industrial society and labour market flexibility. Section 9.4 considers the implications of these changes for trade unions, while the final section of the chapter looks at the emergence of a new labour geography that explores labour as an active participant in the construction of the economic landscape.

9.1 Introduction

For most people regular interaction with the 'economy' takes place through work in its different forms, and our chances of making a decent living are shaped by our relationship to the forms of work and employment that predominate in society at any one point in time and space. In the developed economies of the world, economic well-being is shaped largely by access to paid work or employment. A small minority of people run their own firms or are self-employed but the majority of us make a living through forms of paid employment, where we exchange our human labour in return for a wage. Being unemployed for any length of time is likely to lead to a serious deterioration in living standards. In less-developed countries, much of the population is still engaged in non-capitalist forms of work, linked to subsistence agriculture, although the spread of the global economy often threatens such traditional lifestyles while incorporating a growing number into paid employment. In this respect, the processes of geographical uneven development and economic restructuring, considered in earlier chapters, take on particular meanings in the way that they shape work and people's livelihoods. Often, it is only when a person's job is under threat, through the closure of their office or factory, that they become aware of the geographical web of connections, involving workers in other places, within which they are themselves entangled.

In this chapter, our purpose is to examine the nature of employment change in the contemporary economy, exploring in particular the transformation of work that has occurred since the 1970s and the role played by geography in shaping change. We also emphasize that, unlike other factors of production, labour is not passive to processes of economic restructuring but rather plays a more active role. At both the individual level, and collectively (through organizations such as trade unions), labour helps to shape the changing landscape of capitalism in the same way that capital itself does although, as we have also noted in earlier chapters, labour is often at a strategic disadvantage because of its relative immobility (Herod, 2001).

9.2 Conceptualizing work and employment

9.2.1 Definitions

Human beings have to perform basic work tasks (e.g. hunting and gathering food, finding shelter, making clothes, looking after and raising children) to reproduce daily life. In this sense, **work** is essential to all societies, however primitive or advanced. How work is organized has varied and changed dramatically over time as societies have developed from early nomadic peoples to the more advanced global capitalist society of today. In section 3.3 we examined labour as a basic category under capitalism and developed an understanding of how a more complex **division of labour** emerged with the growth of an industrialized society. The concept of a division of labour allows us to differentiate between different forms of work, some of which are paid and others that receive no financial reward.

The determination of which work or jobs are paid, and which are not, and how much different types of work are paid, relates to the social division of labour (section 3.3.2) Under advanced capitalism, a basic distinction can be made between two very different forms of work. Being in **employment,** selling your labour to work for an employer in the formal economy, is usually paid (though see next section), whereas many forms of **household or domestic labour**, typically undertaken by female family members, are not. Arguably, household work is more valuable to the basic reproduction of society than many forms of employment in the formal economy. The fact that it is (usually) not financially rewarded reflects prevailing values and (gendered) power relations within society. As Pahl notes quite succinctly:

> Someone arriving from another planet might be surprised and puzzled by the way that we distinguish between work and employment and the differential rewards that are paid to employees based on the kind of work they do and the kind of person they are. Interesting, creative and varied employment is highly rewarded; dull, repetitive and routine work is poorly rewarded. Men receive more than women, and this is related to social attitudes and conventions more than the actual amount or quantity of work that the individual or the gender category does.

(Pahl, 1988, p.1)

The broader point is that work is highly differentiated and these differences reflect facets of **social identity** such as gender, ethnicity and age. Up until relatively recently, women were discouraged and actively discriminated against in the labour market (both through government legislation and employer attitudes). Indeed, in most countries wages for women's employment are usually much lower than the equivalent for men. While most advanced industrial economies have laws promoting equal opportunities, employers continue to discriminate on the basis of ethnicity, with white workers typically enjoying better job prospects than non-whites, and age, with older workers and young people often disadvantaged in the labour market.

Geography is critical in understanding how work and employment are organized. As we have noted in past chapters, the economy under capitalism is characterized by spatially uneven development, and the geographical differentiation of work is an important part of this. Through the development of large multiplant corporations and the construction of a **spatial division of labour** (section 4.3.2), a highly varied landscape of employment emerges reflecting differences in the nature and type of work between places. With the emergence of a global economy and through the activities of MNCs this has been translated into a '**new international division of labour**' as we noted in Chapter 7 (section 7.3.4). The important point to note here is that these spatial variations in labour (at national and international levels) are in turn important for future rounds of capital investment, allowing firms to make location decisions based upon the different types of labour (e.g. rates of pay, levels of skill) available across the economic landscape.

Spatial divisions of labour are reflected in variations in employment conditions between places, most obviously in the wide discrepancies in wages between developed and developing countries (Figure 9.1) (section 7.3.4). Such differences also exist within countries. For example, in the US, the average hourly earnings of a manufacturing worker in Dakota were found to be 60 per cent of those in Michigan in the 1990s (Hayter, 1997, p.88). More recent data from the US Bureau of Labor Statistics indicates considerable variation by region in the incidence of unemployment (Figure 9.2). States hit by deindustrialization such as Michigan in the north, or predominantly rural states heavily dependent on agricultural and other traditional sectors such as Mississippi or Kentucky, continue to suffer the highest rates of unemployment. Spatial variations in employment do not just amount to wage or unemployment differentials but also include different cultures of labour associated with particular industries (section 4.3) (see also Peck,

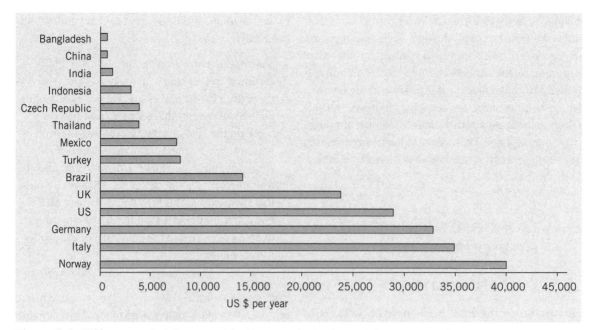

Figure 9.1 Differences in labour cost between selected countries.
Source: Adapted from World Bank, 2006, Table 2.6.

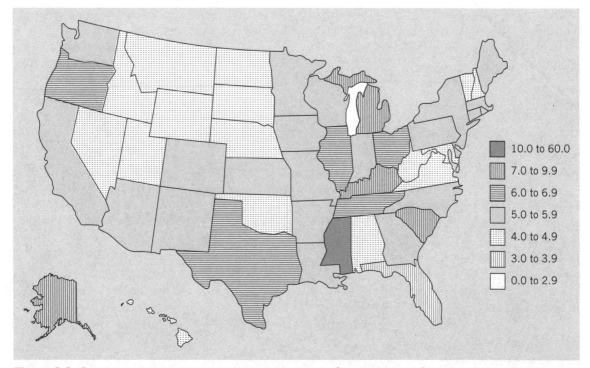

Figure 9.2 Spatial variations in unemployment by state for the United States, January 2006.
Source: Bureau of Labor Statistics, http://data.bls.gov.

1996). These include differences in working practices, levels of unionization and the way **local labour markets** operate (e.g. through different kinds of recruitment and training strategies).

9.2.2 The transition to industrial work

The forms of employment that we have come to associate with the capitalist economy are relatively recent in historical terms. Prior to the development of the industrial revolution in the late eighteenth century, the concept of paid work was relatively marginal to the overall functioning of the economy. The vast majority of the population lived 'off the land' and engaged, to a varying extent, in forms of **subsistence agriculture** (Malcolmson, 1988). Within the pre-industrial economy wage labour was used to supplement other household activities as a way of 'making a living' and would have taken the form of supplying labour to larger farms and estates at particular times of the year, such as during harvesting (Figure 9.4). Some men and women

may also have been employed as maids, cooks and other servants within the household of local ruling elites.

In geographical terms the idea of 'going to work', in the sense of travelling to a dedicated 'place of work', would have been an alien concept. Work centred on the household. Early craft work, for example in textile and weaving activities, would have take place within the home, the origins of the term 'cottage industry'. At a broader geographical scale, work was dispersed across the landscape, in contrast to the massive concentrations of work in large towns and cities that were to develop with the emergence of modern industry and industrial capitalism. In the world's first industrial economy, the UK, there had been only two cities, London and Edinburgh, with a population of over 50,000 in 1750. By 1851, there were nine with two of over 100,000 and over half of the UK's population were now living in urban areas as a result of industrialization (Hobsbawm, 1999, p.64).

The transition to an **industrial society** therefore caused massive spatial and social upheaval, and in the process fundamentally transformed the nature of work

Figure 9.3 Labouring in a pre-modern landscape: 'The Harvesters', by Hans Brasen.
Source: Corbis.

(see also sections 3.2.4 and 3.3.2), as the renowned Marxist historian Eric Hobsbawm puts it:

> It transformed the lives of men beyond recognition. Or, to be more exact, in its [the Industrial Revolution] initial stages it destroyed the old ways of living and left them free to discover or make for themselves new ones.
>
> (Hobsbawm, 1999, p.58)

For the majority of the working population, this transformation meant the destruction of a predominantly subsistence rural lifestyle, largely controlled by the rhythm of the seasons, to forms of routine, regularized paid work under the strict supervision of employers in densely populated towns and cities.

9.2.3 Work and employment in the contemporary global economy

A consequence of the transformation of work was the emergence of a large **industrial working class** whose sole means of 'earning a living' was employment or waged labour. The appalling conditions faced by labour in the early phases of industrialization gave rise to the emergence of **trade unions** in the late nineteenth century as collective organizations geared towards defending the interests of the working class (see Box 9.1). Trade unions reflect the inherent conflict between employees, concerned with earning a living through paid employment, and a capitalist class of employers, dedicated to producing higher profits and reducing costs, for which wages are often the greatest element.

Box 9.1

The emergence of trade unions as key actors in the landscape of capitalism

Harvey's concept of a spatial fix was used in earlier chapters to highlight the role of firms in constructing the geography of the economy. But labour also shapes the economic landscape directly through its own strategies and actions in defending and promoting its interests. The fundamental interests of labour are focused upon securing improved wages and conditions, and the interaction between this and the profit-seeking imperative of capital are of critical importance. This focuses on the organizations set up by workers to promote their interests, principally trade unions. The formation of unions was the result of struggle by groups of workers to defend their interests and improve

their lot, beginning in the industrial heartlands of the UK in the nineteenth century before spreading to continental Europe and the US (Hudson, 2001, p.100). Such struggles were often conducted against employers who were bitterly opposed to trade unions, sometimes supported by the state.

As part of a broader labour movement, unions have been closely associated with social democratic and socialist political parties in many countries. The British Labour Party, for example, has been closely tied to trade unions throughout its history. Unions had considerable success in securing better conditions for their members alongside their political allies over the

middle decades of the twentieth century. By the late 1960s growing union strength (see Figure 9.4) in the manufacturing sector in particular was leading to demands for higher wages, as well as a challenge to management's ability to control the nature and speed of work. Since the late 1970s, however, unions in the advanced industrial countries have faced tougher times, reflected in falling membership, as many of the heavily unionized traditional industries have collapsed and governments in many countries have placed increased legal restrictions on their activities. Despite such reversals, organized labour remains an important force shaping the geography of capitalism.

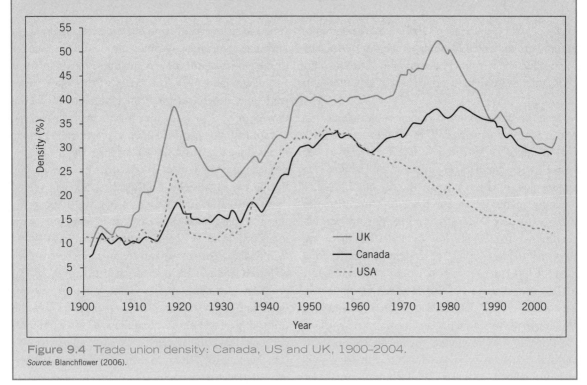

Figure 9.4 Trade union density: Canada, US and UK, 1900–2004.
Source: Blanchflower (2006).

While employment has become the dominant category of work in the global economy, its geographical incidence is highly uneven. Using figures from the World Bank, Castree *et al.* (2004, p.11) estimated that in the year 2000 three-quarters of world employment (waged labour) was located in just 22 countries with 'almost half of the world's labour force located in just four countries: China, India, the United States and Indonesia'. Within the global South, the incidence of wage labour varies considerably, reflecting differences in levels of integration within the global economy. In many parts of Africa, Asia and Latin America forms of subsistence agriculture persist, although attempts to modernize economies (see Chapter 11) have often led to the forced destruction of such traditional ways of 'making a living', without always replacing them with sustainable alternatives. One of the most spectacular examples of such destructive development has been the 'development' project to construct 30 major dams in the Narmada river valley in central India, submerging 248 towns and villages and displacing up to 15 million people (Routledge, 2002, p.321). As noted in Chapter 11, for much of sub-Saharan Africa in particular, integration into the global economy remains dominated by the production of primary commodities such as minerals and agricultural products such as coffee and cotton.

Even in those countries where wage labour has grown with industrialization, the conditions of work are often poor, as evidenced by the difference in labour costs between countries (Figure 9.1). Child labour – the employment of children under the age of 16 – is also common in the global South, more often than not under extremely exploitative and badly paid working conditions. In China, for example, children are employed to make toys for 12 to 16 hours per day for a wage of less than 14 cents per hour (Hudson, 2001, p.245). The ILO estimates that there are approximately 246 million child labourers (between 5 and 17) working in the world, 70 per cent of whom are engaged in hazardous work in mining, chemicals, working with harmful pesticides in agriculture, or with dangerous machinery. Some 73 million child labourers are under the age of 10 (UNICEF, 2006, p.46).

Although it has spread with the development of globalization, it is important to emphasize that only around half the population at any one time are in paid employment. There are also other forms of work that are unpaid, notably housework, **voluntary work** and care work (particularly family members), looking after older people, children and the disabled. The retreat of the state from welfare provision in many countries since the 1980s (section 6.5.4) has meant that unpaid care work has become increasingly critical both to supporting the economy – through the supervision of children and their 'socialization' for future employment – and in providing for the more disadvantaged sections of society. Unpaid housework is still the dominant form of domestic labour in both advanced and less-developed economies, although the situation is slightly complicated by the existence of paid work for some household tasks (e.g. nannying, cleaning), particularly in many Western societies in the US and Western Europe where both members of a household are engaged in full-time employment (Gregson and Lowe, 1994).

Alongside paid work, other categories include the **self-employed**, the definition of which often varies in time and space according to differences in employment laws and regulations. Additionally, a growing number of the global labour force is subject to unemployment or underemployment. The extent of this again varies over time, dependent upon the pendulum of uneven development. A growing number of people in Western Europe have been exposed to unemployment in various forms since the 1970s, while after the collapse of communism unemployment rocketed in Eastern Europe during the early 1990s. During the latter half of the 1990s, unemployment became a massive problem for the previously successful East Asian economies with many people made jobless by the financial crisis.

A final category of work to comment on here is **slave or forced labour**, which remains a feature of the global economy (see Box 9.2). One of the more pernicious outcomes of globalization has been an increase in the trafficking of people in conditions of forced labour, particularly for the sex trade. Illegal immigrants, who lack the status of citizens in their destination countries, are particularly vulnerable to highly exploitative employers, and an estimated two and a half million people are engaged in forced labour as a result of trafficking (ILO, 2005a, p.14).

Box 9.2

Forced labour under capitalism

While wage labour has become the dominant form of employment under capitalism, other more exploitative forms of labour continue to exist. In particular, the slave trade, from the fifteenth to the nineteenth centuries, was critical in the emergence of an international system of capitalism and represents one of the most ignominious episodes in European colonial history. As part of colonial expansion, various European states engaged in the enforced transfer of slaves from the African continent to work in the new colonies in the Americas. Although slavery largely predated formal political colonialism, its effect was arguably as devastating for the countries involved. From the late seventeenth century until the nineteenth, it is estimated that between 8 and 10.5 million slaves were transported from West Africa to the Americas to work primarily in tobacco, sugar and cotton plantations. Apart from the appalling situation of slavery itself, the conditions of transportation were dreadful and inhumane; on an average Dutch slaver in the seventeenth century, for example, 14.8 per cent of slaves would die en route from diseases such as smallpox, dysentery and scurvy (Potter *et al.*, 2004, p.59).

With the abolition of slavery in the nineteenth century, many countries developed systems of 'indentured labour', whereby Asian workers, in particular, were recruited to work in European colonies. Workers would be employed under contract to a single employer for a fixed period (typically between four and seven years) in return for transportation, accommodation and food. In practice, conditions for indentured workers were little better than slavery. Many died during transportation, and workers were unable to terminate their contracts. Employers, however, had the freedom to sell indentured workers on before the end of the contract period.

Despite the near universal rejection of slavery by states in the modern economy, it sadly persists in various guises. The International Labour Office's 1998 *Declaration on Fundamental Principles and Rights at Work* has drawn attention to the continuing use by employers of what it terms 'forced labour' in the global economy. Forced labour is defined as 'all work or service which is exacted from any person under the menace of any penalty and for which the said person has not offered themselves voluntarily' (ILO, 2005a, p.5). Although difficult to accurately measure, for obvious reasons, a conservative estimate is that over 12 million people are in some form of forced labour globally, with Asia and the Pacific region dominating the trade (ibid., p.12). Two million people are in 'state or military imposed' forced labour, such as prisons, while another growing category is sexual exploitation with over one million 'workers' forced to sell their bodies for sex. Forced labour is a highly gendered affair and often involves children: women and girls account for 56 per per cent of total forced labour and 98 per cent of the sex trade.

Reflect

➤ In what ways has the nature of work changed with the transition from a pre-industrial to an industrial society?

➤ How does geography affect different people's experience of work in the global economy?

9.3 Changing forms of employment

The nature of employment has changed through time, reflecting broader changes within the capitalist economy. The shift from a craft-based form of production to large-scale, factory-based production was critical to the reorganization of work and the introduction of a more complex division of labour in the nineteenth century (section 3.3.2). More recently, there has been considerable debate about the changes taking place in the contemporary workplace, particularly since the 1970s. In part, these are linked to MNC restructuring and a developing global division of labour as firms seek out cheaper labour resources in the global South (see Chapter 7), but changes have also been driven by the process of deindustrialization and the emergence of a service-based economy. Restructuring also reflects more deep-seated tensions between employers and employees over both the organization and control of work (**labour process**), and the distribution of the surplus generated by production (i.e. wages versus profits).

Many commentators have viewed the period from the mid-1970s to the present day as a new phase of capitalism, characterized first and foremost by a transformation in the nature of employment. We explore two sets of claims here: arguments about a transition from an industrial to a **post-industrial** workplace; and, debates about the shift from **Fordism** (see section 3.2.2) to more **flexible forms of work** organization. Both have aroused considerable academic controversy and, in evaluating the claims made, it is important to be wary of 'myths at work', and 'unpick and deconstruct such myths to show which aspects of them carry credibility and which do not' (Bradley *et al.*, 2000, p.2).

9.3.1 A post-industrial economy

In the advanced industrial economies, there has been a massive shift in employment from manufacturing to services (section 8.3), prompting a number of well-known theories about work in a post-industrial economy. Many have chosen to characterize this as a dramatic transformation in the whole relationship between work and society (e.g. Bell, 1973; Gorz, 1982; Lash and Urry, 1994), in which the following claims are made.

➤ The growth of services has led to an increasing number of people in middle-class, white-collar occupations. Linked to the emergence of high-tech industries, a growing proportion of the workforce are in more educated and professional forms of employment (Dicken, 2003a, p.526) that involve more autonomy and control in contrast to Fordist mass production.

➤ In the context of a global economy, developed countries should not try to compete in less-skilled activities but should prioritize knowledge-based activities for jobs growth. With the advance of global communications and information technology, more routine activities in both services and manufacturing will be increasingly exposed to low-wage competition. Labour market policy should therefore focus upon skills and training rather than job creation.

➤ Reflecting the increasing interaction with consumers that characterizes much service work, new social relations are being established that involve employees 'performing' scripted roles under the supervision of management (see Box 9.3). Job prospects and career enhancement are increasingly differentiated by the quality of an individual's performance and the ability to internalize management's wishes (du Gay, 1996; McDowell, 1997).

➤ Traditional class identities are becoming eroded with the decline of the industrial working class that formed the basis of trade union strength. Alongside the growth of the middle classes, the erosion of traditional working-class livelihoods has left a residual underclass of male, unskilled workers stripped of their traditional identities (McDowell, 2003).

The idea that the post-industrial economy is knowledge-intensive, and provides plenty of jobs as long as the workforce has the appropriate skills, is a particularly powerful one that has been expounded by key figures in the US government. For example Robert Reich, President Clinton's Secretary for Labor during his first term of office in the 1990s, famously claimed:

The most rapidly growing job categories are knowledge-intensive; I've called them 'symbolic analysts'. Why are they growing so quickly? Why are they paying so well? Because technology is generating all sorts of new possibilities.... The problem is that too many people don't have the right skills.

(Quoted in Henwood, 1998, p.17).

However, the evidence suggests that this is overstated. In both the US and the UK, for example, the fastest-growing job categories are in more menial work that does not require high skill or education levels (Henwood, 2000; Thompson, 2004). The disappearance of relatively well-paid jobs in the manufacturing sector, which also offered the chance for manual workers to progress through craft apprenticeships and vocational

Box 9.3

It's showtime! Philip Crang on workplace geographies of display

Arguments about the emergence of a new kind of post-industrial economy based on the production and consumption of information, knowledge and images often stress the distinctive nature of work in service industries, involving interaction with people rather than machines. This interaction with customers requires a social performance from workers, incorporating their appearance, social skills and personalities. One notable study in economic geography that addresses such questions is Phil Crang's assessment of workplace geographies of display in a restaurant in south-east England. The research is based on Crang's own experiences as a waiter in restaurant, employing a research methodology known as **ethnography** – long employed by anthropologists and sociologists – which involves detailed observation of people in particular settings over a long period of time, aiming to understand and represent their experiences in their own terms.

The restaurant in which Crang worked is called Smoky Joe's (a pseudonym), a South American-style barbecue establishment that occupies

a middle-market position between traditional restaurants and fast-food outlets. The service offered to customers was partly standardized through an 'order of service' that identified 23 different operations to be performed by the waiter (involving 16 separate visits to the table). The level of standardization and managerial surveillance was necessarily limited, however, by the need to offer a distinctive service and experience to customers, requiring staff to exercise a degree of autonomy and spontaneity. The motif of performance was central, explicitly identified by the company through its literature, notices and costume. Such performance is not tightly 'scripted' or 'engineered', Crang argues, involving, instead, a subtler form of performance as interaction, based on social encounters between staff and customers. Waiters attempt to anticipate what the customer wants, often beginning by dividing them into recognized 'types', engage in joint projects such as order taking with them and act as their ally in dealing with other parts of the restaurant (the kitchen or bar) on their behalf.

The encounter between waiters

and customers is underpinned by routine geographies of movement and stopping, requiring the former to both rush around between locations, generating a buzz within the restaurant, and engage with their 'tables' through the right kind of talk. It is also highly embodied and gendered, with waiters required to be attractive by management. They were also subject to being engaged in sexual interactions such as flirting with customers or other staff. Crucially, Crang concludes that the self is vital to this kind of work: the performance of staff draws on their inner emotions and feelings, meaning that such employment is deeply personalised. This offers a contrast with more traditional forms of manual labour in manufacturing industries where work was regarded as separate from the personal realm, involving physical rather than emotional competency (see Box 3.4). Along with other research on contemporary forms of work and identity (Allen and du Gay, 1994), Crang's research does suggest that employment in the service sector is qualitatively different from manufacturing industry.

training, have been replaced by low-paid, low-skilled work in service activities such as cleaning, security, retailing and catering, with limited opportunity for career advancement. In terms of the nature of employment in the new service economy, many of the jobs created are often very similar to the more routine kinds of work associated with manufacturing assembly lines (see Figure 9.5), where work is strictly controlled and monitored by management, often highly intensive, pressurized and monotonous (Beynon, 1997; Henwood, 1998) (see section 8.2.2). In contrast, evidence suggests a continuing decline in 'middle-level, craft and skilled manual employment' (Thompson, 2004, p.30).

Sociologists Philip Brown and Anthony Hesketh's recent study of occupations in the US economy found that only 20 per cent of the workforce were involved in knowledge-intensive activities (that required creative and innovative tasks), with the other 80 per cent involved in 'serving or making things rather than thinking for a living' (ibid., p.55). While knowledge-intensive jobs were likely to increase as a proportion of the workforce over the decade 2000–10, in absolute terms the more routine jobs in production and 'in-person services' were likely to be the fastest growing (Table 9.1). For Brown and Hesketh, therefore, the post-industrial economy is one of increasing **social**

Figure 9.5 Call centres: an 'assembly line in the head'?
Source: Sherwin Crasto/Reuters/Corbis.

Table 9.1 Trends in knowledge v routine jobs in the US economy

Category	2000 (%)	Number (million)	2010 (predicted) (%)	Number (million)
Knowledge work	20	29.1	22	36.9
In-person services/routine production	80	116.5	78	130.8
Total	100	145.6	100	167.7

Source: Derived from Brown and Hesketh, 2004, p.55.

polarization, with the opportunity to use their knowledge and skills being confined to a small minority of workers.

The shift from manufacturing into services has also been accompanied by a growth in female employment relative to male. For some, this reflects the greater personal abilities of women when dealing with customers, which makes them more suitable for service employment. However, it should be noted that many of the jobs created for women are lower paid than those lost to men, while one of the highest growth rates has been recorded in part-time employment (Dex and McCulloch, 1997).

9.3.2 The geography of uneven development under services

The more optimistic proponents of the post-industrial thesis also overlook the geography of manufacturing decline and services growth. Across the advanced industrial economies of Western Europe and North America, the timing and extent of deindustrialization have varied considerably, as we noted in Chapter 5. Germany, France and Japan have been more successful at retaining manufacturing employment than the US and the UK (Table 9.2). Within countries, deindustrialization has affected areas of traditional and heavy industry (e.g. coal mining, steelmaking and shipbuilding, textiles, auto production) particularly hard

(see section 4.3.2). In Western Europe, particular industrial areas experienced a virtual collapse of the local labour market as key industries and employers closed down, some in a dramatically short period of time (Table 9.3).

In the case of the UK, while all regions experienced a more sustained recovery in employment from the mid-1990s onwards, the growth in service employment was highly uneven between regions. For example, the employment performance of London (+623,000 jobs) and south-east England (+510,000) between 1995 and 2001 massively exceeded that of Scotland (+11,000) and north-east England (+10,000) (UK Census of Employment, available at: www.nomisweb.co.uk). At the same time, five of the UK's traditional industrial regions (East Midlands, Yorkshire/Humberside, north-east England, Wales and Scotland) continued to experience a decline in full-time male employment during a period of supposed 'boom' conditions.

The geography of employment in the UK's service economy continues to reflect a pronounced spatial division of labour between London and the South-east, and the rest of the economy. In peripheral regions such as north-east England and Scotland employment growth has been dominated by lower-paid and more routine jobs, while the lion's share of high-status jobs has been concentrated in south-east England. The massive growth in call-centre work, in particular, as larger corporations in the business services sector outsourced more routine work to lower-cost locations in the peripheral regions, has almost directly matched

Table 9.2 Deindustrialization in selected developed economies: an uneven global picture

	Share of industrial employment	
	1990 %	2004 %
France	29.7	23.0
Germany	38.6[a]	31.0
Japan	34.1	28.4
UK	32.3	22.3
US	26.2	20.0

[a] West Germany only.
Source: OECD in Figures, various editions, available at www.oecd.org.

Table 9.3 Job losses in selected West European regions through deindustrialization

Industries suffering losses	Regions/areas	Country	Scale and timing of job losses
Textiles	Nord-Pas-de-Calais	France	100,000 (1945–90)
Steel	Ruhr	Germany	100,000 (1970–80)
Steel	Scotland / Northern England / Wales	UK	152,000 (1971–90)
Heavy engineering, shipbuilding, steel	Teesside	UK	46,000 (1979–82)

Source: Derived from Charrie, 1994, pp.174–9.

earlier phases of spatial reorganization by manufacturing firms (Massey, 1984). Rather than the new environment producing better jobs, for many of the UK's regions it has led to a deskilling and downgrading of work (Hudson, 1989).

9.3.3 The transition from Fordism to flexibility

A second very influential perspective on the changing workplace is associated with the shift towards **post-Fordism** or flexible working (section 4.3.3). Not only have variants of this become dominant with academic and management discourses, but they have also proved very popular with certain politicians, keen to pursue labour market deregulation as a route to competitiveness in the global economy. During the UK's recent presidency of the European Commission, Prime Minister Tony Blair urged his continental counterparts to adopt what he termed the 'Anglo-Saxon model of **labour market flexibility**' (reported in *The Guardian*, 1 July 2005), while in April 2006 the French Prime Minister, Dominique de Villepin, was forced to withdraw a new law proposing greater flexibility for employers to 'hire and fire' younger workers in the face of mass public protests.

Fordism as a concept was derived originally from the mass production system (see sections 3.3.2 and

4.3.2) introduced into the US automobile industry by Henry Ford. As mass production and its principles spread throughout industry in the early part of the twentieth century, the term Fordism became used in a wider sense to encapsulate the system of modern work under capitalism. The Italian Marxist Antonio Gramsci was one of the first to use Fordism in this way, recognizing an innovative US model of capitalism that he saw as replacing more antiquated European forms. Fordism has subsequently been associated with a particular phase of industrial capitalism and work organization, originating in the early 1900s, but having its heyday between 1945 and 1970 when it spread from North America to Western Europe. In employment terms, the key features of Fordism were (see Table 9.4):

➤ a highly detailed division of labour with considerable deskilling and management control over job tasks (i.e. Taylorism);

➤ the development of large corporations with complex and extremely hierarchical 'internal labour markets' through which recruitment and promotion were organized;

➤ relatively high job security centred around the norm of the white full-time male worker;

➤ recognition of trade unions as 'social partners' with government and employers in regulating and man-

Table 9.4 Fordist and post-Fordist labour markets

	Fordist–Keynesian	Post-Fordist
Production organization	Mass production	Flexible production
Labour process	Deskilled and Taylorised, detailed division of labour	Flexible, functional and numerical
Industrial relations	High union densities; strong worker rights; centralized bargaining	Disorganization of unions; individualized employment relations; decentralized bargaining
Labour segmentation	Institutionalized; rigid hierarchies; large, internal labour markets	Fluid; core–periphery divide; breakdown of internal labour markets
Employment norms	Male, full-time workers; occupational stability and job security	Privileging adaptable workers; normalization of employment insecurity
Income distribution	Rising real incomes and declining pay inequality	Polarization of incomes and pay inequality
Labour market policy	Full employment; secure and high level of male employment	Full employability; ensuring workforce adaptability
Scale characteristics	Privileging of national economy for economic management and labour regulation	Deprivileging of national; global economic imperatives; decentralization of labour regulation
Geographical tendencies	Dispersal	Concentration

Source: Peck, 2000, p.139.

aging employment conditions, with high levels of union membership and national-level collective bargaining.

As part of the response to the economic crisis of the 1970s, European and US firms began to rethink their models of employment organization in order to (i) reduce costs and (ii) create more flexible organizational structures that allowed them to better adapt to increasingly unstable markets arising from global competition (see Table 9.4). For our purposes here we can distinguish between two types of effect: changes to the way work is organized (i.e. the labour process), and changes in employers' recruitment strategies (i.e. the labour market).

Labour process

➤ The replacement of systems of mass production and Taylorist work practices with new high-performance forms of production, most often termed 'lean pro-

duction' (Womack *et al.*, 1990) or flexible production (Peck, 2000). Heavily influenced by the success of Japanese auto manufacturers during the late 1970s and 1980s, particularly the example of Toyota in out-competing US firms, lean production methods involve slimming down the workforce, a greater attention to quality and the elimination of waste (through quality circles and total quality control and management). As part of the commitment to more efficient methods just-in-time production is being introduced, whereby manufacturers keep only the bare minimum of components, eliminating the costly investment in huge stocks associated with Fordism. This necessitates greater interaction and proximity (both in terms of geographical and managerial distance) with suppliers.

➤ For employees, these new production concepts require a shift away from the deskilled and intensive division of labour associated with Fordist assembly work towards a more multi-skilled and multi-tasked labour force, more able to be flexibly deployed between different parts of the production process or indeed able to switch to new products as the need arises.

➤ The alienated worker of Fordism, subjected to strong and coercive managerial control, is replaced by a more involved worker, who has greater skills and knowledge of the labour process, and greater autonomy over his/her work.

➤ For some, this brings about the potential for a change in the social relations between management and worker with a culture of confrontation giving way to trust and cooperation (Oliver and Wilkinson, 1992).

Labour market

➤ Accompanying the shift towards lean production, firms are restructuring their workforces, shifting away from internal labour markets where recruitment and selection are organized within the firm

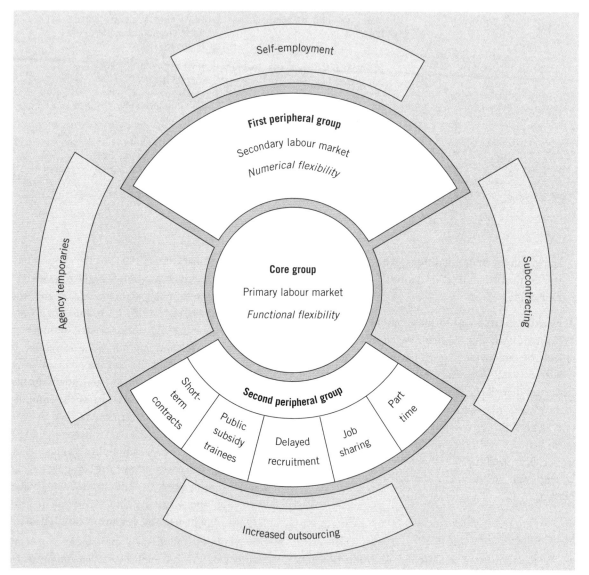

Figure 9.6 The flexible firm model.
Source: Allen *et al.*, 1988, p.202.

towards dual structures where the workforce is divided into 'core' and 'peripheral' elements (see Figure 9.6). Consequently, there has been a decline in permanent full-time work and an increase in non-standard forms of employment such as part-time, temporary, self-employed and agency workers. This increase in non-standard forms of employment provides firms with 'numerical' flexibility, whereby the workforce can be reduced or increased in line with fluctuations in product markets (Atkinson, 1985).

➤ In return for job stability, workers in the core segments of the model agree to greater 'functional' flexibility, whereby managers can move them between job tasks to a greater extent than hitherto. This involves agreeing to perform a wider range of tasks as part of the job in the short term, or even agreeing to retrain or move to another job within the firm in the longer term.

➤ The introduction of a range of non-standard forms of work also provides firms with increased temporal flexibility. By being able to draw upon part-time labour, for example, firms can vary shift patterns and times, or even introduce extra shifts into a working day.

➤ Finally, the transfer of much of the workforce to non-permanent status often means that firms can reduce the non-wage labour costs, such as the payment of redundancy, health and social insurance, and holiday pay, providing them with considerable 'financial' flexibility.

9.3.4 The limits to flexibility for workers

While these changes to the nature of employment have been perceived as important by politicians and businesses to enhance competitiveness in the context of a global economy, it is clear that for individual employees the effects can be more pernicious, depending on which side of the core–peripheral divide you fall. Various commentators have catalogued how the processes of 'downsizing' and rationalization that have accompanied the search for greater employment flexibility have led to the shedding of hundreds of thousands of jobs

within large corporations (e.g. Gordon, 1996; Harrison, 1994).

Even for those remaining in work, there is little impression – contrary to lean production rhetoric – that the work environment has been enhanced through skills upgrading, greater empowerment in decision-making or an increased spirit of collaboration and team working. Rather, research has suggested an intensification of work, increased monitoring and control of job tasks and growing employee dissatisfaction and insecurity (Beynon et al., 2001; Green 2001, 2004; Grimshaw et al., 2001; Thompson, 2003). Indeed, one group of critics has even suggested that the much-vaunted Japanese production methods used in the automobile industry have more in common with Fordism than many would admit, being first and foremost about the speeding up and intensification of work, though made more efficient through added attention to quality and stock control (Williams et al., 1992).

Rather than a more egalitarian workplace, the evidence suggests the opposite. For example, in the UK there is a growing gap between the salaries of top managers and the average worker. A recent report suggested that 20 years ago, the typical chief executive of a FTSE-100 company earned some 25 times the pay of the average worker; today it is close to 120 times (The Guardian, 1 April 2006). Similarly, research by the US trade union confederation, the AFL-CIO, found that the ratio of chief executive pay to the average worker had risen from 1:41 in 1980 to 1:431 by 2004 (see http://www.aflcio.org/corporatewatch/paywatch/pay/index.cfm).

9.3.5 The variable geography of flexibility in a global economy

Although these trends are evident in most advanced economies (Standing, 1999), the use of flexibility has been greater in the Anglo-American world, in countries such as the UK, the US and Australia where successive governments have pursued policies of labour market deregulation and where job protection legislation is minimal. Firms find it easier therefore to 'hire and fire' workers, compared with many countries in continental

Table 9.5 Average job tenure and tenure distribution for selected OECD countries

	Average tenure (years)	Workers with < 1 year tenure (%)	Workers with > 1 year tenure (%)
Greece	13.6	9.8	52.1
Japan	12.2	8.3	43.1
Italy	12.2	10.8	49.3
France	11.2	15.3	44.2
Germany	10.6	14.3	41.7
Denmark	8.3	20.9	31.5
UK	8.2	19.1	32.1
US	6.6	24.5	26.2

Source: *World Employment Report 2004–05*, p.191, Table 4.1. Copyright © International Labour Organization.

Europe where employment legislation means that firms have to pay much higher redundancy and other social costs. For example, when the French automobile manufacturer Peugeot decided to close its plant near Coventry in the English Midlands rather than one of its domestic plants, trade unionists at the plant contrasted the £50,000 it cost to lay off a British worker with the £140,000 for a French worker (*The Guardian*, 20 April 2006).

Recent research by the International Labour Office confirmed that, despite increasing global economic integration, there remain wide variations in labour market flexibility. This is reflected in the average length of time that workers spend in a particular job. Once again, the main contrasts exist between a continental European labour market model and the Anglo-American with, for example, the average tenure of American workers being almost half that of larger European countries such as Italy and France (Table 9.5). The differences are even greater when we consider the proportion of the workforce with under a year's tenure: 24 per cent in the US, compared with 11 per cent for Italy. Japan is the other country in the developed world with a tradition of more stable employment, where employers in the large corporate sector have traditionally provided lifetime employment guarantees in return for functional flexibility from employees.

Advocates of **labour market flexibility** claim that it helps to reduce unemployment, by encouraging labour to move more rapidly from sectors and regions that are in decline to newer growth sectors and regions. The US labour market is held up as the model here, because it has consistently outperformed that of the EU over the past decade with an average unemployment rate of 5.6 per cent compared with 9.4 per cent for the EU (Sorrentino and Moy, 2002, p.19). The UK, as another model of labour market flexibility, has also outperformed its European neighbours, but it should be pointed out that there are wide variations in experience between European countries with similarly 'rigid' systems of employment protection, while no study has been able to find a decisive link between economic performance and labour market flexibility (ILO, 2005b). Additionally, many of the jobs created in the UK and the US are in contingent or temporary work that is highly exploitative (see Box 9.4)

There has been surprisingly little research into the geography of flexibility at the subnational level between regions, although studies of flexibility in particular regions have cast doubt upon the extent to which multi-skilled forms of flexible working have been introduced either in older industrial areas (e.g. Cumbers, 1996; Hudson, 1989) or newer service spaces in areas such as the south-east of England (Pinch and Storey,

Box 9.4

The growth of the temporary agency as a labour market mechanism under post-Fordism

One of the most tangible signs of the new flexible labour market has been the growth of the temporary agency. In the US, it is estimated that one in five jobs created since 1984 has been through a temping agency (Peck and Theodore, 2001, p.475). The number of temporary workers in the US grew from 250,000 in 1973 (0.3 per cent of the workforce) to 4.4 million by 1999 (4.3 per cent). Temporary agencies have traditionally been associated with supplying office workers, particularly women, to lower-level administrative positions within firms. However, recent research indicates that they have greatly expanded their remit, both spatially and sectorally. Some temporary agencies are now global transnationals in their own right. The two leading agencies – Adecco and Manpower – both operate in over 60 countries, with combined sales of over $30 billion per annum and the placement of over two million workers (Ward, 2004, p.252).

The greater use of temporary workers by firms and other organizations has meant that the use of agencies has expanded beyond the office to the factory, the warehouse and even the public sector (e.g. in schools and hospitals). In these conditions, Peck and Theodore's study of

temporary agencies in Chicago found emerging divisions within the temporary workforce (Peck and Theodore, 1998). For some workers (typically highly experienced administrative workers or secretaries working in the corporate sector) with highly prized skills and attributes, the flexibility provided by temporary work can be empowering. During periods of low unemployment it allows them to choose between employers and make considerable demands in terms of pay and conditions. In the words of one agency manager:

> If you're really good, most companies will hire you on. If [good temps] tell me Friday they want to change placements, Monday I will have a job for them. They're that good. I have to keep them busy or their other agency will take them.
> (Manager, small office placement)

However, for the majority of workers in the temping sector, who lack the skills to differentiate them from other workers, the experience is more negative. At the bottom end of the labour market (in areas such as warehouse work, cleaning and construction) many temporary workers are hired, by the day, in impersonal hiring halls,

encapsulated by the phrase 'warm bodies delivered on time'. The benefits to the employer of this system is that the disciplining of workers is also outsourced to agencies who are asked to filter out workers with the 'wrong attitude'. In addition, the cost benefits are enormous, revealed in the following quote from an agency manager:

> They'd have to pay $8–9 an hour with benefits, vacation time, sick days, payroll taxes, they'd have to hire someone to pay them... And their orders fluctuate so much... Now they can cut 50–100 people in a day. All they have to do is call us.
> (Peck and Theodore, 1998, p.661)

From a firm's point of view, using temporary agencies to recruit labour reduces the non-wage costs and increases their numerical flexibility. In this way, labour is subject to a process of 'hyper-commodification' (ibid., p.661) in the sense that employees really are treated as no more than commodities whose labour can be bought and sold on a daily basis. At the same time, the scope for forming unions and organizing for better conditions is greatly reduced by the precarious and ephemeral nature of employment.

1992). Rather, research has tended to highlight flexibility being related to increased casualization and job security for lower-status workers in otherwise prosperous local economies (Allen and Henry, 1997). Numerical flexibility in this sense seems to be more prevalent than functional flexibility in the new forms of work emerging at the regional level.

Focusing upon the UK, Mohan (1999, pp.84–100) noted that while a small increase in part-time working

is notable across the country, from 20.6 per cent of the labour force in 1981 to 24.9 per cent by 1996, the most significant increases have tended to be in rural areas where employment is dominated by tourism and agriculture whose seasonal nature requires more temporal flexibility. One of the few pieces of recent research in this area is by Monastiriotis (2005), who noted a North–South divide in the UK in the incidence of flexibility. While functional flexibility was more prevalent in the North, perhaps suggesting a link to the establishment of post-Fordism in manufacturing, financial and numerical flexibility tended to be greater in the South. Monastiriotis suggests that this may in turn reflect the South's greater exposure to global market pressures, where employers need to vary employment levels in line with changes in product markets.

Reflect

> To what extent is the shift towards a post-industrial economy improving conditions for the average worker?
> What are the main elements of labour market flexibility and how does its take up vary across space?

9.3.6 Historical precedents for labour flexibility

At a more conceptual level, some critical commentators have pointed out that in historical terms employers' search for flexibility from their workers is nothing new and indeed is as old as capitalism itself (Pollert, 1988). Additionally, employers have always found it useful to use labour strategies that structure the labour force into different segments, privileging some workers over others, as a means of 'divide and rule' (Gordon *et al.*, 1982). The imperative to find new ways of working that reduce costs is particularly acute when competition is intensifying and when markets are relatively stagnant and mature. These conditions have been present for many workers and firms in the developed world as a result of increased global competition from less-developed economies since the 1970s, although, as we noted in Chapter 5, it is important not to exaggerate the threat of globalization. From a management point of view, the rhetoric of labour market flexibility can be seen as a useful device to discipline workers and reduce costs, but from the point of view of workers it has some serious shortcomings (Bradley *et al.*, 2000).

9.4 A crisis of trade unionism?

9.4.1 The decline of the mass collective worker

Taken together, the shift towards a post-industrial economy and a more flexible workforce are seen as heralding the decline and perhaps ultimate demise of trade unions and the 'mass collective worker' (Hudson, 1997). Five themes in particular run through these arguments.

> In the industrial economy, social identities were tied to and constructed through the workplace (e.g. car worker, shipbuilder, steelmaker). Workers tended to remain in the same occupation for most of their working lives and therefore work became an important 'fixing' element in their social consciousness. By contrast, employment in services – for less-skilled workers – tends to be more fluid and casualized. People move between jobs more frequently and social identity is less associated with work but more with consumption activities (Lash and Urry, 1994, p.57).

> Subsequently, workers are becoming more individualistic in attitude and less inclined to join trade unions. The fragmented nature of the service economy with large factories giving way to smaller and more dispersed offices and retail workplaces make it difficult for unions to organize.

> Trends towards employment flexibility also weaken the position of trade unions. The slimming down of the core, unionized workforce and the creation of more temporary and agency workers undermines traditional union strength in the large single workplaces of Fordism.

> Additionally, increased outsourcing and geographical relocation of work to non-union plants in areas without a culture of strong union representation further weakens the ability to organize, as well as providing employers with increased opportunities to play off employees in one location against those in another.

Table 9.6 Geographical variations in trade union decline for selected OECD countries (union density = percentage of workforce in union)

	Union density	
	1980 %	2003 %
Australia	49.5	22.9
Canada	34.7	28.4
France	18.3	8.3
Germany	34.9	22.6
Italy	49.6	33.7
Netherlands	34.8	22.3
New Zealand	69.1	22.1 (2002)
Norway	58.3	53.3
Sweden	78.0	78.0
UK	50.7	29.2
USA	19.5	12.4

Source: Blanchflower, 2006, p.30, Table 1.

➤ Finally, the increased 'feminization' of the workforce also undermines trade unions as women are less likely than men to join trade unions.

Undoubtedly, union membership has declined considerably in most advanced industrial economies since the heyday of the 1970s but the extent of its decline is often exaggerated (Table 9.6). A factor has been the decline of manufacturing, where union strength has traditionally been high, and the growth of service-related employment, where unions have traditionally been weaker. But other factors are also at play. If we consider the international comparative evidence of union decline, what is significant is that decline has been greatest in those countries where employment regulation is weakest and governments have been hostile to unions (see Table 9.6). Union membership in the UK declined faster than in most other countries in the 1980s and 1990s but this was in the context of Conservative governments passing anti-union laws restricting the right to strike and encouraging employers to derecognize trade unions. US, Australian and New Zealand unions similarly have seen their influence declining in the context of neoliberal econ-

omic policies (section 6.5.1), hostile governments and anti-union employment legislation.

In other countries, where governments have been less willing to challenge trade unions, they remain a stronger presence in the labour market, particularly in some northern European countries such as Norway, Sweden, Germany and the Netherlands. This does not mean that unions have not been on the back foot in these countries, however. Indeed, the evidence from Germany suggests that employers have been attempting to withdraw from national systems of national collective bargaining to more decentralized plant and company-based systems (Berndt, 2000; Zeller, 2000). However, critically, here, as in other Western European countries, political support for the idea of **social partnership** still persists, even if employers are demanding more labour flexibility (Jeffreys, 2001).

It is also worth noting that the situation facing trade unions is different in much of the global South. In South Africa, trade unions experienced considerable growth in the period between the mid 1980s and 1990s, both in membership and prestige, due to their wider role in campaigns for democratic change alongside their ability to organize in the context of rapid

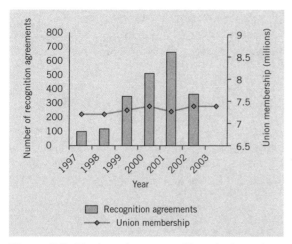

Figure 9.7 Trade union recognition deals and membership levels in the UK, 1997–2003.
Source: Adapted from Labour Force Survey TUC, and Cumbers, 2005.

industrialization (Adler and Webster, 1999). Similarly, the trade union movement in Brazil expanded during the 1980s associated with the struggle for democracy, although it has subsequently fallen back in recent years as employers seek to subcontract work away from the formal unionized sectors of the economy to the small-firm-dominated informal sectors (Ramalho, 1999). Indeed, in many developing countries, during the 1990s, unions started to come under pressure from processes of corporate restructuring and similar patterns of outsourcing by employers to non-union workplaces as those described for unions in the global North.

9.4.2 The end of trade unions or a crisis of a particular form?

The tendency to write off trade unions as dinosaurs whose days are numbered has been criticized by many commentators who point to the continuing importance of labour action and struggle in the global economy (Herod, 2001). Not only are there very different experiences between countries, but even in those places where unions have suffered the worst setbacks there are signs of revival in recent years. Notably, in the UK the election of a more sympathetic Labour government in 1997 and the passing of new laws providing unions with improved, though still restricted, rights to

organize has resulted in a dramatic upsurge in agreements with firms to bargain collectively, helping stem the decline in the number of members (Figure 9.7). Significantly, while male trade union numbers have continued to decline in the 2000s, female membership has increased, suggesting the potential for trade unionism to renew itself for the demands of the post-industrial workplace.

Indeed, more perceptive commentators point to contemporary union decline as representing the crisis of a particular form of trade unionism in time and space, one that derived its strength from the economic conditions of the post-1945 period, namely the development of large-manufacturing-based workforces, nationally regulated economies and systems of collective bargaining, and the social model of a **male breadwinner** in permanent employment (Munck, 1999). As these realities have changed, trade unions need to 'change their spots', making themselves more relevant to the dispersed and fragmented economy of services by appealing to non-traditional constituencies such as women and those in non-standard work who have little social protection. Some have also called for unions to become less narrowly oriented to workplace issues, but instead to work for broader social and political goals, forging alliances with other social movements within local communities on issues such as housing and combating poverty (Wills, 2003), what has been termed a **new social movement unionism** (Moody, 1997). Others argue for a **new labour internationalism**, capable of facing up to the emerging global networks of TNCs by unions themselves becoming less nationally oriented and more capable of operating globally (Waterman, 2000).

Perhaps the most serious issue facing trade unions worldwide is the growth of a massive non-unionized and highly exploited labour force in the global South. The ILO recently noted that the fastest-growing area of employment in both Latin America and Africa was in the urban informal economy, accounting for 90 per cent of all jobs in Africa during the 1990s (ILO, 2005b, p.6). Significantly, this sector includes a growing number of 'unregistered' workers engaged in outsourced work linked into global production networks, but with minimal social protection and employment regulation. Not only are such workers subject to considerable exploitation by managers and key customers, but their presence also drives down the wages and conditions of workers in formal employment. Tackling the unorganized workforce is now a common problem for unionized workers in the North and South, which could become a rallying cry for a new labour internationalism.

Reflect

➤ What are the main challenges facing unions in stemming membership decline?

9.5 From a geography of labour to a labour geography

9.5.1 Constructing labour's spatial fix

In the past, geographers have considered the distribution and organization of labour across space and how this shapes the geography of the economy (Massey, 1984). More recently, however, a new generation of economic geographers has emphasized the active role that labour plays in constructing the global economy. The persistence of measures of employment protection and regulation in all advanced economies (even in the Anglo-American world) and the growth of trade unions in some of the most important economies of the global South suggests a need to understand how workers, as well as employers, help to organize and regulate the changing landscape of capitalism. As one of the most prominent of the new **labour geographers**, Andrew Herod, notes:

the production of the geography of capitalism is not the sole prerogative of capital. Understanding only how capital is structured and operates is not sufficient to understand the making of the geography of capitalism. For sure, this does not mean that labor is free to construct landscapes as it pleases, for its agency is restricted just as is capital's – by history, geography, by structures that it cannot control, and by the actions of its opponents. But it

does mean that a more active conception of workers' geographical agency must be incorporated into explanations of how economic landscapes come to look and function the way they do.

(Herod, 2001, p.34)

While there is a long-established tradition within Marxist geography of studies that explore how workers seek to 'defend place' in the threat of plant closure and industrial restructuring (e.g. Hudson and Sadler, 1986), Herod and others have been pursuing research that highlights the more proactive role of trade unions in creating their own 'spatial fixes'. Critical here in the context of globalization is the need for trade unions to rethink their own geographical strategies to respond to the global production networks being developed by MNCs.

9.5.2 The 'rescaling' of trade union action

The need for trade unions and workers to develop new spatial strategies to deal with the changing economic landscape of globalization has been termed **rescaling** by labour geographers (e.g. Herod *et al.*, 2003). Three themes are evident in the research that is under way on the subject:

➤ Studies exploring attempts by trade unions to develop more effective global solidarity networks to challenge TNCs and the global organization of production (e.g. Cumbers, 2004; Sadler, 2004; Willis, 2002). Different spatial strategies have been advocated. On the one hand, many in the trade union hierarchy have prioritized the importance of developing a set of minimum global labour standards (with regard to pay and conditions, gender equality, the right to join unions and the end to child and forced labour) by lobbying governments and key global institutions such as the World Bank and IMF. Others have prioritized the signing of collective agreements with MNCs as a means of improving workers' rights (see Box 9.5). In opposition to these 'top-down' internationalist strategies, others have championed a more grass-roots approach that seeks to develop independent networks of workers within global production chains (Waterman, 2000). This

Box 9.5

Constructing labour action at the global scale through the signing of frame agreements

The most significant recent manifestation of trade union action in the global economy has been the signing of 39 **global frame agreements**, in the period between 1994 and September 2005, between global union federations and TNCs. While agreements vary in scope, they usually involve the following commitment by TNCs in their global operations:

➤ to pay a decent living wage

Table 9.7 Global framework agreements concluded between MNCs and global union federations, as at September 2005

Company	Employees	Country	Sector	GUF	Year
Danone	100,000	France	Food processing	IUF	1988
Accor	147,000	France	Hotels	IUF	1995
IKEA	70,000	Sweden	Furniture	IFBWW	1998
Statoil	16,000	Norway	Oil industry	ICEM	1998
Faber-Castell	6,000	Germany	Office materials	IFBWW	1999
Freudenberg	27,500	Germany	Chemical industry	ICEM	2000
Hochtief	37,000	Germany	Construction	IFBWW	2000
Carrefour	383,000	France	Retail industry	UNI	2001
Chiquita	26,000	US	Agriculture	IUF	2001
OTE Telecom	18,500	Greece	Telecommunications	UNI	2001
Skanska	79,000	Sweden	Construction	IFBWW	2001
Telefonica	161,500	Spain	Telecommunications	UNI	2001
Merloni	20,000	Italy	Metal industry	IMF	2002
Endesa	13,600	Spain	Power industry	ICEM	2002
Ballast Nedam	7,800	Netherlands	Construction	IFBWW	2002
Fonterra	20,000	New Zealand	Dairy industry	IUF	2002
Volkswagen	325,000	Germany	Auto industry	IMF	2002
Norske Skog	11,000	Norway	Paper	ICEM	2002
AngloGold	64,900	South Africa	Mining	ICEM	2002
DaimlerChrysler	372,500	Germany	Auto industry	IMF	2002
ENI	70,000	Italy	Energy/oil	ICEM	2002
Leoni	18,000	Germany	Electrical/automotive	IMF	2003
ISS	280,000	Denmark	Cleaning/maintenance	UNI	2003
GEA	14,000	Germany	Engineering	IMF	2003
SKF	39,000	Sweden	Ball bearings	IMF	2003

Box 9.5 (continued)

Table 9.7 continued

Company	Employees	Country	Sector	GUF	Year
Rheinmetall	25,950	Germany	Defence/auto/electron.	IMF	2003
H&M	40,000	Sweden	Retail	UNI	2004
Bosch	225,900	Germany	Automotive/electronics	IMF	2004
Prym	4,000	Germany	Metal manufacturing	IMF	2004
SCA	46,000	Sweden	Paper industry	ICEM	2004
Lukoil	150,000	Russia	Energy/oil	ICEM	2004
Renault	130,700	France	Auto industry	IMF	2004
Impregilo	13,000	Italy	Construction	IFBWW	2004
Electricité de France (EDF)	167,000	France	Energy sector	ICEM/PSI	2005
Rhodia	20,000	France	Chemical industry	ICEM	2005
Veidekke	5,000	Norway	Construction	IFBWW	2005
BMW	106,000	Germany	Auto industry	IMF	2005
EADS	110,000	Netherlands	Aerospace	IMF	2005
Röchling	8000	Germany	Auto industry/plastics	IMF	2005

Source: Gibb, 2005, Appendix 1. Global Labour Institute, www.global-labour.org

(usually above local averages) in every country of operation;

➤ to recognize the rights of workers to form and organize trade unions;

➤ to provide time and space for union representatives from different companies to meet;

➤ to pursue equal opportunities policies in relation to gender and race.

In many ways, GFAs have their limitations and are a long way from genuine global collective bargaining agreements, that give unions the right to bargain over detailed pay and conditions, that they enjoy at national level. It is also worth noting that most agreements only cover workers within company boundaries and therefore outsourced work is typically exempt. Additionally, a review of the GFAs signed so far reveals an uneven geography, with most of the agreements between unions and European MNCs (Table 9.7). There is a marked absence of MNCs from the Anglo-American world in particular, suggesting that anti-trade union attitudes are being scaled up from national labour market cultures to the global level.

Within the trade union movement, debate continues about their effectiveness in organizing within global corporate networks. Clearly they rely upon employer cooperation in the first instance, which limits the capacity to organize in hostile MNCs. Additionally, even in companies that do sign up to the idea of social partnership, the difficulties of policing agreements remains. Considerable resources are required for unions to monitor conditions across dispersed global production networks at a time when union strength is declining in most countries. Despite these problems, GFAs are undoubtedly a step forward for the trade union movement in combating the effects of globalization, affording the opportunity for workers from different countries and backgrounds to compare experiences and develop new solidarity networks.

approach tends to be more militant, viewing attempts by union leaders to forge social partnerships at an international level with governments and employers as inevitably compromised. In contrast to these opposing positions, some labour geographers have tended to advocate **multi-scalar** approaches (e.g. Cumbers, 2004), in other words strategies that coordinate action at a number of different geographical scales simultaneously. For example, global labour standards agreed by international bodies are likely to founder without the involvement of grass-roots union members capable of implementing, monitoring and policing them at the level of individual workplaces. By the same token, grass-roots initiatives between workers of different countries campaigning against individual MNCs may result in short-term victories but not lead to more sustainable transnational solidarity without effective support from national and international leaderships.

➤ A focus upon creative ways that unions are responding to the spatial reorganization of production (Herod, 2001; Holmes, 2004). For example, instead of assuming that post-Fordist geographies of production will inevitably undermine the position of trade unions, researchers have pointed out that much depends upon the strategies developed by unions themselves. Rutherford and Gertler (2002) have observed that one effect of employers introducing more flexible production that requires a skilled labour force will mean a greater dependency upon the workers who have key skills and knowledge about the production process. Workers located in strategic hubs within global production systems can be considerably empowered, particularly where MNCs are operating global just-in-time strategies that are vulnerable to stoppages (Herod, 2001).

➤ Attempts to organize workers locally in the dispersed and fragmented landscape of the post-industrial service-based economy (Savage, 1998; Walsh, 2000; Wills, 2005). As labour geographers have been quick to point out, not all capital is mobile. Much service work in particular is tied to place. Restaurants and hotels, cleaning and security work and employment in public services all have to locate where the customers are, and in particular within large urban areas. Nevertheless, as we have already documented, such work is often low paid and non-unionized. The geographical dilemmas facing unions in this context are about developing more effective strategies for organizing across a city in multiple workplaces. One of the most successful

examples was the 1990s 'Justice for Janitors' campaign in Los Angeles, whereby the union, the Service Employees International Union, won a campaign for recognition and better wages for hotel cleaners by focusing upon particular districts in the city. Because the workforce was highly fragmented, distributed among hundreds of different employers, a district-wide strategy was more effective than bargaining with individual employers (Herod, 2001, pp.262–3). Across the US, coalitions of trade unions, church groups and other social activists have achieved considerable success in recent years, in an otherwise hostile industrial relations climate, in winning '**living-wage campaigns**' at the local level (Box 9.6).

9.5.3 Global networks of labour

Alongside the collective agency of labour, it is also important to highlight the way workers at an individual level help to shape the global economy. The most obvious manifestation of this is through the decision to move in search of a better life. In the global economy an estimated 100 million workers are now living away from their countries of origin (Castree *et al.*, 2004, p.189). In contrast to previous phases of international migration, what is notable about the current period is the complexity of linkages, although in broad terms movements are from the global South to the North, reflecting the different global geography of income opportunities (see Figure 9.8).

Different kinds of labour migration can also be identified that reflect the nature of underlying power relations within the capitalist economy and the divisions that exist within the global workforce. On the one hand, we can identify a small group of highly skilled international migrants who form a **transnational capitalist elite**, working in the areas of global finance and management. This privileged group of 'workers' operate in a global space of flows (Castells, 2000), located in and moving freely between the headquarters of the world's most powerful TNCs in global cities such as London, New York, Tokyo and Paris (Beaverstock, 2002). They exercise considerable power

Box 9.6

Organizing city-wide: the 'living-wage movement' in the US

The 'living-wage movement' in the US has been described by celebrated columnist Robert Kuttner as 'the most interesting (and underreported) grassroots enterprise to emerge since the civil rights movement' (www.livingwagecampaign.org). Living-wage campaigns show that local interventions can still be highly effective in a global economy. They involve the passing of laws by local or city authorities, guaranteeing decent wage levels to all workers employed either directly in the public services, or by private employers contracted to the public sector. Campaigns in the US have cleverly linked pay to public spending, suggesting that the latter should not be used to subsidize 'poverty wages', and their success can be measured by the fact that Republican as well as Democrat city authorities have signed living-wage ordinances. In some cities, a living-wage ordinance has been used as a

local marketing strategy, in effect using a 'progressive localism' (Castree et al., 2004, pp.179–80) to attract more socially concerned investors.

The first living-wage coalition was launched in Baltimore in 1994 when, in the context of deindustrialization and a process of gentrified waterfront regeneration that had done little to deal with the problems of poverty and alienation affecting the traditional workforce, an alliance between trade unionists and religious leaders forced the city government to pass a law requiring a 'local living wage' for public-sector workers. Subsequently, 122 cities across the US followed suit and passed living-wage ordinances.

While all living-wage ordinances guarantee wages above the US national minimum wage (currently a paltry $5.15), rates vary widely, reflecting local labour market conditions. Prosperous Fairfax, in the San

Francisco Bay region of California, has a rate of $13 whereas Milwaukee in Wisconsin has a rate of $5.70. There are also variations in terms of non-wage benefits, with some ordinances including health benefits, vacation days and even acceptance of union rights to organize (see www.livingwagecampaign.org).

There are signs that the living-wage movement might be gaining ground in other countries. A campaign has recently been launched in the UK by community and trade union groups in London with an agreement that the 2012 London Olympics will adopt living-wage principles in its procurement strategy. Additionally, Queen Mary College has become the first British university to agree a living wage, with a minimum hourly rate for all staff (including contractors) of £6.70 (about $10) and a minimum of 28 days holiday (*The Guardian*, 11 April 2006).

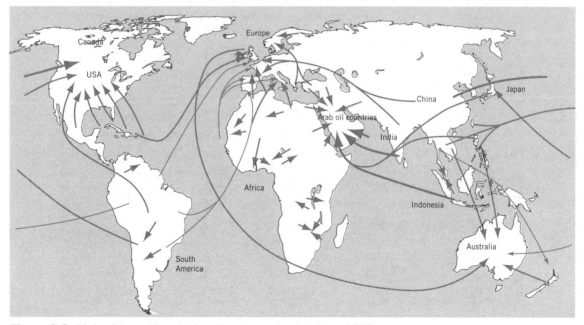

Figure 9.8 Major international migration movements since 1973.

Source: Castles, S. and Miller, M., *The Age of Migration*, 2nd edn, 1998, Macmillan, reproduced with permission of Palgrave Macmillan.

as the 'movers and shakers' of the global economy whose decisions affect the lives of most of the rest of the world. On the other hand, the vast majority of labour migrants have a less-exalted status and are differentiated variously by the extent to which they are temporary or permanent, skilled or unskilled, and voluntary as opposed to 'forced' (see Box 9.1), and legal or illegal.

As Castree *et al.* (2004, p.191) note, despite the rhetoric of globalization and a borderless world, 'national governments actively regulate the international migration of workers', filtering, in the same way that firms do, workers on the basis of their 'desirability'. Not only are migrants increasingly assessed in terms of education and skill levels, but racial and ethnic characteristics typically still inform immigration policy. Racist discourses and nationalistic stances towards immigration often lead to policies that are contradictory from the point of view of capital; for example, tight immigration controls operating during periods of low unemployment and economic boom, when there is a high demand for foreign labour. Recent evidence from the US suggests that immigration actually fuelled economic growth during the 1990s when around 13.5 million people entered the country (*The Guardian*, 3 December 2002), while another report on

the UK estimated that the economy needed 150,000 migrants per year to maintain the labour force at current levels in the face of an ageing population (*The Observer*, 8 December 2002). Despite these facts, official policy still tends towards ever stricter immigration controls (see Figure 9.9).

Although international labour migration can be taken as evidence of worker agency in moving to find work, requiring considerable resourcefulness, ingenuity and courage, migrant workers continue to be among the most exploited groups in the global economy. Undocumented or illegal workers are routinely abused by unscrupulous employers, often having their lives endangered as well as suffering appalling working conditions. One of the most dreadful incidents in recent times involved the death by drowning of 23

Figure 9.9 Illegal immigrants crossing the Mexican–US border.
Source: © Danny Lehman/Corbis.

Chinese migrant workers in the UK in the Morecambe Bay area of Lancashire while being employed as cockle pickers (see http://news.bbc.co.uk/1/hi/england/lancashire/3827623.stm).

9.6 Summary

The nature of employment has changed dramatically in the period since the 1970s through processes of deindustrialization and the reorganization of work from Fordist to flexible workplaces. This has spawned a number of important debates about the magnitude of the changes taking place and their geographical effects. In this chapter we have critically reviewed these claims, drawing upon both data and counter-arguments. While we find considerable evidence of a changing workplace, particularly a shift to greater flexibility and insecurity, there is less evidence that these changes have brought positive benefits for the majority of the workforce. There continue to be wide spatial variations and divisions within the workforce, in terms of pay and conditions, and little evidence of a more egalitarian workplace that some commentators predicted with post-Fordism.

We have also highlighted the importance of geography. Continuing spatial variations in employment conditions, and in the culture of the workplace, in turn help to shape new forms of work and employment to develop forms that are place-specific. Different political and institutional frameworks between regions and countries also lead to variations in how local and national labour markets are organized. As part of the recognition of the role of geography, we have also demonstrated the importance of workers and their unions in shaping the global economy and highlighted the emerging field of labour geography. While labour has been on the back foot in processes of global restructuring, due to the increasing mobility of capital, we have highlighted the continuing possibilities for labour action at a number of different scales in the global economy. Some of the more interesting research in economic geography in recent years has been examining the new spatial strategies being developed by unions in response to globalization.

Exercises

Using Table 9.5 as a starting point and employment tenure as a measure of labour market flexibility, compare the economic performance of two 'flexible economies' with those of less flexible ones.

1. How do the countries compare in terms of job creation, unemployment rates and economic growth?

2. What is the evidence that labour market flexibility leads to better economic performance and social wellbeing?

3. What other indicators might be used to assess the social impact of flexibility?

4. Does the use of such statistics provide us with adequate knowledge of geographical variations in the performance of labour markets?

5. What are the limitations of using statistical data and what other methods might be used to understand the different experience of flexibility for individual workers?

Key readings

Bradley, H., Erickson, M., Stephenson, C. and Williams, S. (2000) *Myths at Work*, Cambridge: Polity.
A thorough and critical analysis of various theories about changes in the contemporary workplace. As the title suggests, the book usefully unpacks the various myths about employment transition, with particularly valuable chapters on flexibility and the 'end of trade unionism'.

Castree, N., Coe, N., Ward, K. and Samers, M. (2004) *Spaces of Work: Global Capitalism and Geographies of Labour*, London: Sage.
An excellent general text on labour geography, with the first two chapters providing useful introductions to the spatial variations in the experience of work in the global economy and key concepts in understanding employment relations. In later chapters there is a wealth of empirical data and maps to help illustrate essays.

Pahl, R. (ed.) (1988) *On Work: Historical, Comparative and Theoretical Approaches*, Oxford: Blackwell.
A little dated but contains a number of very useful chapters about the history of work from the pre-industrial period

through to the industrial. It is particularly good at understanding the changing social and technical divisions of labour through time, evaluating changes in gender and class relations.

Peck, J. and Theodore, N. (2001) 'Contingent Chicago: restructuring the spaces of temporary labor', *International Journal of Urban and Regional* Research 25 (3): 471–96.
A detailed empirical case study of the workings of the flexible labour market for those at the sharp end of the US economy. The paper offers real insights into how the labour market is being restructured and how the meaning of flexibility is very different for firms and employees.

Wills, J. (2002) 'Bargaining for the space to organize in the global economy: A review of the Accor–IUF trade union rights agreement', *Review of International Political Economy*, 9: 675–700.
This is a well-balanced assessment of the potential and the limitations for unions in developing global frame agreements. It contains a detailed case study on one particular agreement between the international foodworkers union and the French catering chain Accor.

Useful websites

www.//ilo.org
The International Labour Organization's website: the UN body that carries out research into global labour issues. Produces data and research papers on a range of labour issues from trade union membership to trends in labour flexibility.

www.icftu.org
The International Confederation of Free Trade Unions. The main website for the umbrella body that represents the international trade union movement.

http://www.aflcio.org
The website for the main US trade union federation.

www.global-labour.org
Website for the independent Global Labour Institute, which tends to be more radical and militant than the official trade unions.

www.//tuc.org.uk
British trade union confederation site.

http://www.bls.gov/
US Bureau of Labor Statistics.
www.//nomisweb.co.uk
UK official labour market statistics.

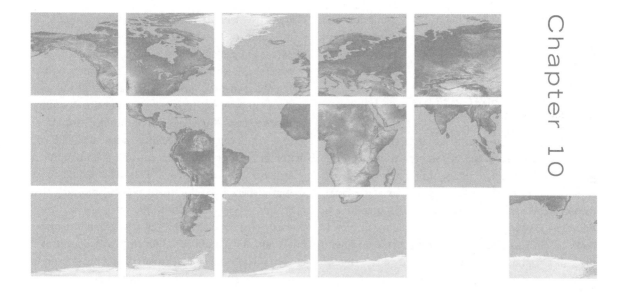

Chapter 10

Towards a knowledge-based economy: innovation, learning and clusters

Topics covered in this chapter

➤ Globalization and local and regional economic development.

➤ The current emphasis on knowledge and innovation as key drivers of economic development.

➤ Contemporary theories of the spatial agglomeration (concentration) of economic activity adopted by economic geographers and others.

➤ The role of industrial clusters in supporting local and regional growth.

➤ The concept of 'learning regions' based on innovation and knowledge generation.

➤ The significance of wider global linkages for the development of clusters and learning regions.

Chapter map

The aim of this chapter is to assess the processes that lead to the **spatial agglomeration** (concentration) of economic activity at the local and regional scales. In the introduction, we stress the importance of the geographical concentration of industries in the context of globalization and the notion of the knowledge-based economy. Section 10.2 outlines contemporary notions of knowledge, innovation and learning, focusing on the level of the firm in particular. We then consider traditional theories of agglomeration, relating these to the contemporary approaches assessed in the remainder of the chapter. Section 10.4 examines the concept of industrial clusters, highlighting its influence on regional development policy. This is followed by a consideration of the related idea of learning regions. Section 10.6 provides a critical assessment of these concepts, focusing on the underlying notion of competitiveness and questioning the assumption that innovation and learning are products of geographical proximity per se.

10.1 Introduction

A resurgence of interest in localities and regions as economic units has been apparent within economic geography since the late 1980s. At first sight, this seems paradoxical, given the prevailing emphasis on **globalization** as perhaps *the* political and economic force of the last 15 or so years. As we argued in Chapter 5, however, **globalization** is an uneven process, leading to the concentration of economic activity in particular places and creating increasingly close linkages between the local/regional and global scales of activity. National economic coherence has been undermined since the 1970s as states have lost control over increasingly globalized flows of investment. The abandonment of Keynesian policies of demand management and full employment has exposed regions to the effects of international competition (section 6.5.4.). This has focused attention on the need for regional-level action if regions are to be able to shape their own development prospects in a climate of rapid technological change and increased capital mobility (Amin and Thrift, 1994).

The revival of the region as a focus of interest has breathed new life into the topic of **spatial agglomeration** or concentration, referring to the tendency for industries to cluster in particular places (section 4.3). This has been an issue of recurring interest for economic geographers and spatial economists since the late nineteenth century. Agglomeration can be contrasted with the opposite process of **spatial dispersal**, where firms move out of existing centres into other, often less-developed regions. The balance between concentration and dispersal processes will vary across different economic sectors, with some exhibiting a highly concentrated pattern while others are more evenly dispersed across the economic landscape. The two processes operate at different geographical scales, for example the global, regional and local, creating distinctive patterns of **uneven development** (section 1.2.2). The balance between agglomeration and dispersal also

changes over time, shaped by the development of technology and organizational structures (e.g. the growth of MNCs) in particular.

Several influential economists and business writers have argued that the growth of advanced information and communications technologies will bring about the 'end of geography' or the death of distance, since businesses can locate their facilities anywhere and still maintain close contact with customers and suppliers (Cairncross, 1997; O'Brien, 1992). This would make **spatial dispersal** the dominant process, meaning that questions of agglomeration are of only historical interest. In reality, however, geography remains important and a number of industries are concentrated in particular locations (e.g. financial and business services in world cities, section 8.5). Indeed, the concept of the **knowledge-based economy** has fostered renewed interest in the role of spatial proximity in facilitating processes of innovation and learning (Malmberg and Maskell, 2002). From this perspective, the capacity to generate new ideas, products and services has been identified as a key source of **competitive advantage** for both firms and regions in the global economy (Cooke and Morgan, 1998).

10.2 Knowledge, learning and innovation: key concepts

10.2.1 Codified and tacit knowledge

Over the last decade or so, the growing importance of knowledge in supporting and driving economic growth has been emphasized by academics, business commentators and policy-makers in developed countries (Castells, 1996; Leadbetter, 1999). According to the Scandinavian economist Lundvall, capitalism has entered a new stage in which 'knowledge is the most important resource and learning the most important process' (Lundvall, 1994). A key point to note is that the importance of knowledge and learning is not confined to a new set of advanced, high-technology industries such as software or biotechnology. Instead, knowledge is developed and applied across all sectors of the

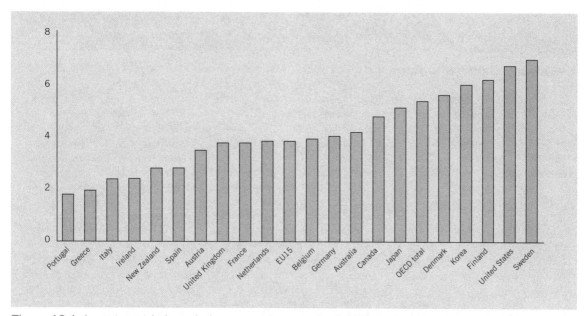

Figure 10.1 Investment in knowledge as a percentage of GDP among OECD countries, 2002.
Source: 'Investment in knowledge', *OECD Factbook 2006: Economic, Environmental and Social Statistics*, © OECD, 2006.

Figure 10.2 Research and development.
Source: Jim Craigmyle/Corbis.

economy, being embodied in work practices and accessed through information technology (Henry and Pollard, 2000, p.v).

A key contextual factor behind the focus on innovation and learning as the basis of economic growth in developed countries is globalization. Instead of following a 'low-road' strategy of trying to compete with low-wage economies on the basis of costs, policymakers have advocated a 'high-road' strategy of specializing in advanced, high-value activities, with competitiveness seen to depend on the generation and commercialization of knowledge (Figure 10.1). This

Box 10.1

A smart, successful Scotland

Since the introduction of UK devolution in 1999, Scotland has led the
way in the development of innovative
economic development polices within
the UK. In particular, the Scottish
Executive (the devolved government)
has embraced the knowledge
economy agenda. A key development
here was the publication of the
Smart, Successful Scotland strategy
in 2001. This identified three main
priorities: 'growing businesses',
'global connections' and 'learning
and skills', becoming popularized as
the 'science and skills' agenda.
Particular importance is attached to
the need to commercialize Scotland's
strong research base by creating
'pipelines of support that take new
ideas all the way from the lab to our
high-growth businesses of tomorrow'

(Scottish Executive, 2002), together
with encouraging a culture of lifelong
learning throughout Scotland. As part
of this agenda, Scottish Enterprise
(the regional development agency for
lowland Scotland) has developed
large-scale initiatives such as Project
Atlas, which aims to upgrade the
telecommunications infrastructure;
the Scottish Co-Investment Fund,
designed to provide venture capital –
in partnership with private-sector
investors – to innovative and dynamic
firms; and the creation of three intermediate technology institutes (ITIs)
to help to commercialize university
research, focusing on energy, life sciences and communications and
digital media (Scottish Enterprise,
2003).
Given the prevailing emphasis since

the 1960s on creating new jobs by
attracting inward investment, the
'science and skills' agenda represents
a significant departure for economic
development in Scotland. It has
gained broad support among policy-
makers and economic commentators
as an appropriate strategy for an
ambitious region aiming to foster
an innovative, knowledge-based
economy. In particular, the
Executive's approach signals a shift
from trying to compete for large-scale
manufacturing employment on the
basis of costs to a focus on the generation of knowledge and skills,
competing on the basis of quality. In
this way, it is representative of the
'high-road' strategies being evolved
by a range of developed regions in
recent years.

agenda has been translated into policy through a focus
on research and development (Figure 10.2), the commercialization of scientific knowledge and the
promotion of business clusters at local and regional
level (e.g. DTI, 1998) (Box 10.1).

Knowledge can be defined as a framework or
structure in which information is stored, processed
and understood (Howells, 2002, p.872). Existing
stocks of knowledge shape how people respond to
particular events and developments, a process that
itself generates new knowledge, enhancing or transforming previous understandings. As this suggests,
knowledge is distinct from information, referring to
the broader frameworks of meaning through that
information or data about real-world events and
trends is processed and understood. As Nonaka *et al.*
(2001, p.15) – a group of influential management
theorists – put it, 'Information becomes knowledge
when it is interpreted by individuals and given a
context and anchored in the beliefs and commitments
of individuals.'

A distinction is often drawn between codified and
tacit forms of knowledge (p.15). **Codified** or **explicit
knowledge** refers to formal, systematic knowledge that
can be conveyed in written form through, for example,
programmes or operating manuals. **Tacit knowledge**,
on the other hand, refers to direct experience and
expertise, which is not communicable through written
documents. It is a form of practical know-how
embodied in the skills and work practices of individuals or organizations. Traditionally, in industries such
as construction, practical skills were acquired through
apprenticeships where new entrants learnt on the job
by shadowing and assisting established tradesmen.

Computer packages provide a good example of the
distinction between codified and tacit knowledge. The
operating manual or programme informs the user
about the operations and capabilities of the package,
telling them what procedures to follow, but it is only by
gaining actual experience of operating the package
(tacit knowledge) that you become a proficient user of
it. As we shall see, much of the economic geography

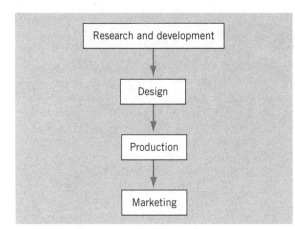

Figure 10.3 The linear model of innovation.
Source: D. Mackinnon.

literature on learning and innovation is based on the assumption that codified knowledge has become increasingly global in organization and reach, while tacit knowledge remains local, relying on geographical proximity to foster communication and interaction between firms in specialist industrial clusters (Maskell and Malmberg, 1999).

10.2.2 The firm as a laboratory: new models of innovation

Innovation can be defined as the creation of new products and services or the modification of existing ones to gain a competitive advantage in the market. The commercial exploitation of ideas is crucial, distinguishing innovation from invention. Traditionally, a distinction was drawn between product and process innovations, with the former referring to new outputs (e.g. mobile phones) and the latter to new methods of making or doing things (e.g. the moving assembly line). This was derived from studies of manufacturing industries, however, with **innovation** in services industries often involving a particular service being delivered or packaged in new ways. The utilization and development of new technology is often central to innovation in services, with the Internet becoming very important in recent years. The online bookstore, Amazon, for example, offers customers easy access to a wide range of goods, having moved into electronic goods and used books, and enabling it to maintain low inventory and overheads costs compared with conventional bookstores (*The Economist*, 2001a). The Internet auction site eBay has also grown rapidly since the late 1990s, with $34 billion of goods being exchanged in 2004 (Waters, 2005), effectively creating a market where none had existed before by acting as an intermediary that brings together buyers and sellers of goods.

The traditional **linear model of innovation** was focused on large corporations, breaking it down into a series of well-defined stages running from the research laboratory to the production line, marketing department and retail outlet (Figure 10.3). It emphasizes

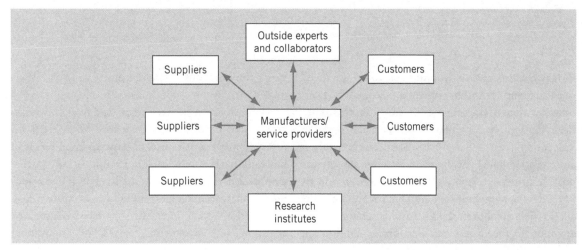

Figure 10.4 The interactive model of innovation.
Source: D. Mackinnon.

Box 10.2

The development of Apple's iPod

Apple's iPod has rapidly become the most popular and fashionable digital music player in the marketplace, following its launch in October 2001 (Figure 10.5). It is based on the recombination of several existing components, involving collaboration between a number of different companies. Rather than stemming from a revolutionary new invention, the iPod began with a sense in early 2001 that Apple could develop a better product than any of its rivals within the emerging market for MP3 players (Hardagon, 2005). The original idea came from an independent contractor, Tony Fadell, who was hired by Apple to develop the product. The platform design came from Portal Player and the operating system from Pixo (both Silicon Valley start-ups), while the hard disk was developed in collaboration with Toshiba and the lithium battery was obtained from Sony (Nambisan, 2005). By integrating these diverse components during an intensive eight-month design period, Apple was able to produce the most portable, user-friendly and fashionable digital music player, which combines small size and ease of use with a large storage capacity (holding over a 1,000 songs) thanks to the hard drive developed by Toshiba. Sales grew rapidly, equalling Apple's computer sales in two years and reaching 14 million in the three months up to 31 December 2005, as the iPod became a 'must-have' Christmas gift for many people (*The Economist*, 2006).

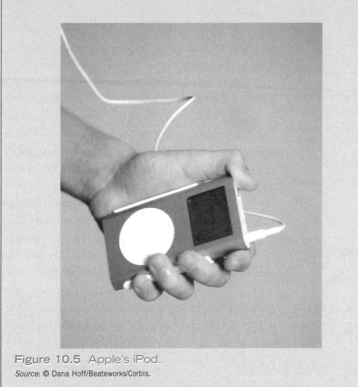

Figure 10.5 Apple's iPod.
Source: © Dana Hoff/Beateworks/Corbis.

formal research and development based on advanced scientists and engineers operating separately from other divisions of the company. In recent years, this has been replaced by an **interactive approach**, viewing **innovation** as a circular process based on cooperation and collaboration between manufacturers or services providers, users (customers), suppliers, research institutes, development agencies, etc. (Figure 10.4 and Box 10.2). The metaphor of the firm as a laboratory draws attention to the experimental nature of innovation, which is often based on trial and error, involving the adoption of existing practices and the trying out of new combinations (Cooke and Morgan, 1998, pp.47–53).

The distinction between linear and interactive conceptions of innovation is associated with a parallel distinction between **radical** and **incremental forms of innovation** (Freeman, 1994). Traditional approaches emphasized radical innovations such as the development of new technologies and industries (section 3.5). By contrast, recent interactive models have focused attention on incremental forms of innovation, involving relatively small improvements in the design and operation of particular products and services.

Innovation, in this sense, can be seen as a continuous process of technical improvement or learning, akin to the 'kaizen' concept derived from Japanese management practice. Indeed, the success of the Japanese model of corporate organization in the late 1980s and early to mid-1990s was important in directing attention towards incremental forms of innovation (Cooke and Morgan, 1998, pp.43–4). The concept of **innovation** as an interactive and incremental process has broadened the scope of research, incorporating small firms and peripheral regions in addition to large corporations and core regions.

Firms are key agents of learning and innovation in the economy. Much of the work on knowledge creation within the firm is informed by a **competence-** or **resource-based theory of the firm** derived from the economist Edith Penrose (section 3.2.2). This stresses the importance of the competences, capabilities or resources of individual workers and managers and the 'bundles' of customs, routines, procedures and practices that define the firm (Gertler, 2003, p.136). From this perspective, the firm is a repository of knowledge. By its nature, learning is an evolutionary process, based on developing, processing and absorbing knowledge.

Probably the best-known model of knowledge gen-eration within the firm is the four-stage **SECI approach** developed by Japanese management theorists, drawing on the practices employed by Japanese corporations in the 1980s and 1990s (Nonaka and Takeuchi, 1995). The four stages are socialization, externalization, combination and internalization, viewed as a spiral, based on continuous and dynamic interaction between **tacit** and **codified knowledge** (Figure 10.6). The model is illustrated with reference to the development of a new car by Honda in the late 1970s (ibid., pp.11–12).

➤ Socialization involves the articulation and exchange of tacit knowledge, requiring face-to-face interaction between experts within a firm. The key problem here is that of tapping into tacit knowledge that already exists within the firm, as expressed in the famous statement by Lew Platt, the former chief executive of Hewlett Packard, 'if only HP *knew* what HP knows, we would be three times as profitable' (quoted in Morgan, 2004a, p.7, emphasis in original). Such knowledge is 'sticky', making it difficult to harness and move to where it is most required within the firm. Efforts to develop a new product need to overcome this problem. The first stage is often to create a 'field' of interaction whereby experts from different

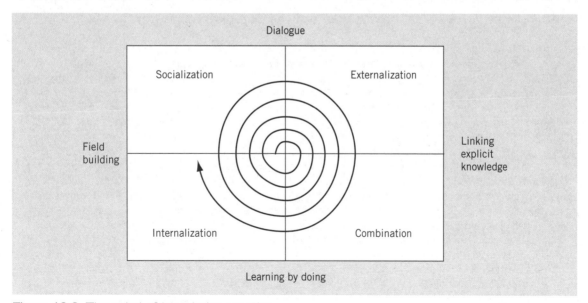

Figure 10.6 The spiral of knowledge creation.
Source: The Knowledge-creating Company: How Japanese Companies Create the Dynamics of Innovation (Nonaka, I. and Taekuchi, H., 1995). By permission of Oxford University Press, Inc.

departments get together to identify and present the different types of relevant knowledge. This is often done in rounds of structured discussion and brainstorming (Bathelt *et al.*, 2004, p.35). When Honda decided to develop a new car in 1978 under the slogan 'Let's gamble', a product development team of young engineers and designers was formed who shared their ideas and skills in discussing what an ideal car should look like in a series of brainstorming sessions (Nonaka and Takeuchi, 1995, pp.11–12).

➤ Externalization is based on the transformation of this tacit knowledge into codified form. It becomes formal and systematic, allowing it be shared with others. This is not a simple task, requiring the development of models, languages and metaphors (Hudson, 2005, p.58). The work of the Honda team was shaped by the metaphor of 'automobile evolution', viewing the car as an organism and seeking to identify its ultimate form (Nonaka and Takeuchi, 1995, p.65). This fed into the concept of a car that was simultaneously tall and short, offering maximum comfort and room for the passenger while taking up least space on the road. The team's

concept of the 'tall boy' eventually led to the Honda City, the company's distinctive urban car of the 1980s.

➤ Combination refers to the combination of different bodies of codified knowledge into more complex and integrated systems. This involves the reconfiguration of existing knowledge through sorting, adding, combining and categorizing it, often involving middle managers breaking down and operationalizing a corporate vision, business model or product concepts (ibid., pp.68–9). Indeed, it is through such processes that new product ideas and models are linked to a company's overall business strategy or vision. The Honda City design team's revolutionary concept of the 'tall boy' – which contradicted the conventional wisdom at the time of long, low sedans – fed back into Honda's senior management sense that a radical change in style was required, with existing models becoming too familiar and routine, helping to launch a new generation of Japanese cars.

➤ Finally, internalization is the process by which firms embody codified knowledge in the skills of workers and the routines and work practices of the firm,

Box 10.3

'Fumbling the future': Xerox, Apple and the personal computer

A group of scientists working at Xerox's Palo Alto Research Centre in Silicon Valley discovered key elements of the personal computer (PC) in the 1970s. This included not only the processing equipment that sits beneath the desk, but also the screen desktop of icons, folders and menus that make up systems such as Windows, Macintosh and the World Wide Web. Since the 1970s, of course, this technology has been transmitted from a small group of experts to hundreds of millions of ordinary users in offices and homes across the world.

Yet the development of the PC did not benefit the firm that initially created the technology. This was

because of divisions within Xerox between scientists based in Palo Alto, development engineers in Dallas, Texas, and the company management in Stanford, California. Not only were the different groups separated geographically, there was also an absence of the common language and outlook required to exploit emerging technologies commercially. The engineers found the scientists naive and unrealistic while the scientists viewed employees in other divisions as 'toner heads' who were interested only in photocopiers (Brown and Duguid, 2000, p.151).

Such internal divisions meant that the knowledge embodied in the emerging computer technology

leaked out of Xerox, to be exploited by another company, Apple, based nearby in Cupertino. In what has become a well-known story in business folklore, Steve Jobs, one of the founders of Apple, visited the Xerox plant in 1979. After Jobs recognized the potential of the technology being developed there – something that the management of Xerox had failed to do – Apple copied and exploited it, licensing some aspects and replicating others. Thus, 'the knowledge that stuck within Xerox leaked readily through its front door' (ibid., p.151). In this way, the company 'fumbled the future', losing out to one of its main competitors.

turning it back into tacit knowledge. It is closely related to 'learning by doing', enabling new forms of knowledge to be absorbed into individuals' tacit knowledge base in the form of shared concepts, technical know-how and practical skills (ibid., p.69). Members of the Honda City team internalized the knowledge generated through the development of the new product, enabling them to play a leading role in other R & D projects. The actual production of the new car saw a large number of workers throughout the company becoming familiar with the concept. A key challenge in this internalization stage is to successfully contain the new knowledge within the firm, ensuring that it does leak out to competitors.

A key factor in determining a firm's success in innovation is its '**absorptive capacity**', referring to the ability to recognize, assimilate and exploit knowledge, derived from either internal or external sources (Cooke and Morgan, 1998, p.16). The absorption of knowledge, in turn, depends upon the existence of a common corporate culture and language, meaning that everybody shares the same broad outlook and sense of the company's overall purpose and objectives. This common culture allows knowledge and meaning to be generated and communicated between groups of workers and managers. In the absence of a common language and outlook, it becomes far more difficult to exploit tacit knowledge without it leaking out to competitors (Box 10.3). The task is often complicated by size as knowledge becomes increasingly fragmented between different divisions.

Reflect

> What measures do you think that firms should adopt to stimulate learning, particularly in terms of encouraging communication between departments?

10.3 Agglomeration and learning

10.3.1 The importance of agglomeration

The main question that arises for economic geographers from the above discussion concerns the significance of geographical proximity in facilitating learning and innovation within and between firms. Nonaka and Takeuchi's SECI model stresses the importance of the context in which knowledge is shared, created and utilized, using the Japanese term *ba* to emphasize the importance of context. While not necessarily corresponding to a particular physical space, *ba* does refer to a specific time and space shaped by the interaction between individuals involved in the process of knowledge creation (Nonaka *et al.*, 2001, p.22). As such, it overlaps significantly with the conceptions of space held by geographers studying processes of **spatial agglomeration** or clustering. Since the late 1980s, a profusion of case studies of innovation and learning in dynamic, successful regions such as Silicon Valley, Hollywood, Boston, parts of northern, eastern and central Italy (the so-called 'Third Italy', Box 1.4), Baden Wurttemberg, Cambridge and Motor Sport Valley in south-east England have been published. This reflects the fact that agglomeration remains an important feature of economic life (Box 10.4), alongside processes of dispersal such as the movement of call centres to developing countries like India (section 8.6).

10.3.2 Marshall's theory of agglomeration

The writings of Alfred Marshall, the renowned Cambridge economist, in the late nineteenth and early twentieth centuries are a key influence on contemporary theories of **spatial agglomeration**. Marshall's ideas were drawn from his observation of specialized **industrial districts** in Britain such as the Sheffield steel industry, emphasizing their distinctive 'industrial atmosphere'. More specifically, traditional explanations of the spatial agglomeration of industries, derived from Marshall, emphasize three main factors (compare Box 4.2):

➤ the growth of various intermediate and subsidiary industries that provide specialized inputs. This refers to the development of close linkages between manufacturers and suppliers of particular components and services, with their co-location serving to reduce transportation costs.

➤ the development of a pool of skilled labour as workers acquire the skills required by local industry. Thus, workers can readily find suitable employment and employers can find skilled labour locally, reducing the search costs for both parties.

➤ the establishment of a dedicated infrastructure and other collective resources (Malmberg and Maskell, 2002, p.432). This includes aspects of the built environment such as the transport and communication system, industrial property, basic services such as sewage and electricity as well as education and training programmes. Once again, the existence of these common facilities serves to reduce costs for individual firms.

These three factors generate **agglomeration economies,** which can be defined as cost advantages that accrue to individual firms because of their location within a cluster of industrial growth (Knox *et al.*, 2003, p.242). These advantages are sometimes also known as external economies because they stem from circumstances beyond a firm's own practices, reflecting broader features of the local environment. Agglomeration economies can be divided into **localization economies** stemming from the concentration of firms in the *same* industry, and **urbanization economies**, derived from the concentration of firms in *different* industries in large urban areas. The basic factors identified by Marshall have remained prominent in much subsequent work, emphasizing the advantages of agglomeration in minimizing costs for firms.

10.3.3 Agglomeration and learning

In the 1980s, Marshall's work was rediscovered by a new generation of economic geographers and economists concerned with the growth of 'new industrial spaces' in Europe and North America. Studies of **flexible specialization** and the growth of craft-based

---

Box 10.4

How common is the spatial agglomeration of industries?

Much of the literature on local and regional growth is concerned with the agglomeration of economic activity in particular places. Most research, however, has taken the form of individual case studies, supplemented by a smaller number of comparative studies (see Saxenian, 1994). One key question that arises from this concerns the actual significance of spatial clustering as a process. How common is it? While this question has been surprisingly neglected by economic geographers, there are a small number of studies that provide some evidence.

In the UK, a study undertaken for the DTI identified 154 clusters across different regions (Trends Business Research, 2001). A relatively larger number of these were located in London and the South-east, with those in other regions smaller in size and 'shallower' in terms of their internal linkages (Figure 10.7). The percentage of regional employment in clusters averaged 23 per cent, ranging from 34 per cent in London to 15 per cent in the North-west. In Scandinavia, research has found not only that manufacturing industries show a significant degree of agglomeration, but also that such concentration has increased over a 20-year period for the majority of industries (Malmberg and Maskell, 1997).

An analysis of the distribution of 106 industries across the US found that many were strongly agglomerated at state level (Krugman, 1991). More recently, Porter (2003) identified 41 'traded' – referring to those industries that were not resource-dependent and exported good and services to other regions or countries – clusters in the US economy, employing some 35 million workers. He found that regional economic performance was strongly determined by the strength of clusters and the degree and range of innovation occurring within them. Alongside the proliferation of case studies, the results of this more extensive, quantitative research clearly indicate that clustering is an important phenomenon worthy of further study (Malmberg and Maskell, 2002).

Box 10.4 (continued)

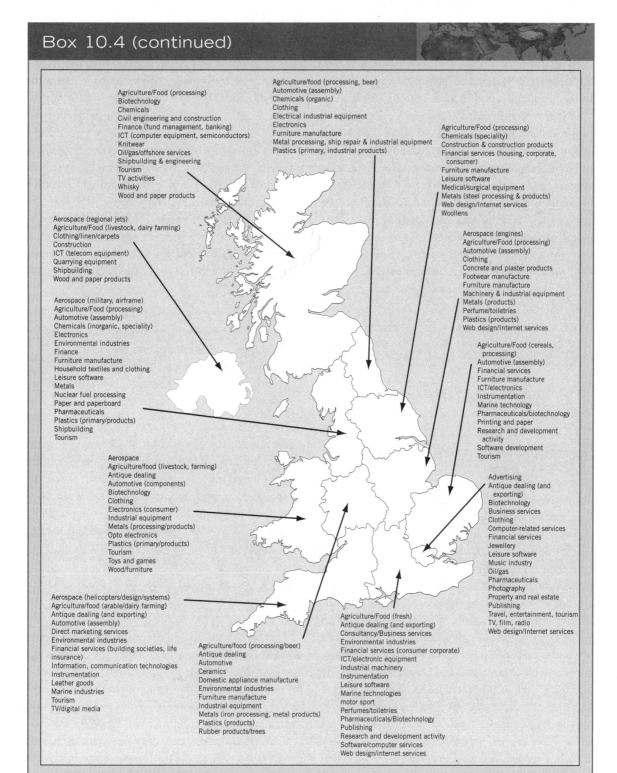

Agriculture/Food (processing)
Biotechnology
Chemicals
Civil engineering and construction
Finance (fund management, banking)
ICT (computer equipment, semiconductors)
Knitwear
Oil/gas/offshore services
Shipbuilding & engineering
Tourism
TV activities
Whisky
Wood and paper products

Aerospace (regional jets)
Agriculture/Food (livestock, dairy farming)
Clothing/linen/carpets
Construction
ICT (telecom equipment)
Quarrying equipment
Shipbuilding
Wood and paper products

Aerospace (military, airframe)
Agriculture/Food (processing)
Automotive (assembly)
Chemicals (inorganic, speciality)
Electronics
Environmental industries
Finance
Furniture manufacture
Household textiles and clothing
Leisure software
Metals
Nuclear fuel processing
Paper and paperboard
Pharmaceuticals
Plastics (primary/products)
Shipbuilding
Tourism

Aerospace
Agriculture/food (livestock, farming)
Antique dealing
Automotive (components)
Biotechnology
Clothing
Electronics (consumer)
Industrial equipment
Metals (processing/products)
Opto electronics
Plastics (primary/products)
Tourism
Toys and games
Wood/furniture

Aerospace (helicopters/design/systems)
Agriculture/food (arable/dairy farming)
Antique dealing (and exporting)
Automotive (assembly)
Direct marketing services
Environmental industries
Financial services (building societies, life insurance)
Information, communication technologies
Instrumentation
Leather goods
Marine industries
Tourism
TV/digital media

Agriculture/food (processing/beer)
Antique dealing
Automotive
Ceramics
Domestic appliance manufacture
Environmental industries
Furniture manufacture
Industrial equipment
Metals (iron processing, metal products)
Plastics (products)
Rubber products/trees

Agriculture/food (processing, beer)
Automotive (assembly)
Chemicals (organic)
Clothing
Electrical industrial equipment
Electronics
Furniture manufacture
Metal processing, ship repair & industrial equipment
Plastics (primary, industrial products)

Agriculture/Food (fresh)
Antique dealing (and exporting)
Consultancy/Business services
Environmental industries
Financial services (consumer corporate)
ICT/electronic equipment
Industrial machinery
Instrumentation
Leisure software
Marine technologies
motor sport
Perfumes/toiletries
Pharmaceuticals/Biotechnology
Publishing
Research and development activity
Software/computer services
Web design/internet services

Agriculture/Food (processing)
Chemicals (speciality)
Construction & construction products
Financial services (housing, corporate, consumer)
Furniture manufacture
Leisure software
Medical/surgical equipment
Metals (steel processing & products)
Web design/Internet services
Woollens

Aerospace (engines)
Agriculture/Food (processing)
Automotive (assembly)
Clothing
Concrete and plaster products
Footwear manufacture
Furniture manufacture
Machinery & industrial equipment
Metals (products)
Perfume/toiletries
Plastics (products)
Web design/Internet services

Agriculture/Food (cereals, processing)
Automotive (assembly)
Financial services
Furniture manufacture
ICT/electronics
Instrumentation
Marine technology
Pharmaceuticals/biotechnology
Printing and paper
Research and development activity
Software development
Tourism

Advertising
Antique dealing (and exporting)
Biotechnology
Business services
Clothing
Computer-related services
Financial services
Jewellery
Leisure software
Music industry
Oil/gas
Pharmaceuticals
Photography
Property and real estate
Publishing
Travel, entertainment, tourism
TV, film, radio
Web design/Internet services

Figure 10.7 Map of UK clusters.
Source: Trends Business Research, 2001, p.21.

industrial districts in particular were directly informed by Marshall's writings (Box 1.4). Since the mid-1990s, however, the focus of agglomeration has moved from the traditional Marshallian notion of cost reduction through **agglomeration** economies to an emphasis on the dynamic benefits of clustering in facilitating innovation and learning processes (Malmberg and Maskell, 2002). A key claim is that geographical proximity between firms allows them to create and share information and knowledge, stimulating processes of adaptation, learning and innovation. The geographically 'sticky' nature of **tacit knowledge** in particular is crucial, and co-location allows firms to share experience through collaborative projects. Local industrial cultures and institutions are held to be important in encouraging learning and collaboration between firms (Box 1.4).

10.4 Clusters, innovation and competitiveness

10.4.1 The focus on clusters

One of the most influential recent models of spatial agglomeration or concentration is Michael Porter's theory of business **clusters**. Porter, a Harvard business economist, has been highly active in promoting his clusters concept among academics, policymakers and consultants. These efforts have met with considerable success and cluster-based policies have been adopted by a range of organizations, including local and regional development agencies, national government units such as the Department of Trade and Industry (DTI) in the UK and supranational bodies such as the Organization for Economic Cooperation and Development (OECD) and the European Commission. As advocates of clusters stress, the concept is not confined to dynamic new clusters in knowledge-economy sectors like biotechnology, but can be applied in a variety of contexts, including mature industries in peripheral regions such as chemicals in northern England or furniture in Scandinavia. The model is based on the assumption

that the prosperity of high-income clusters is a result of innovation and learning processes that add value to existing products and services, rather than involving a strategy of cost reduction. This serves to reinforce the underlying idea that firms and regions in developed countries now compete on the basis of quality, not price.

Porter's writings stress the importance of clusters as a general phenomenon and outline how the operation of particular clusters promotes growth and competitiveness. Porter was originally concerned with the external conditions that support firms' competitiveness at the national scale. More recently, he has focused on the role of sub-national clusters in enhancing innovation and productivity. For Porter, clusters are defined as:

> geographical concentrations of interconnected companies, specialized suppliers, service providers, firms in related industries, and associated institutions (for example, universities, standards agencies and trade associations) that compete but also co-operate.
>
> (Porter, 1998, p.197)

There are two key elements in this definition (Martin and Sunley, 2003, p.10). First, the firms in the cluster must be linked in some way: for example, by the supply of specialized inputs or services. Second, a cluster is defined by geographical concentration, with proximity creating a commonality of interest between firms and encouraging frequent interaction (Box 10.5).

10.4.2 Porter's 'diamond' model

Porter argues that geographical concentration encourages processes of interaction within the 'competitive diamond' by increasing the productivity of constituent firms or industries, stimulating higher rates of innovation and encouraging high rates of business formation. The **diamond model** is based on the contention that clusters enhance **competitiveness** and productivity by fostering the interaction between four sets of factors (Figure 10.8). Demand conditions refer to the tendency for successful clusters to generally serve

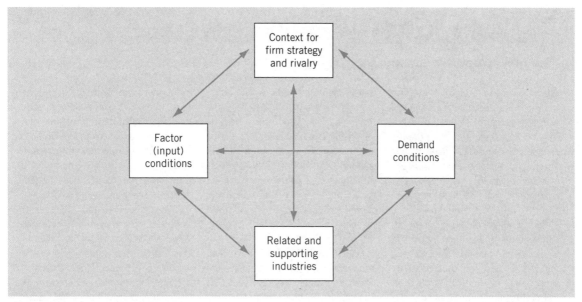

Figure 10.8 Porter's diamond model.
Source: 'Locations, clusters and company strategy', in *The Oxford Handbook of Economic Geography*, edited by Clark, G.L., Feldman, M. and Gertler, M. (Porter, M.E., 2000). By permission of Oxford University Press.

global markets with 'leading-edge' local customers – firms who sell to global markets – playing a key role in encouraging innovation among suppliers. Such exports represent the main external link between clusters and the wider global economy emphasized in Porter's model.

Second, supporting and related industries refer to firms that supply inputs or simply industries located in the same area. Close links between customers and suppliers enable complex communication and interaction to occur, acting as key channels of learning and knowledge transfer, while a concentration of other industries creates the critical mass to support advanced skills, training and infrastructure (Box 10.5).

The third dimension of the model relates to factor conditions, referring to the main factors of production – land, capital, labour and knowledge. The availability of capital to fund investment and growth is important, with access to venture capital having played a key part in the development of successful clusters such as Silicon Valley in California. Advanced clusters in developed countries are generally characterized by high labour costs in global terms, meaning that the presence of a skilled workforce is crucial. Such skills need to be con-

tinually updated and enhanced through training and education programmes. Knowledge is closely related to skills with high labour costs, again placing a premium on the ability to generate and transfer knowledge as the basis of competitiveness. Here, clusters need to support innovation, based on close links between universities, research institutes and leading firms, ensuring that scientific research is adequately funded and commercialized.

The final element in the model is firm strategy, structure and rivalry. Since any particular industry will normally comprise a number of firms that compete, with some succeeding and others losing out, maintaining a substantial number and range of firms is crucial. As such, new-firm formation is important, with processes of corporate spin-off a key mechanism for this. At the same time, rivalry between co-located firms encourages investment and innovation in order to keep up and remain competitive. Proximity enables firms to monitor the activities of local rivals, providing them with access to information and making it difficult to ignore new developments. Apple's poaching of PC technology from Xerox provides an excellent example of this effect (Box 10.3).

Box 10.5

The Hollywood film production cluster

The US motion picture or film industry has become synonymous with the district of Hollywood in Los Angeles, southern California. Hollywood represents a distinctive geographical phenomenon, composed of a dense network of film production companies and service providers (Scott, 2002). As such it represents a striking example of the geographical concentration of a cultural industry in a particular location. Most of the industry is clustered in a relatively small area centred on Hollywood itself, stretching from Burbank in the east through to Beverley Hills and

Santa Monica in the west (ibid., p.965) (Figure 10.9). At the same time as Hollywood has become increasingly globalized, production has remained concentrated in southern California. This presents something of a paradox, at least for those who believe that electronic communications and computerized technologies herald the 'end of geography' (O'Brien, 1992).

Hollywood has historically been dominated by the seven 'majors' – the large corporate studios such as Paramount, 20th Century-Fox and Warner – who have controlled film

production and distribution. In the 'old' system, these majors directly controlled most aspects of film production, distribution and exhibition, owning their own theatre (cinema) chains across the US (Storper and Christopherson, 1987). This system began to break up in the 1950s and 1960s following the Paramount antitrust decision by the US courts in 1948 and the growth of television in the 1950s. The former forced the majors to get rid of their extensive cinema chains while the latter shrunk the audience for films. The result was increased competition and the

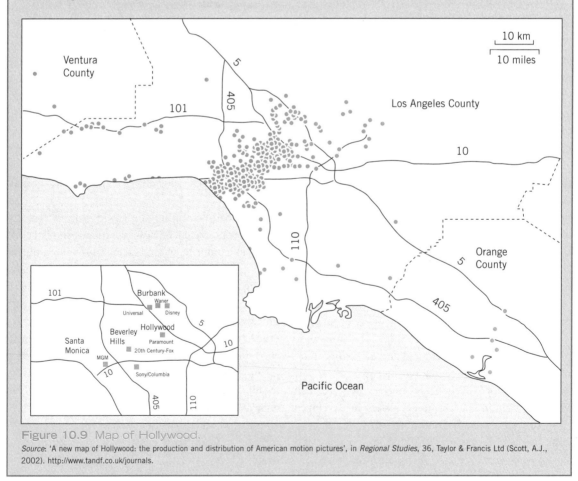

Figure 10.9 Map of Hollywood.

Source: 'A new map of Hollywood: the production and distribution of American motion pictures', in *Regional Studies*, 36, Taylor & Francis Ltd (Scott, A.J., 2002). http://www.tandf.co.uk/journals.

Box 10.5 (continued)

creation of a climate of uncertainty and instability. The majors responded by divesting themselves of many activities, out-sourcing these to small suppliers and becoming 'nerve centres' of production, exercising overall coordination and control. The number of small firms providing specialized services and inputs, such as script writing or film editing, grew markedly. At the same time, the majors have maintained their economic power, particularly in terms of financing, deal-making and distribution.

The Hollywood film production complex is characterized by four key features (Scott, 2002, p.965) that explain the pattern of geographical concentration:

➤ A series of overlapping production networks encompassing majors, independents and providers of spe-

cialized services. The frequency, complexity and volume of transactions (contacts) between firms and individuals fosters a need for face-to-face communication (Storper and Christopherson, 1987, p.112).

➤ The local labour market supplies a large number of individuals with the requisite skills and experiences, constantly replenished by the arrival of new talent from the rest of North America and the world.

➤ The institutional environment consists of many organizations and associations who represent the interest of firms, workers and government agencies.

➤ The broader regional environment – produced in part through the geographical concentration of the industry there over time – provides

a range of crucial resources, ranging from a cinematic tradition to background landscapes and proximity to other cultural industries.

The industry has experienced substantial growth in recent decades, with employment in motion-picture production and services expanding by 194 per cent between 1980 and 1997, maintaining its high level of geographical concentration. Some dispersal to satellite sites has occurred in the search for new locations and reduced costs, particularly for film-shooting, mainly to Canada. While such trends are likely to continue, there seems little immediate prospect of them undermining the fundamental basis of Hollywood, which is rooted in a powerful combination of production networks, skills and creativity, institutional support and a wider regional culture (Scott, 2002).

10.4.3 Clusters policy

The identification and promotion of clusters has become a major component of regional development policy over the last decade or so (Benneworth *et al.*, 2003). In contrast to old-style regional policy, which targeted investment on depressed regions through financial incentives, contemporary approaches aim to foster innovation, learning and entrepreneurship in all regions (section 6.5.4). **Cluster initiatives** fit well with this agenda, promising to increase productivity and growth in a selected range of key industries. In particular, the promotion of clusters is bound up with an emphasis on increasing competitiveness as the economic challenge facing countries and regions (see section 10.6). Nations and regions that have adopted clusters policies include several prominent US states, a number of developing countries and UK regions following the establishment of Regional Development Agencies across the 'standard' regions of England in 1999. Specific forms of support associated with clusters

policy include R & D assistance, training and skills development, the provision of venture capital and efforts to build a sense of cluster identity among constituent firms.

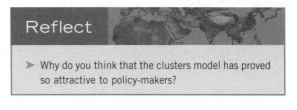

Reflect

➤ Why do you think that the clusters model has proved so attractive to policy-makers?

10.5 Learning regions

Recent work on **learning regions** also explains agglomeration in terms of the generation and circulation of knowledge rather than conventional notions of cost minimization. Compared with the clusters concept, however, which is derived from the work of Porter in a very direct fashion, the notion of learning regions is more diffuse in its origins and development. One source is work on national systems of innovation by industrial economists and others that emphasized important differences in national environments for supporting innovation. These differences were understood in terms of the main relationships between key institutions such as government agencies, large corporations, small firms, university research, technology institutes and training organizations (Cooke and Morgan, 1998, pp.24-6). The insights derived from research on national systems of innovation were transferred to the regional level, informing the learning regions concept. While it is probably fair to say that learning regions are less clearly specified than clusters in much of the literature, the following propositions can be identified.

10.5.1 The importance of tacit knowledge

First, rather than leading to the annihilation of geography, **globalization** is associated with the emergence of new forms of agglomeration based upon knowledge creation (Storper, 1997). In an increasingly globalized economy, firms can access standard factors of production from virtually any location, and access to a growing stock of **codified knowledge** has increased greatly through advanced communication technologies such as the Internet. This means that those places that are able to generate and sustain tacit, specialized local knowledge will gain a competitive advantage (Maskell *et al.*, 1998, pp.62–3). This claim is underpinned by the model of interactive innovation we described earlier, where proximity helps firms to work together in addressing particular problems, and sharing information and knowledge. Second, and implicit in the above, is the claim that **tacit knowledge** is geographically 'sticky', making it difficult to transmit across space, compared with codified knowledge (section 10.2). Transmission of such knowledge is best achieved by regular face-to-face communication, requiring spatial proximity (see section 8.5). Thus, tacit knowledge is regarded as local and codified knowledge as global.

10.5.2 Informal relations and conventions

Third, new forms of agglomeration are underpinned by informal relations and conventions that link firms and organizations in particular places. The informal and often intangible nature of these relations distinguishes them from the traditional emphasis on material linkages, referring to exchanges of goods and services between firms (for example, suppliers and manufacturers). While the latter represent 'traded interdependencies', the economic geographer Michael Storper (1995) emphasizes the importance of what he calls '**untraded interdependencies**'. These are informal linkages and relationships that tie firms together. They are made up of an intangible set of skills, attitudes, habits and understandings that become associated with specialized production. An example would be the complex of information technology skills, entrepreneurial attitudes and networking habits that help to define Silicon Valley as a distinctive 'world of production'. Such linkages are 'untraded', being transmitted through routine forms of communication and interaction between firms. Particular mechanisms of transmission include the labour market, social contacts between managers/executives (e.g. golf clubs), and organizations like development agencies, chambers of commerce and trade associations.

10.5.3 Local 'buzz'

Fourth, learning regions are characterized by a particular form of internal vibrancy that Storper and the economist Anthony Venables term **local 'buzz'** (Storper and Venables, 2002). Reminiscent of Marshall's 'industrial atmosphere', this intangible dimension of the agglomeration process can be identified as a form of **urbanization economy** derived from the concentration of a large number of firms in a location. The term conveys the vibrancy and excitement of daily life within a cluster with a proliferation of activities and events occurring simultaneously, generating information of significant interest to local actors. In this sense, 'buzz' refers to the information and communication ecology created by face-to-face contacts, co-presence and co-location of people and

Box 10.6

Innovation and learning in Motor Sport Valley (MSV)

The economic geographers Nick Henry and Stephen Pinch's work on the British motor sport industry is an excellent example of recent economic geography research on innovation and knowledge circulation. Their study sought to apply Storper's notion of 'untraded interdependencies' (Storper, 1997), conceiving of Motor Sport Valley (MSV) – the 'cluster' of car-producing and supplying firms located in Oxfordshire and adjacent counties in southern England – as a distinctive 'world of production'. It represents a classic example of a world-leading concentration of small firms. The industry employs over 30,000 people, consisting of scores of small and medium-sized firms located within a 50-mile radius of Oxfordshire (Figure 10.10). 'Approximately 75 per cent of the world's single-seater racing cars are designed and assembled in this region', including the majority of Formula 1 cars (Henry and Pinch, 2000, p.192). MSV has taken over from northern Italy as the centre of the world's motor sport industry since the 1960s, reflecting its success in capitalizing on the industry's move to a technological system based on aerospace technologies (Pinch and Henry, 1999).

In their research, Henry and Pinch (2000, p.192) focus on the 'knowledge community' of MSV, defining this as 'a group of people (principally designers, managers and engineers in this case) often in separate organizations but united by a common set of norms, values and understandings'. Based on this notion, their study actually tracked 'knowledge processes', as distinct from many articles that have merely provided a theoretical discussion of these. Henry and Pinch examine the mechanisms through which knowledge is circulated within MSV, including staff turnover between firms, the formation of new firms as 'spin-offs' from existing organizations, and informal exchanges based on 'gossip' and 'rumour' between individuals within

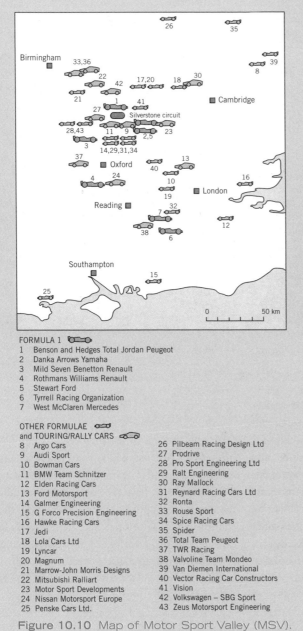

FORMULA 1
1   Benson and Hedges Total Jordan Peugeot
2   Danka Arrows Yamaha
3   Mild Seven Benetton Renault
4   Rothmans Williams Renault
5   Stewart Ford
6   Tyrrell Racing Organization
7   West McClaren Mercedes

OTHER FORMULAE
and TOURING/RALLY CARS
8   Argo Cars
9   Audi Sport
10 Bowman Cars
11 BMW Team Schnitzer
12 Elden Racing Cars
13 Ford Motorsport
14 Galmer Engineering
15 G Forco Precision Engineering
16 Hawke Racing Cars
17 Jedi
18 Lola Cars Ltd
19 Lyncar
20 Magnum
21 Marrow-John Morris Designs
22 Mitsubishi Ralliart
23 Motor Sport Developments
24 Nissan Motorsport Europe
25 Penske Cars Ltd.
26 Pilbeam Racing Design Ltd
27 Prodrive
28 Pro Sport Engineering Ltd
29 Ralt Engineering
30 Ray Mallock
31 Reynard Racing Cars Ltd
32 Ronta
33 Rouse Sport
34 Spice Racing Cars
35 Spider
36 Total Team Peugeot
37 TWR Racing
38 Valvoline Team Mondeo
39 Van Diemen International
40 Vector Racing Car Constructors
41 Vision
42 Volkswagen – SBG Sport
43 Zeus Motorsport Engineering

Figure 10.10 Map of Motor Sport Valley (MSV).
Source: Henry and Pinch, 2000, p.192.

Box 10.6 (continued)

the industry. Career histories and biographies of key individuals are provided along with diagrams of linkages between individuals and 'family trees' of 'parent' and spin-off companies (Figure 10.11).

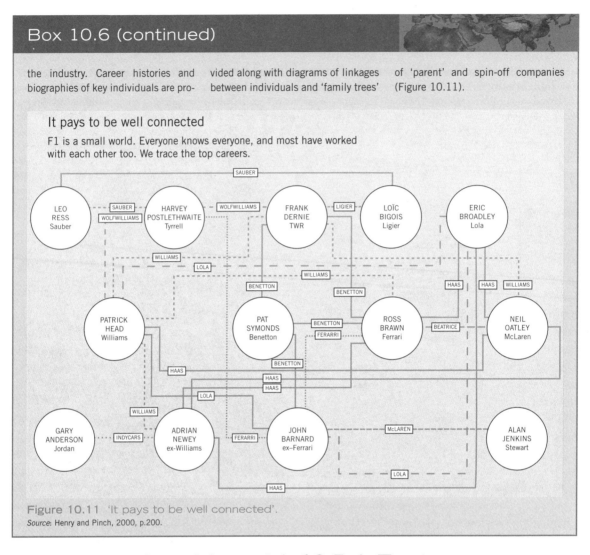

It pays to be well connected

F1 is a small world. Everyone knows everyone, and most have worked with each other too. We trace the top careers.

Figure 10.11 'It pays to be well connected'.
Source: Henry and Pinch, 2000, p.200.

firms within the same industry and place or region' (Bathelt *et al.*, 2004, p.38). It operates through different modes of communication such as chatting, gossiping, brainstorming, in-depth discussions and problem analysis. Firms receive information automatically from their presence in the region and participation in its economic and social activities. Through frequent communication and interaction, firms develop a shared language, attitudes towards technology and interpretative schemes that are used to process new knowledge (ibid., p.38). A study of the knowledge-based cluster of Motor Sport Valley in southern England has illustrated how such internal processes of learning and interaction actually operate (Box 10.6).

10.5.4 Trust

Fifth, the concept of learning regions emphasizes the importance of trust, which is an essential support for collaboration and learning between firms. At a minimal level, firms must have the confidence that other firms, for example suppliers or customers, will meet their commitments, a **'competence' trust** that is based on expectations of performance derived from reputation. A deeper sense of trust is that of **'goodwill' trust**, associated with collaborative ventures where firms are satisfied that the commitment of partners goes beyond explicit contracts and agreements (Lazaric and Lorenz, 1998). As such, the former seems conducive to the transmission of **codified knowledge** and the latter to the exchange of **tacit knowledge**. The key claim made in the literature is that geographical proximity between associated firms and producers is more likely to generate high levels of 'goodwill' trust in particular than more dispersed relations. Proximity enables frequent face-to-face contact between economic actors, allowing for the development of reciprocity and 'goodwill' trust, something that would be more difficult with longer-distance relationships where contact is likely to be periodic and more formal (Morgan, 2004a).

10.5.5 Regional development policy

Finally, at the level of policy, work on learning regions fits with the general emphasis on 'high-road' strategies of specializing in more advanced, high-value activities, compared with a 'low-road' strategy of trying to compete with low-cost developing countries. In particular, regional **competitiveness** is seen to depend on the generation and commercialization of knowledge. Moreover, the concept of learning regions suggests that the impetus for development comes from within a region. Thus, regional development is a 'bottom-up' process, requiring the construction of local capacities to support learning and innovation (see below). Economic underdevelopment is not about regions being exploited by MNCs, but reflects poor learning characteristics that are internal to regions themselves. The task for less-favoured regions particularly is to develop better learning capabilities, focusing on the 'softer' intangible factors of interaction, networking and collaboration, what has been called the 'infostructure' of regional development policy (Morgan, 1997, pp.498–9), alongside the traditional emphasis on inward investment and physical infrastructure. An example of a specific policy initiative that sought to build such capacities was the EU's Regional Innovation Strategies programme (Box 6.6), targeting on less-favoured regions such as Wales (UK), Lorraine (France), Saxony-Anhalt (Germany) and Limburg (Netherlands). It was designed to encourage interaction and collaboration between different interest groups and organizations, including business, government agencies, universities, vocational colleges, local government, and enterprise and development agencies (Morgan, 2004b, pp.12–17).

Reflect

➤ Do you think that policies encouraging inter-firm collaboration and learning – derived from successful regions such as Silicon Valley and the 'Third Italy' – offer a solution to the problems of less-favoured regions? Justify your answer.

10.6 Clusters and learning regions: an appraisal

10.6.1 Learning and competitiveness

A basic question concerns the issue of who learns. This focuses attention on the model of learning and competition that underpins the literature. On a very simple level, it is individuals and firms that learn, not regions. While this may seem rather obvious and pedantic, claims about learning have been extended from firms to regions in a rather loose and imprecise manner. The idea that regions are the main agents of learning is erroneous, reflecting, perhaps, how the concept has emerged from pressing policy concerns about regional development, representing a form of 'theory led by policy' (Lovering, 1999). The whole competitiveness agenda, which underpins the clusters and learning literature, has been subject to similar criticism (Box 10.7).

In reality, of course, regions are not single entities, but territorial aggregations of individuals and social groups. The learning process in which different individuals and firms are engaged will vary in terms of both quantity and quality. There is no single regional voice or interest, for conflicts can sometimes develop between different groups. For example, tensions between large inward investors and small local firms are not uncommon, with the latter often complaining that MNCs are overly prominent in shaping the regional development agenda and claiming funds. Regional development priorities tend to be set by development agencies, local authorities and business leaders who claim to speak on behalf of the region in negotiating

Box 10.7

Competitiveness: a dangerous obsession?

Since the early 1990s, the notion of regional competitiveness has gained hegemonic status within economic policy debates in developed countries (section 6.5.4) (Bristow, 2005, p.285). This means that it has become widely accepted, being regarded as simple common sense. It refers to the notion that regions compete with one another directly for market share in the global economy, emphasizing the need to promote learning and innovation in order to gain a competitive edge on rival regions. As well as using the notion to develop specific polices such as the identification of clusters, the commercialization of research and the promotion of exports, government bodies have attempted to measure and model competitiveness. The UK, for example, has published regional competitiveness indicators since 1995, while the European Commission has also sought to measure regional economic performance in this way (ibid., p.285).

'Benchmarking' studies where regions track and measure their economic performance against other regions have become particularly prominent.

For critics such as the economist Paul Krugman, the focus on competitiveness is unfortunate since it is fundamentally erroneous, representing a 'dangerous obsession' among policy-makers and commentators. Krugman argues that competitiveness is based on a false analogy between the corporation and the national or regional economy, involving 'a set of crude misconceptions, presented as if they were sophisticated insights' (Krugman, 1996, p.256). When applied to firms, competitiveness has a clear meaning, referring to their capacity to grow and be profitable in the marketplace, and can be measured in terms of output and profitability (Bristow, 2005, p.287). While regions or nations may compete to attract particular investments such as a new car factory or government research facility, the idea

that they compete for market share in a similar fashion to firms is highly simplistic. In reality, much economic activity remains domestic in nature, particularly in large economies like the US. Furthermore, international trade is not the zero sum game as the competitiveness industry portrays it (section 4.2); instead, it generates additional income for both parties, with the demand for imports by country B boosting the export industries of country A and vice versa. Krugman suggests three reasons for the popularity of competitiveness rhetoric: it is exciting, portraying economic development as a contest between competing territories; it makes economic problems seem easier to solve, targeting supply-side measures to boost innovation and learning; and it is politically useful, justifying tough policies (e.g. welfare cuts or wage restraints) as a necessary response to the pressures of global competition (Coe and Kelly, 2002).

with prospective investors and lobbying government or the EU for additional funds (MacLeod, 1999). In this sense, work on clusters and learning regions tends to ignore the politics of regional development, assuming that regions are coherent identities defined by a shared interest in enhancing their competitiveness.

Krugman's critique of the competitiveness agenda is forceful and compelling (Box 10.7), offering a necessary rejoinder to what we might call the 'hyper-competitors' (following our discussion of globalization, section 5.2). Yet Krugman's arguments lack much geographical sensibility, providing little sense of the nature of regions or nations as social entities. A substantial volume of research shows, however, that aspects of the

national and regional environment shape the competitiveness of firms. Advantages derived from the spatial concentration of economic activity in particular places (e.g. close links with suppliers, a pool of skilled labour and shared infrastructures; section 10.3) are important in underpinning the success of firms within dynamic clusters. Thus, a modified version of regional competitiveness that stresses the role of the regional environment in shaping firms' growth remains useful (Kitson et al., 2004). This can be seen as a middle-ground perspective, between the 'extremes' of regions competing directly for market share and Krugman's 'sceptical' response, which dismisses the concept outright.

10.6.2 The effects of spatial proximity and distance

A key characteristic of research on clusters and learning regions is an over-emphasis on interfirm links and networks within regions, as opposed to connections between local firms and external organizations and actors. Such external links are rarely considered in such studies (Markusen, 1996), except in terms of the link between local producers and global markets emphasized by Porter and others. Recently, however, the emphasis on local linkages has been questioned, with critics suggesting that proximity is not only a spatial phenomenon, but can also take social and organizational forms. As individuals, we may have close relations with people located in other places, relatives or close friends. 'Close' long-distance relationships are also possible both within and between firms, although the former are likely to be more common. The main target of these critiques is the assumption that **tacit knowledge** is invariably local and **codified knowledge** global. Instead, the geographer John Allen (2000) implies that tacit knowledge can flow across long distances, arguing that 'thick relationships' and 'shared understandings' depend on the quality of the inter-personal and organizational relations involved, not their geography.

This emphasis on the exchange and sharing of knowledge through spatially distant relationships is consistent with a pattern of dispersal where firms are located in different places. Processes of dispersal have long been apparent at the local scale in particular, with firms moving certain administrative and back-office functions out of major metropolitan centres like London and New York to the surrounding suburbs where land costs (rents) in particular are lower (section 8.5). The growth of MNCs and advanced communication technologies such as the Internet create additional opportunities for dispersal on a global scale, as highlighted by the relocation of call centres to countries such as India. The argument that knowledge can be generated and exchanged through spatially distant but organizationally close relationships often relies on the notion of **'communities of practice'**, derived from organizational studies

(Amin and Cohendet, 1999). This emphasizes the close informal contact and ties built between individuals working together in the same field. Increasingly, it is argued, such communities of practice can be developed and maintained through networks that operate across distance, using technologies such as email, conference calls and videoconferencing. As such, the notion of communities of practice is often opposed to the traditional emphasis on geographical communities.

It has become increasingly apparent, however, that the communities of practice argument is itself questionable, offering little real evidence in support. Merely asserting that the local does not matter in the face of corporate networks hardly represents real progress over the opposite claim (Gertler, 2003, p.138). In a recent response to critics of the spatial proximity thesis, Kevin Morgan (2004a, pp.12–13) – one of its main proponents in economic geography – identifies three key flaws in their arguments. First, rather ironically, the tendency to counterpose social and organizational proximity to geographical proximity is itself based on an impoverished notion of geography. Assuming that the two are completely separate succeeds only in reducing geography to its narrow, physical dimension. In reality, geographical proximity works through the social and organizational relations that actually connect firms together in a particular place. Second, the sceptics tend to exaggerate the possibilities of close interaction over distance. It seems unlikely that this can offer the same scope for reciprocity, serendipity and trust as face-to-face contact, structured by body language and other forms of non-verbal communication (Brown and Duguid, 2000). Finally, so-called 'communities of practice' within firms (as opposed to geographical communities) do not seem to offer the same opportunities for learning as local clusters where dense webs of linkages generate a real sense of 'buzz' and vibrancy (see Box 10.6). While organizational proximity may well have an important role to play, it offers only a partial substitute for geographical proximity. The value of the latter is demonstrated by real-world examples of firms with many MNCs retaining their R & D functions in their home countries, often concentrated in one particular location (Box 10.8).

Box 10.8

Proximity and learning: BMW's Research and Engineering Centre

In order to ensure that the product development process is as integrated and 'seamless' as possible, BMW – the German luxury car maker – has embarked upon a radical experiment. This involves the co-location of some 6,000 professional and technical staff at its Research and Engineering Centre near Munich in what is believed to be the largest concentration of vehicle engineering expertise in Europe. The architecture of the centre is designed to encourage collaboration, being planned to ensure that no one has to walk more than 50 metres to meet a colleague, reflecting the belief that R & D is facilitated by face-to-face interaction (Cooke and Morgan, 1998, pp.45-6). The centre brings together a much broader range of skills than conventional R & D facilities, including research, design, development, manufacturing, personnel, procurement and patents.

The overriding goal is to reduce the development cycles of new models. It is in the sphere of product development that bringing together different disciplines – normally divided into separate departments based in different locations – is paramount. As such, the BMW centre reflects the emphasis on interactive forms of innovation, compared with the linear model of old (section 10.2.2). In particular, co-location is designed to overcome the problem of tapping into existing pools of tacit knowledge, enabling the firm to harness and share such knowledge. Face-to-face interaction is regarded as conducive to the development of the shared language that facilitates the process of learning and innovation (compare Box 10.3).

10.6.3 Local buzz and global pipelines

While the balance of evidence indicates to us that spatial proximity holds advantages over more distant relationships, we know little about how local and global linkages actually shape the development of clusters. The idea of 'global pipelines', developed by a group of Scandinavian researchers, represents a promising line of enquiry (Bathelt et al., 2004). The key claim is that, in addition to engaging in processes of localized learning with a cluster, firms seek to build channels of communication or pipelines with selected partners outside the cluster (Figure 10.12). Such strategic partnerships offer access to knowledge and assets not available locally, although their number and scope is limited by the cost and time involved in building them. Successful establishment of global pipelines requires firms to develop a shared organizational context that enables them to learn and solve problems together.

At the same time, such relationships complement and enhance local linkages, rather than being a substitute for them (ibid., pp.41–2). While a firm's **embeddedness** in a cluster provides automatic and routine access to a range of information and knowledge, requiring little investment in scanning the environment for particular kinds of information, pipelines provide access to more specialized forms of knowledge that are not locally available. This specialized knowledge may relate to the development of new technologies or new market opportunities. The role of external linkages in providing access to scarce knowledge provides a basis for the hypothesis that 'the more developed the pipelines between the cluster and distant sites of knowledge the higher the quality (and value) of the local buzz benefiting all firms in the local cluster' (ibid., p.46).

Bathelt et al. (2004, p.41) suggest that wider links are particularly important during the early stages of cluster formation, providing access to markets and knowledge before critical mass is achieved locally. Maintaining such links as clusters mature is also seen as important to avoid introversion as local linkages become too close and rigid, leading to 'lock in' as firms fail to respond to change. Successful renewal of a cluster is likely to occur through the adaptation or upgrading of its industrial base, or diversification into new markets or technologies. As such, it will require external links, providing information on market opportunities, new technologies, regulatory changes, etc. In certain

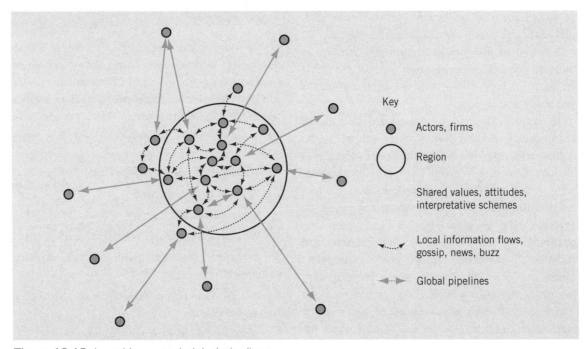

Key

○ Actors, firms

◯ Region

Shared values, attitudes, interpretative schemes

Local information flows, gossip, news, buzz

Global pipelines

Figure 10.12 Local buzz and global pipelines.
Source: Bathelt *et al.*, 2004, p.46.

circumstances, a conflict is likely to arise between the need to a) protect and maintain local knowledge and b) to access and plug into non-local networks.

If the latter becomes important, it can lead to a leakage of local tacit knowledge that is assumed to act as the source of competitive advantage for a cluster. It is difficult to see how this can be prevented, something that would require very strong identity and loyalty from key firms that ultimately have to pursue their own interests in generating profits for shareholders. As such, individual firms may follow these links outside the cluster and relocate, particularly if it starts to decline. Even locally derived capital will tend to flow to areas that offer higher profits once clusters start to decline, as indicated by the fate of 'old' industrial regions in the UK (Box 3.1).

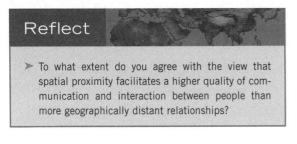

Reflect

➤ To what extent do you agree with the view that spatial proximity facilitates a higher quality of communication and interaction between people than more geographically distant relationships?

10.7 Summary

Over the last decade or so, the notion of the knowledge-based economy has shaped economic policy in developed countries. While knowledge has always been important to economic development, the pressures of global competition mean that the ability to generate and disseminate knowledge has become a crucial source of **competitive advantage**. Traditional linear models of innovation have been revised to emphasize the interactive nature of the innovation process, based on relationships between manufacturers/service providers, users, suppliers, universities and research institutes. The focus on knowledge and innovation has revived interest in the topic of spatial agglomeration, which is of enduring interest for economic geographers and spatial economists. In particular, knowledge-based aspects of agglomeration have been stressed over the traditional emphasis on cost minimization. The concepts of clusters and learning regions are underpinned by the key claim that spatial proximity allows firms to engage in processes of interactive learning involving the exchange of tacit knowledge. The latter is seen as geographically fixed and 'sticky' compared with codified knowledge, which is globally mobile. The idea that regions should promote innovation and learning in order to enhance their competitiveness in the global economy has been embraced by academic researchers and policy-makers alike.

In the previous section, we appraised the value of clusters and learning regions as theories in relation to notions of **competitiveness** and the role of wider global linkages. Clustering remains an important phenomenon and the notion of **agglomeration economies** (section 10.3) provides a basis for understanding the process. These traditional models need to be supplemented by an emphasis on the creation and exchange of knowledge through processes of innovation and learning. This is where contemporary theories of clusters and learning regions make a useful contribution. **Porter's 'diamond model'** identifies the main advantages of geographical concentration for individual firms, including enhancing productivity by offering access to specialized inputs, skills, capital and infrastructures; providing knowledge about buyer needs and the competitive environment; and encouraging new firm formation through processes of 'spin-off' in particular. At the same time, it is important to appreciate that wider global linkages – beyond those connecting firms to product markets – play an important role in the development of clusters. As the emphasis on both **local buzz** and **global pipelines** suggests, localization and globalization can be complementary rather than contradictory processes. Clustering is a dynamic process, however, and variations in the opportunities for profits offered by different locations mean that the balance between agglomeration and dispersal will change over time. The mobility of capital as it 'see-saws' between places means that some established clusters experience decline (automobiles in Detroit or shipbuilding in north-east England), while others continue to attract investment (world cities like London and New York) and new ones emerge (Silicon Valley, Cambridge).

Exercises

Select a long-established cluster of economic activity and review its development over time. Appropriate resources to draw on here are academic articles, reports by regional development agencies or local authorities, economic statistics, materials or websites produced by business associations, newspaper articles and media reports.

1. What were the origins of the cluster?

2. Why did it develop there?

3. What markets did it serve?

4. What were its sources of competitive advantage?

5. Are these consistent with Porter's diamond model?

6. Was it based on a few large firms or a network of smaller firms?

7. What were the key supporting institutions?

8. What was the role of local linkages between firms and institutions?

9. What were the main external linkages (global pipelines)?

10. Did the cluster remain competitive or experience a process of decline as it matured?

11. How would you explain this outcome?

12. How important were internal conditions (capital, skills, knowledge, infrastructure, attitudes/practices) and external factors (competition, technology, markets)?

13. If the cluster remained competitive, what processes of adjustment were involved?

14. If it experienced decline, why didn't adjustment occur.

Key reading

Cooke, P. and Morgan, K. (1998) *The Associational Economy: Firms, Regions, and Innovation,* Oxford: Oxford University Press, pp.35–59.
Written by two leading researchers, this text provides an excellent introduction to recent theories of innovation and learning regions. Chapter 2 provides an accessible account of interactive processes of innovation within and around corporations, based on the metaphor of the firm as a laboratory.

Gertler, M. (2003) 'A cultural economic geography of production', in Anderson, K., Domosh, M., Pile, S. and Thrift, N.J. (eds) *The Handbook of Cultural Geography,* London: Sage, pp.131–46.
A very insightful review of debates about cultures of learning and innovation in economic geography. In particular, Gertler argues that recent research lacks a convincing understanding of the firm as key learning agent. As a result, key questions regarding the importance of local linkages and global networks respectively in facilitating learning and knowledge transfer remain unanswered.

Henry, N. and Pinch, S. (2000) 'Spatialising knowledge: placing the knowledge community of Motor Sport Valley', *Geoforum* 31: 191–208.
One of the best examples of research on the circulation of knowledge within a particular cluster. Employing some novel techniques such as firm histories and career biographies, Henry and Pinch 'track' the movement of knowledge within Motor Sport Valley (Box 10.6).

Morgan, K. (1997) 'The learning region: institutions, innovation and regional renewal', *Regional Studies* 31: 491–504.
A key statement setting out the concept of learning regions in a clear and engaging fashion. Morgan pulls together ideas from innovation studies and research in economic geography to develop the central idea. Focusing directly on policy, he illustrates his argument with reference to EU regional policy and the implementation of a regional innovation initiative in Wales.

Morgan, K. (2004a) 'The exaggerated death of geography; learning, proximity and territorial innovation systems', *Journal of Economic Geography* 4: 3–21
This article is a response to criticisms of the key claim that the exchange of tacit knowledge depends upon spatial proximity. Morgan argues that face-to-face interaction holds benefits over more distant interaction, identifying specific flaws in the argument of the critics, informed by relevant examples from the corporate world.

Porter, M.E. (2000) 'Locations, clusters, and company strategy', in Clark, G.L., Feldman, M. and Gertler, M. (eds) *The Oxford Handbook of Economic Geography,* Oxford: Oxford University Press, pp.253–74.
An accessible account of the clusters concept from its founder. Porter defines the concept and explains the key mechanisms of growth within clusters, drawing on a range of examples. The discussion is set within a framework of firm strategy, focusing on the notion of competitiveness that he has done so much to popularize.

Useful websites

http://www.isc.hbs.edu/
The website of the Institute for Strategy and Competitiveness at Harvard Business School, led by Michael Porter. Contains a wealth of information on clusters, competitiveness and business strategy.

http://www.dti.gov.uk/regional/clusters/index.html
UK government's Department of Trade and Industry website on clusters policy. Contains a number of policy documents and links.

http://www.compete.org/nri/ncric.asp
The website of the US Council of Competitiveness – a partnership between industry, universities and trade unions – covering issues of competitiveness and regional innovation.

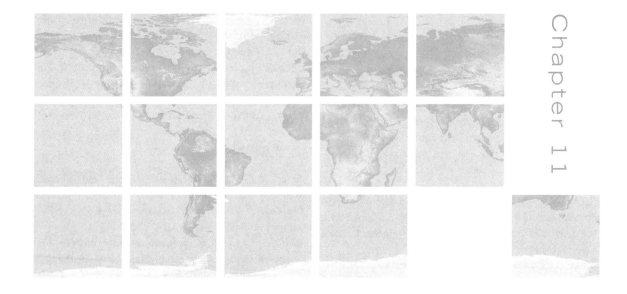

Chapter 11

Geographies of development

Topics covered in this chapter

➤ The meaning and purpose of development as a political and economic project.

➤ Key theories of development that have influenced policy and practice.

➤ The policies of the World Bank and IMF.

➤ The impact of the debt crisis and structural adjustment programmes.

➤ Global inequalities between rich and poor countries.

➤ Patterns of uneven development across the developing world.

➤ Examples of local social movements established to resist particular development projects.

➤ Contemporary development challenges and issues, focusing on trade, debt and aid.

Chapter map

After a brief introduction to the issue of development, we consider the meaning of development as a term and its history as a concept in section 11.2. This is followed by a review of the main theories of development that have influenced policy and practice. We focus on the modernization school of the 1950s and 1960s, the dependency theories of the 1960s and the neoliberal model of recent decades. Section 11.4 focuses on changing patterns of development, emphasizing the increased economic divergence between different parts of the developing world that has become apparent in recent years. We then review the growth of social movements in developing countries, highlighting examples of local groups resisting particular projects. Current development issues and challenges are discussed in section 11.6, focusing on trade, debt relief and aid. This is followed by a brief summary that highlights the main points covered in the chapter.

11.1 Introduction

Development is defined by the *Concise Oxford English Dictionary* as a 'Gradual unfolding, fuller working out; growth; evolution ... ; a well-grown state, stage of advancement; product; more elaborate form ...' (quoted in Potter *et al.*, 2004, p.4). In the context of economic policy, the term conveys a sense of positive change over time, applied to a particular country or region. Such change involves growth or progress as countries become more prosperous and advanced. As an economic and social policy, development has been directed at those 'underdeveloped areas' of the world that require economic growth and modernization. The 'underdeveloped world', which we are concerned with in this chapter, is made up of Africa, Latin America and Asia, representing the main focus of international development policy since the 1950s. The division between this periphery and the core of developed countries in Europe, North America, Japan and Australasia can be expressed as a geographical divide between the global North and South (Figure 11.1)

The problem of developing countries in the global South has attracted increased attention in recent years. The July 2005 G8 summit in Gleneagles focused on the problem of global poverty, symbolized by the plight of Africa in particular (Box 1.2). This focus on global poverty and the need for development reflects the impact of campaign groups such as Jubilee 2000 and 'Make Poverty History', prompting leading Western politicians such as Tony Blair, George Bush and Jacques Chirac to address the issue in the face of concerns about the uneven impact of globalization and the growth of international terrorism. The promotion of development as the solution to the problem of large-scale poverty in Africa, Latin America and Asia is not new, however. During the **cold war** (late 1940s to late 1980s), Western capitalist countries, led by the US, established large-scale development programmes in the newly liberated colonies to 'save' them from the 'evils' of communism, stressing the need for economic progress and modernization. While post-war development was driven by cold war geopolitics, the current interest in global poverty and development is shaped by opposing perspectives on globalization. For pro-globalists, globalization is a long-term process that will

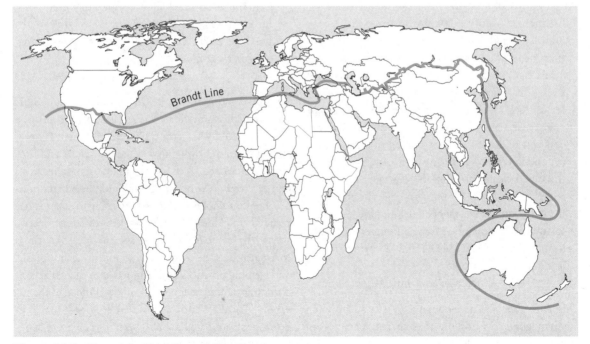

Figure 11.1 The global North and South.
Source: Adapted from Knox *et al.*, 2003, p.24.

eventually lift all people out of poverty as long as their governments pursue market-oriented policies. For 'anti-globalization' groups, however, neoliberal globalization is the problem, not the solution, and the global economic system requires major reform if not outright revolution (sections 5.2, 5.4).

11.2 The project of development

The countries of Africa, Latin America and Asia were designated as the 'underdeveloped world' in the 1950s in a context of cold war geopolitics and decolonization. As part of its efforts to break the old imperial trading blocs and create a world open for international trade and investment, the US sought to encourage economic growth and modernization in the newly liberated colonies. In his inaugural address in 1949, President Truman emphasized the need to tackle poverty through the promotion of development:

We must embark on a bold new program for making the benefits of our scientific advances and industrial progress available for the improvement and growth of underdeveloped areas ... For the first time in history, humanity possesses the knowledge and the skill to relieve the suffering of these people ... I believe that we should make available to peace-loving peoples the benefits of our store of technical knowledge in order to help them realize their aspirations for a better life ... What we envisage is a program of development based on the concepts of democratic fair-dealing ... Greater production is the key to prosperity and peace. And the key to greater production is a wider and more vigorous application of modern scientific and technical knowledge. Only by helping the least fortunate of its members to help themselves can the human family achieve the decent, satisfying life that is the right of all people.

(http://www.bartleby.com/124/pres53.html)

This 'program of development' was not only an inherent good, overcoming the 'ancient enemies' of hunger, misery and despair, it would also provide a necessary bulwark against communism, viewed as a

disease to which those living in poverty were particularly prone. As such, the former colonies in Africa, Latin America and Asia became key sites of the cold war struggle as the US and Soviet blocs competed for influence and power.

The notion of the '**Third World**' was invented as a label for the 'underdeveloped areas', in contrast to the 'First World' of Western democracies and the 'Second World' of communist states in the USSR and Eastern Europe. The term was applied to vast areas of the globe, reflecting a kind of negative labelling of them as backward and in need of assistance from outside. It is a view that has been perpetuated over the decades, supported by powerful images of poverty and underdevelopment. As the development geographer Morag Bell remarks, however, this dramatic image masks a less clear-cut reality:

What is the geography of the Third World? Certain common features come to mind: poverty, famine, environmental disaster and degradation, political instabilities, regional inequalities and so on. A powerful and negative image is created that has coherence, resolution and definition. But behind this tragic stereotype there is an alternative geography, one which demonstrates that the introduction of development into the countries of the Third World has been a protracted, painstaking and fiercely contested process.

(Bell, 1994, p.175)

The presentation of the 'Third World' as a monolithic bloc is highly simplistic and misleading, ignoring a complex and diverse geography. Development is never straightforward, involving a range of policies and programmes devised by main development organizations and received or 'consumed' by the nations, communities and households that they are aiming to assist. It has resulted in different outcomes in different places, shaped by a range of locally specific factors such as social attitudes, environmental conditions, labour skills and farming practices. It is important to recognize that economic development policy is not confined to the so-called 'Third World', as underdeveloped areas in developed countries have also been the subject of state-sponsored programmes and initiatives (sections 6.2, 6.5.4. 10.4.3, 10.5.5).

Following Truman's crucial speech and the hardening of cold war divisions, a development 'industry' emerged in the 1950s and 1960s, funded by the US and other Western capitalist countries. This was defined by a shared belief in economic planning, modern technology and outside investment as the key drivers of change. Academic disciplines, particularly economics, played a significant role in providing expert knowledge about the process of development and the conditions shaping it. International organizations such as the World Bank, International Monetary Fund (IMF) and United Nations (UN), established as part of the new post-war order, were charged with the responsibility of promoting development in poor countries. Driven by an ideology of modernization and development, such organizations sought to introduce modern knowledge and investment in order to overcome the 'ancient enemies' that held back progress in the underdeveloped lands. The UN, for instance, designated the 1960s as the 'decade of development'. Other key players in the development field included the governments of developing countries and **non-government organizations** (NGOs) working in the development field. NGOs are organizations, often of a voluntary or charitable nature, which

that make up the so-called 'third sector' belonging to neither the private nor public sectors. Oxfam is a good example of an NGO focused on development (Box 11.1). Others include Christian Aid, Save the Children and the Red Cross.

Economic aspects of development were heavily stressed in the 1960s and 1970s as policy focused on the need to foster investment and growth at the national level, assuming that this would generate increased incomes and employment opportunities for individuals. Development indicators concentrated on the material well-being of a country or region, with gross domestic product (GDP) and gross national product (GNP) by far the most commonly used measures of development. This corresponds to the quantitative change section of Figure 11.2. Since the late 1970s, however, the notion of development has broadened to include more qualitative aspects of change, encompassing broader social and political goals such as quality of life, choice, empowerment and human rights (Potter *et al.*, 2004, pp.11–12). The work of the renowned Indian-born economist Amartya Sen, winner of the Nobel Prize for Economic Science in 1998, has been particularly important in advancing the

Box 11.1

The development of Oxfam

Oxfam has grown to become one the most important NGOs operating in the field of development. It is an independent British organization, registered as a charity, working with partners, volunteers, supporters and staff of many nationalities. It employs about 4,700 staff working overseas and 1,700 staff in Great Britain, as well as 23,000 volunteers working in more than 750 shops.

Oxfam's origins lie in the Second World War when the Oxford Committee for Famine Relief was set up in response to famine in Greece in 1942. The committee remained in existence after the war, focusing on 'the relief of suffering in consequence

of the war' in Europe in the late 1940s. After 1949, the scope of its operations expanded to encompass the world. In the 1960s, Oxfam's income trebled, reflecting growing concern for the world's poor. Support for self-help schemes, whereby communities improved their own farming practices, water supplies and health provision became the major focus of work. Community involvement and control remain key principles of Oxfam's work.

As well as famine relief and self-help schemes, Oxfam campaigns to raise awareness of global poverty and its structural causes in terms of the debt burden, unfair terms of trade and

inappropriate agricultural policies. In 1995, it opened an office in Washington DC in order to lobby the World Bank, the IMF and UN more effectively. Its strategic review identified four key priorities for the 2003–6 period: sustainable livelihoods, allowing poor people to achieve food and income security and secure employment; basic social services, with access to education, especially for girls, emphasized particularly strongly; life and security, focusing on providing an improved response to conflict and natural disasters; and gender and diversity, integrating gender issues into all aspects of operations.

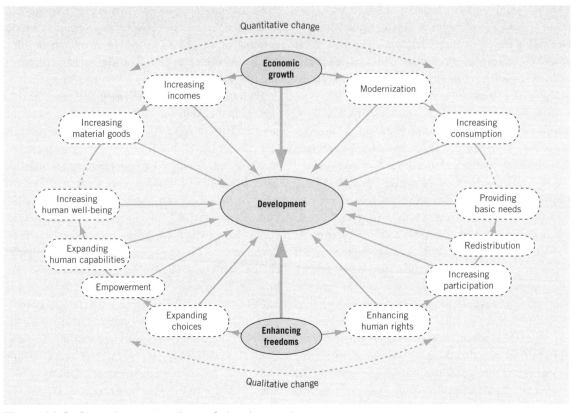

Figure 11.2 Changing conceptions of development.
Source: Geographies of Development, 2nd edn, Potter, R.B., Binns, T., Elliott, J.A. and Smith, D., 2004, p.16. Pearson Education Limited.

idea of 'development as freedom'. Through factors such as better education, increased political participation and free speech, working alongside the process of economic growth, people are liberated from 'unfreedoms' such as starvation, undernourishment, oppression, disease and illiteracy. This broader conception of development is reflected in the evolution of development indicators, with more emphasis now focused on assessing social and political aspects of development alongside the economic dimension (Box 11.2).

Box 11.2

Measuring development

As indicated above, the main economic measures of development are GDP and GNP, usually expressed on a per capita basis. GDP is a measure of the total value of goods and services produced within a country while GNP also includes income generated from investments abroad, but excludes profits repatriated by foreign MNCs to their home countries. GDP and GNP were the key measures employed by international development agencies in the post-war era, but began to attract criticism from development activists and analysts in the late 1960s and 1970s for neglecting social aspects of development. They remain important, however, providing a useful summary measure of development, emphasizing the divide between the global North and South, and growing divergence between the regions of the South (section 6.4).

A wide range of social indicators of development were published in the 1970s and 1980s, focusing on issues such as poverty, education, health and gender. The proliferation of these social indicators threatened to generate considerable confusion,

Box 11.2 (continued)

however, as different measures could be used to show different things, and it was almost always possible to find some statistics to 'prove' a particular argument (Potter *et al.*, 2004 p.10). What seemed to be required was some kind of summary measure constructed out of key economic and social indicators.

The development of the **Human** **Development Index** (HDI) by the United Nations Development Programme (UNDP) – published annually since 1990 in the *Human Development Report* – has meet this need, becoming widely used and adopted. The HDI 'measures the overall achievement of a country in three basic dimensions of human development – longevity, knowledge and a decent standard of living' (UNDP, 2001, p.14). The specific measures used are life expectancy, educational attainment (adult literacy and combined primary, secondary and tertiary enrolment) and GDP per capita in US dollars (Figure 11.3). A separate ratio is calculated for each of the three dimensions, where the actual value is divided by the maximum possible one

Dimension	A long and healthy life	Knowledge		A decent standard of living
Indicator	Life expectancy at birth	Adult literacy rate Adult literacy index	Gross enrolment ratio (GER) GER index	GDP per capita (PPP US$)
Dimension index	Life expectancy index	Education index		GDP index
		Human development index (HDI)		

Figure 11.3 Calculating the human development index (HDI).
Source: UNDP, 2005, p.340.

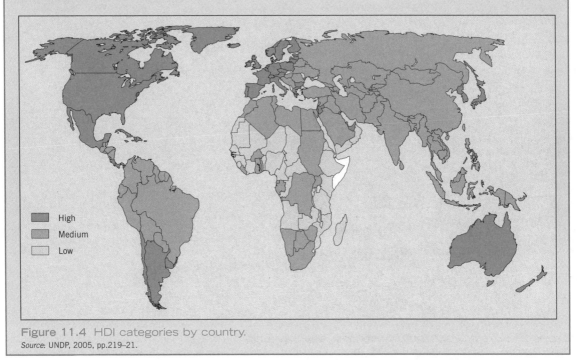

Figure 11.4 HDI categories by country.
Source: UNDP, 2005, pp.219–21.

High
Medium
Low

Box 11.2 (continued)

(for life expectancy, 85; for GDP per capita US$ 40,000), giving a value between 0 and 1. The HDI is then calculated as a simple average of the three dimension indices (UNDP, 2005, p.340). In recent years, is has been supplemented by the development of other related indexes, including the Human Poverty Index, the Gender-related Development Index and the Gender Empowerment Measure (GEM). Countries with HDI scores above 0.8 are classified as high human development; ones with scores between 0.5 and 0.8 as medium human development; and ones with scores below 0.5 as low human development (Figure 11.4).

Reflect

➤ To what extent should development be led by external Northern agencies such as the World Bank, IMF and NGOs?

11.3 Theories of development

11.3.1 The modernization school.

This approach was dominant in the 1950s and 1960s, shaping and informing the efforts of development planners and agencies (Table 11.1). It emerged from the writings of theorists and experts based in the global

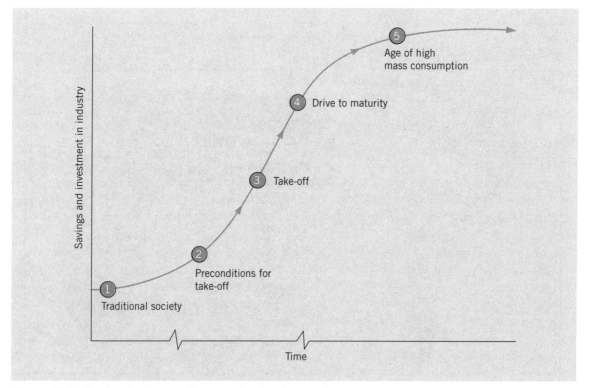

Figure 11.5 Rostow's stages of economic development.
Source: Geographies of Development, 2nd edn, Potter, R.B., Binns, T., Elliott, J.A. and Smith, D., 2004, p.91. Pearson Education Limited.

Table 11.1 Key theories of development

Theory	Key theorists	Time frame	Main argument	Recommendations	Problems
Modernization theory	Rostow.	1950s and 1960s	Development occurs in distinct stages; developing countries undergo linear process of modernization, akin to developed countries in the 19th century.	Need for external funds and expertise, along with modern planning and investment methods, to generate growth.	Eurocentric; ignores structure of world economy; growth fails to alleviate poverty.
Structuralism and dependency theory	Prebisch, Frank.	1960s and 1970s	Metropolitan core of world economy exploits 'satellites'. Development and underdevelopment are opposite sides of the same coin.	Import substitution or withdrawal from world economy.	Crude and static view of global relationships. Undermined by experience of East Asian NICs.
Neoliberalism	Lal and Balassa in development circles.	1980s and 1990s	Developing countries should reduce state intervention in the economy and embrace the free market.	Economic reform through SAPs (structural adjustment programmes), free trade, promotion of exports.	Focus on growth failed to alleviate poverty; SAPs and privatization led to cuts in public services, and the introduction of user charges.
Grass-roots approach	Emerged from activities of development NGOs and activists.	1970s onwards	Development agencies should focus on meeting the everyday needs of the poor.	Local participation and self-help should be encouraged.	Limited funds. Tends to alleviate the symptoms of poverty rather than addressing the causes.

North, notably the US, which was playing a key role in funding and supporting the project of development as the leading economic power in the Western capitalist bloc. The work of the US economist Walt Rostow was particularly influential, setting out a model of the **stages of economic growth** that was applied to the situation of developing countries in the 1960s (Figure 11.5). The process of take-off is triggered by a combination of external and internal factors such as technological innovation, increased rates of investment and saving, the development of modern banks and manufacturing industry. The metaphor invoked here is that of an aeroplane accelerating along a runway before gathering power to become airborne (M. Power, 2005, p.192).

The basic idea is that developing countries undergo a linear process of transformation (modernization),

analogous to the changes experienced by developed countries in the nineteenth century following the Industrial Revolution. The assumption was that developing countries would simply follow this existing model, with the outcomes of modernization being the consolidation of Western norms of economic and social organization (e.g. a large manufacturing sector, commercialized agriculture, the importance of class groups rather than family or tribal structures, and governments based on democratic election rather than tribal or religious loyalties (Willis, 2005, p.189)). The process of modernization would be driven by the injection of external funds and expertise, coupled with national government intervention and planning to mobilize resources and stimulate investment. Economic growth was paramount, generating increased income and employment opportunities that, it was assumed, would 'trickle down' to the poorest groups in society.

From this perspective, the problems of underdevelopment are internal to the countries concerned, reflecting the attachment to traditional values and the absence of modern technology and scientific knowledge. As such, the solution was to import these factors from outside, giving Western experts and agencies a key role in assisting development. When linked to initiatives and reforms carried out by developing country governments, this would generate the momentum required for take-off. The **Eurocentric** assumption that Western values and methods are always superior to those found in developing countries has been criticized as arrogant and condescending.

The focus on internal factors means that the structure of the world economy, particularly in terms of the relationships between richer and poorer countries, was neglected by modernization theorists. It is simply assumed that it is just as easy for developing countries to develop as it was for developed countries in the nineteenth century. This ignores the fact that the latter were not required to compete with a group of already industrialized countries that dominated world production of manufactured goods and services. As such, the policies followed by Europe and North America were not necessarily appropriate to the problems confronting developing countries in the 1960s and 1970s.

11.3.2 Structuralism and dependency theory

The idea that the existing structure of the world economy was impeding the development of 'Third World' countries was a key point of departure for

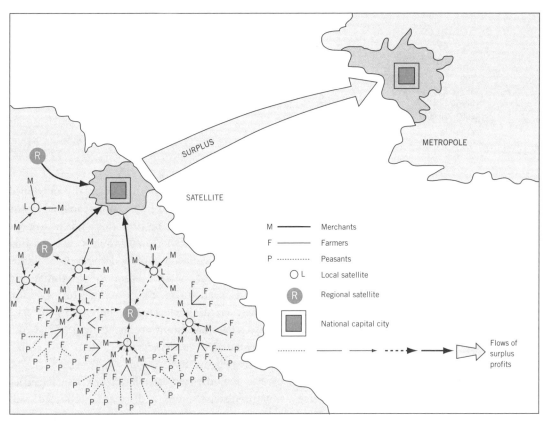

Figure 11.6 Dependency theory.

Source: Geographies of Development, 2nd edn, Potter, R.B., Binns, T., Elliott, J.A. and Smith, D., 2004, p.111. Pearson Education Limited. Adapted from R.B. Potter, *Urbanisation in the Third World* (OUP, 1992). By permission of Oxford University Press.

radical critics of modernization theory in the 1950s and 1960s. These critics adopted a Marxist approach, focusing on the political economy of economic development. In particular, they sought to apply Marx's insights about the historical and geographical unevenness of capitalist development to the experience of developing countries in the post-war period (M. Power, 2005, p.193). In contrast to the modernization school, **structuralist or dependency theories** emanated from the global South, being particularly associated with a group of theorists and activists based in Latin America.

This approach can be termed structuralist for its characteristic focus on the structure of the world economy, particularly the relationship between developed countries and developing countries (see Table 11.1). It focused on the mode of incorporation of individual countries into the world economy, viewing this as a key source of exploitation. From this perspective, the causes of poverty and underdevelopment are external to developing countries, stemming from the relationship between them and the wider world economy. According to Andre Gunder Frank, the most influential of the dependency theorists in the 1960s, the metropolitan core exploits its 'satellites', extracting profits (surplus) for investment elsewhere (Figure 11.6). Colonialism was a key force here, creating unequal economic relations that were then perpetuated by the more informal imperialism characteristic of the post-war period.

Many developing countries have found it difficult to overcome the legacy of colonialism, involving them in exporting primary commodities (agricultural goods, minerals, fuels) while the European powers exported manufactured goods. Over time, the price of the former have tended to fall relative to the latter, reducing the export earnings of developing countries and making them less able to pay for imports. At the same time, MNCs based in Europe and North America have been able to repatriate the profits from their plantations and factories in developing countries to their home countries, paying low wages and having very few links with the host economy in which they are operating.

The implications of dependency theory are that developing countries should protect themselves against external forces or even withdraw from the world economy altogether. Development and underdevelopment are opposite sides of the same coin; closer links between core and periphery merely widen the economic gap between them. The milder form of structuralism associated with Raul Prebisch and the UN Commission for Latin America in the 1950s advocated that countries should follow protectionist policies rather than simply embracing external assistance (Willis, 2005, p.191). In particular, they should focus on **import substitution industrialization** (ISI), seeking to develop domestic industries to produce goods that are currently imported and using import tariffs to protect them against competition from the established industries of Europe and North America (section 6.4.1). For the radical version of dependency developed by Frank, however, the solution was withdrawal from the global economy and the creation of alternative forms of society based on socialism (ibid., p.193).

Dependency theories have been heavily criticized, relying, again, on a simple divide between 'core' and 'satellite' economies. Their view of global relationships is static and crude, assuming that the patterns established under colonialism will inevitably persist. This assumption was undermined by the experience of rapid economic growth in the emerging **newly industrialized countries (NICs)** of East Asia such as South Korea, Taiwan and Singapore in the late 1960s and 1970s (Box 5.3). The success of these countries was based on policies of **export-oriented industrialization** (EOI), where industrial development is based on serving international rather than domestic markets. As such it seemed to undermine some of the key foundations of dependency theory, indicating that countries could overcome the legacy of colonialism. By contrast, the notion of withdrawal from global markets seemed impractical, promising only further economic stagnation and decay.

11.3.3 Neoliberalism

Since the late 1980s, development policy has been shaped by **neoliberal** (free market) theories (section 6.5.1). These emphasize the need to reduce government intervention in the economy, encouraging the development of private enterprise and competition. The neoliberal principles developed by key theorists

Box 11.3

The debt crisis

The **debt crisis** has been a key factor shaping North–South relations over the past 25 years, with many developing countries struggling to service loans originally taken out in the 1970s. It first came to attention when Mexico defaulted on its loans in August 1982, with other Latin American countries such as Brazil and Argentina also experiencing major problems. Much of sub-Saharan Africa and parts of Asia have also been severely affected. The debt crisis has not only seriously undermined development efforts, requiring developing countries to spend their limited export earnings on servicing debts; it has also threatened the viability of the world financial system, forcing Northern governments and bodies such as the IMF and World Bank to intervene. The impact of debt on development has been empha-

sized by development activists and groups such as Jubilee 2000 over the last decade or so and feeding into the 'Make Poverty History' and 'Live 8' campaigns, which brought the issue into the public consciousness.

The origins of the debt crisis lie in the interactions between three sets of factors:

➤ The borrowing of large sums by developing countries from Northern banks and institutions in the 1970s. Following the sharp rise in oil prices that occurred in 1973, the world economy was awash with funds invested by the oil-exporting companies. These so-called 'petrodollars' were lent by leading banks and government bodies to developing countries to pay for oil imports and to finance large-scale industrial development programmes.

➤ Most of the loans were made for a period of 5–7 years, denominated in US dollars and subject to floating interest rates (Corbridge, 2002, p.477). In the late 1970s and early 1980s, interest rates rose markedly, following the introduction of monetarist policies in the US and the UK particularly. For example, the London Interbank Offered Rate, the main index of the price of an international loan, rose from an average of 9.2 per cent in 1978 to 16.63 in 1981 (ibid., p.478). As a result, developing countries were faced with hugely increased debt repayments and the world economy was plunged into a recession.

➤ The collapse of commodity prices in the early 1980s, 'such that in 1993 prices were 32 per cent lower than in 1980; and in

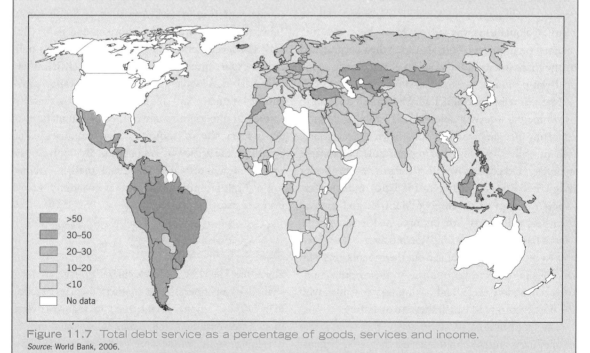

>50

30–50

20–30

10–20

<10

No data

Figure 11.7 Total debt service as a percentage of goods, services and income.
Source: World Bank, 2006.

Box 11.3 (continued)

relation to the price of manufactured goods, they were 55 per
cent lower than in 1960' (Potter
et al., 2004, p.355). This
amounted to a serious deterioration in the terms of trade for
developing countries, reducing
their export revenues relative to
the price of imported goods. As a
result, they were faced with a
'scissors' crisis of declining export
revenues and mounting debt
repayments in a strengthening
dollar (Corbridge, 2002, p.478).
This is what led to a number of

countries defaulting, threatening
the viability of the international
financial system.

The scale of the debt crisis seemed
greater in Latin America in the
1980s, but the difficulties facing
developing countries in sub-Saharan
Africa particularly have become
apparent since the early 1990s.
Figure 11.7 shows debt levels relative
to the size of a country's economy,
represented by GDP, indicating its
capacity to service its debt. While
Latin American and Asian countries

often have the largest absolute debts,
it is African countries that are worst
affected in terms of the relationship
between debt and GDP. Africa's
indebtedness is estimated to have
trebled to around $US 216 billion
between 1980 and 1999, increasing
from approximately 28 per cent of its
GDP to 72 per cent, compared with
40 per cent for Latin America (Potter
et al., 2004, p.356). In some cases,
annual repayments to creditors –
around $15 billion every year – out-
weigh expenditure on education and
health.

---

such as Friedmann and Hayek in North America and
Western Europe were translated into development
thinking in the early 1980s, following the counter-
revolution in development economics when neoliberal
ideas overturned the Keynesian orthodoxies of the
1960s and 1970s. Key figures included Depak Lal and
Bela Balassa, both based at leading US universities, who
argued in favour of free trade and the application of
standard economic (neoclassical) principles to the
developing world (Peet and Hardwick, 1999,
pp.49–50).

Neoliberal principles have underpinned the so-
called **Washington Consensus**, which has structured
economic development policy since the early 1990s
(section 6.5.2). Key strands include a focus on low
inflation, the reduction of barriers to trade, openness
towards foreign direct investment (FDI), the liberalization of the financial sector and the privatization of
state enterprises. The World Bank and IMF have played
a crucial role in requiring developing countries to sign
up to this policy agenda. In the context of the debt
crisis, developing countries were often in desperate
need of further financial assistance from those organizations (Box 11.3). This allowed the World Bank and
IMF to set conditions requiring developing countries to
reform their economies. These reforms have become
known as **structural adjustment programmes** (SAPs),

encompassing a range of measures requiring countries
to open up to trade and investment and to reduce
public expenditure (Box 11.4). Developing countries
should seek to compete in the global market through
the development of competitive export sectors, with
the experience of the East Asian NICs often cited in
support of this argument. In a similar fashion to modernization theory, neoliberalism is based on the
imposition of a set of externally derived solutions from
the global North through mechanisms such as SAPs
(Willis, 2005, p.194).

SAPs were refined in the late 1980s and 1990s, taking
better account of local needs and circumstances, with
the longer-term adjustment element rebranded as
economic recovery plans (ERPs) (p.89). Since 1999,
SAPs have been replaced by **poverty reduction strategies** (PRSs), which require national governments to
produce a comprehensive plan for reducing poverty,
requiring consultation with the World Bank and IMF,
NGOs and local communities. The World Bank has
stated its commitment to work more closely with
NGOs to foster democracy and local empowerment
within developing countries (Potter *et al.*, 2004, p.294).
While this has prompted talk of a 'new post-
Washington poverty consensus' (Maxwell, 2005),
economic and political reforms based on neoliberal
principles are still seen as central to the process, based

Box 11.4

The development of structural adjustment programmes (SAPs)

The central aim of SAPs is debt reduction through reform packages designed to enable countries to pay their debts while maintaining economic growth and stability. Agreement to introduce a SAP became an essential condition for developing countries to obtain finance from the World Bank, IMF and other private donors, a condition that few countries were in a position to refuse. As a result, they rapidly spread across the global South. The first SAP was introduced in Turkey in 1980 and 187 had been negotiated for some 64 developing countries by the end of the decade (Potter *et al.*, 2004, p.290).

SAPs were based on four main objectives (Simon, 2002, p.88):

➤ The mobilization of local resources to foster development.

➤ Policy reform to increase economic efficiency.

➤ The generation of foreign revenue through exports, involving diversification into new products as well as expansion of established ones.

➤ Reducing the economic role of the state and ensuring low inflation.

The measures required to achieve these objectives are generally divided into two types (ibid., p.88.)

1. **Stabilization measures**, which were immediate steps designed to address the economic difficulties facing developing countries in the short term, providing a foundation for longer-term measures.

   ➤ A public sector wage freeze – to reduce wage inflation and the government's salary bill.

   ➤ Reduced subsidies on basic foods and other commodities, and on health and education – to lower government expenditure

   ➤ Devaluation of the currency – to make exports cheaper and more competitive, and to deter imports.

2. **Adjustment measures**, which were to be implemented as a second phase, having a longer-term impact. Their objective was to ensure the structural adjustment of the economy, creating a platform for future growth.

   ➤ Export promotion through incentives for enterprise (including increased revenues and access to foreign currency) and diversification.

   ➤ Downsizing the civil service through the retrenchments following a programme of rationalization to reduce 'overstaffing', duplication, inefficiency and cronyism.

   ➤ Economic liberalization – relaxing and removing regulations and restrictions on economic activities. Examples include import tariffs and quotas, import licences, state monopolies, price fixing, subsidies and restrictions on the repatriation of profits by overseas firms.

   ➤ Privatization through the selling off of state enterprises and corporations.

   ➤ Tax reductions to create stronger incentives for individuals and firms to save and invest.

SAPs proved highly controversial. Their impacts have often been harsh, with ordinary people rather than elites bearing the brunt of the adjustment costs (Simon, 2002, p.89). In general, large traders, merchants and rural agricultural producers have benefited from increased export opportunities, often at the expense of the urban poor, who have suffered from the abolition of food subsidies and reductions in public expenditure. The effects of SAPs have been felt through cuts in public services, privatization and the introduction of user charges for services. In Ghana, for instance – regarded as a 'star pupil' by the World Bank and IMF – the introduction of user charges for health services in 1985, in a climate of failing wages and increased poverty, led to a drop in hospital and clinic visits of 25–50 per cent in urban Accra and of 45–80 per cent in certain rural areas (Konadu-Agyemang, 2000, p.474). The privatization of water in Ghana has led to higher water rates in a situation where 35 per cent of the population already lack access to safe water. There is also evidence that economic reform has benefited foreign interests over domestic producers. Large-scale privatization and liberalization of Ghana's lucrative gold mining sector has benefited outside MNCs, attracted by tax breaks and incentives, with 70–85 per cent of the industry now foreign-owned. Minimal environmental regulations have led to widespread contamination and degradation, while 30,000 people were displaced during 1990–98 (Ismi, 2004, p.17). According to many critics, SAPs have reinforced inequality and poverty by reducing access to crucial social services and opening up developing economies to outside interests such as MNCs, although some local traders and farmers have benefited (Simon, 2002).

on the assumption that such reforms will eventually lead to a reduction in poverty (Simon, 2002). Kenya was one of the first countries to experience the shift from SAPs to PRSs after IMF support was suspended in 1998 because of the government's failure to adhere to the terms of its SAP. The resumption of support was made dependent on the production of poverty reduction strategy papers by the government during 2000, with the emphasis on economic reform prompting widespread criticism from development NGOs and activists in Kenya (ibid., pp.90–1). As this indicates, central elements of the neoliberal model remain in place, continuing to shape the policies of the World Bank, IMF and other organizations such as the EU into the twenty-first century.

11.3.4 Grass-roots development

This approach is rather different in nature and orientation from the other theories discussed above. Instead of advancing a set of overarching prescriptions about development policy, based on abstract economic analyses, it is directly concerned with the practical problems and needs of poor people in developing countries, the ultimate targets of development aid (see Table 11.1). While the previous models are designed and implemented in a top-down fashion, focusing on the national scale, grass-roots development, as the term implies, is focused on the local level. In addressing local need in a 'bottom-up' manner, this approach is based on the observation that increased economic growth does not necessarily reduce poverty; in some cases, growth may widen inequalities within society, benefiting some groups at the expense of others. NGOs are closely associated with grassroots strategies, often working in partnership with local agencies and groups (Townsend *et al.*, 2004).

This approach to development has evolved from the **'basic needs'** strategies of the 1960s and 1970s. These focused attention on the everyday lives of the poor, reflecting concerns that they were being neglected by orthodox modernization policies. Recent work shares the same concern for identifying and meeting such needs in relation to food, shelter, employment, education, health, etc. The ethos is one of 'helping people to help themselves', involving small-scale projects that directly benefit individuals, families and households, supporting local services and livelihoods. In urban areas, they are often based on the large informal sector

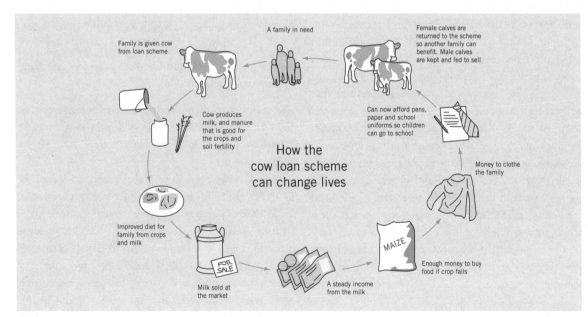

Figure 11.8 Oxfam cow loan scheme.
Source: Willis, 2005, p.196.

of the economy, supporting the efforts of small-scale producers and traders to survive. An example of an urban grass-roots initiative is the partnership of an NGO and two community-based organizations in Mumbai, India, known as The Alliance, which aims to organize and mobilize poor people in slum areas to pursue better housing and sanitation facilities (McFarlane, 2004, pp.891–4). In rural areas, farming is usually the focus of attention, with the establishment of schemes offering targeted assistance and support. An example is Oxfam's cow loan scheme whereby families are given cows from the loan scheme, fertilizing crops and providing milk, which not only provides sustenance for the family, but can be sold to generate an income, allowing them to purchase food and clothes (Figure 11.8) (Willis, 2005, p.196). Calves are returned to the scheme to be loaned out to other families or sold.

The main constraint on grass-roots development is the limited resources available to NGOs and other bodies. While the number of NGOs has increased significantly since the 1970s, their resources are still outweighed by the scale of problems confronting them. Furthermore, as much of their money is derived from donations by individuals and governments, projects often reflect the concerns of donors rather than the priorities of local people. At the same time, the proliferation of small-scale local projects can sometimes be associated with a lack of collaboration and integration between agencies and initiatives. Grass-roots development could be said to focus on alleviating the symptoms of poverty rather than addressing the underlying causes. This helps to explain some of the World Bank's and IMF's reluctance to embrace such approaches, concentrating instead on economic reform programmes. Such a view is rather harsh, however, with NGOs and others doing much valuable work in meeting local needs and highlighting the failure of mainstream initiatives to reach those most in need. The adoption of explicit poverty reduction strategies by the World Bank and others in recent years suggests that such concerns have filtered into the mainstream agenda.

> **Reflect**
>
> ➤ Which of the theories offers the most appropriate model for developing countries? Do elements of different theories need to be combined? If so, which ones? Justify your answers.

11.4 Patterns of development

11.4.1 An unequal world

Since the 1970s, despite many setbacks and difficulties, some modest progress has been made in improving conditions in poorer countries. While these improvements are overshadowed by the sheer magnitude of the gap between rich and poor countries, their significance should not be minimized. Key outcomes include (UNDP, 2001, p.10):

➤ Average incomes in developing countries almost doubled, from US$1,300 to US$2,500 between 1975 and 1998.

➤ Life expectancy increased by 8 years between 1970 and 2000.

➤ The adult literacy rate increased from an estimated 47 per cent in the 1970s to 73 per cent in 1999.

➤ The share of rural families with access to safe water grew more than fivefold.

The number of people living in extreme poverty (less than $1 a day) fell from 1.276 billion in 1990 to 1.072 billion in 2003 (28 to 21 per cent) (UNDP, 2005, pp.20, 42).

At the same time, huge global inequalities remain. GDP per capita, for instance, ranges from an average of US$29,898 (purchasing power parities) for high-income economies to US$2,168 for low-income economies in 2003, using the World Bank's categories (ibid., p.222). The gradual decline in child mortality rates has slowed since 1990, and the gap between rich and poor countries is widening (ibid., p.28). Convergence in life expectancy has ceased since 1990 in sub-Saharan Africa, due to the impact of HIV and

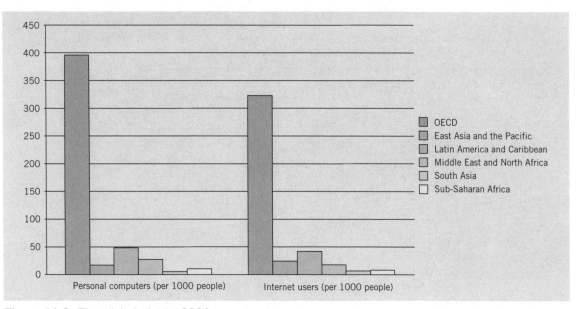

Figure 11.9 The digital divide, 2000.
Source: World Bank, 2006.

AIDS, and the former Soviet Union experiencing actual reversals (World Bank, 2005, p.6). International inequality has declined since the 1950s, thanks largely to rapid growth in populous countries such as China and India since the 1970s (ibid., p.62). This measure takes no account of the distribution of income within countries, however, which seems to have become more unequal in recent decades (ibid., pp.63, 68).

One aspect of global inequalities that has received much attention in recent years is the so-called **digital divide**, with developing countries having very limited access to ICTs compared with developed countries. This is of particular concern since the concentration of new technologies in richer countries is likely to accentuate existing patterns of uneven development. In the late 1990s, there were more telephone lines in Manhattan than in the whole of sub-Saharan Africa (Potter *et al.*, 2004, pp.140–1). Figure 11.9 shows the number of personal computers and Internet users in 2000, emphasizing the massive gap between the OECD and the developing world. Among developing regions, South Asia and sub-Saharan Africa are particularly badly off. The scale of these inequalities is confounding optimistic arguments about the growth of new technologies such as the Internet encouraging economic

development and education in poorer countries. Instead, access to such technologies is conditioned by pre-existing levels of income, effectively reproducing and accentuating established forms of inequality and marginalization in the world economy (Leyshon, 1995).

11.4.2 Divergence between developing regions

It is important to recognize that not all developing countries are equal in terms of economic development. Since 1960 there has been increased divergence between the major regions of the developing world, symbolized by the contrasting performance of East Asia and sub-Saharan Africa (Figure 11.10). The different regions were close together in 1960, with their income standing at about one-ninth/one-tenth of the high-income economies belonging to the Organization for Economic Cooperation and Development (OECD). The main exception was Latin America and the Caribbean, which enjoyed significantly higher income levels, at about one-third/one-half of the OECD level. East Asia and the Pacific experienced strong growth from the early 1970s, progressing to almost one-fifth of

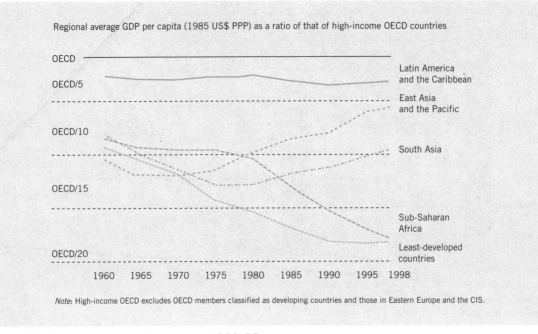

Figure 11.10 Income trends by region, 1960–98.
Source: UNDP, 2001, p.16.

OECD income levels. The position of South Asia worsened in the 1960s and 1970s, before improving from the early 2000s, bringing it back up to about one-tenth of the OECD average.

At the same time, a dramatic decline occurred in sub-Saharan Africa, which plunged from about one-ninth of OECD countries' income in 1960 to one-eighteenth in 1998. Although this is not shown in Fig 11.10, the position of many transition economies in the former Soviet bloc has also worsened since the late 1980s. In absolute terms, the GDP per capita of sub-Saharan Africa grew by 36 per cent between 1960 and 1980 but fell by 15 per cent in the 1980–1990 period (Ismi, 2004, p.11). The divergence between the performance of East Asia and Africa is highlighted by the contrasting experience of Ghana and Malaysia (Box 11.5).

Whereas the economic growth in East Asia was explored in sections 5.4 and 6.4, the decline of sub-Saharan Africa since 1980 requires some explanation. While the efforts of the World Bank and IMF have focused on internal inefficiencies and failures, external relationships are also crucial. The region's decline certainly does not reflect its isolation from the world economy; instead, trade expanded considerably in the

1990s, as did FDI, reflecting the impact of SAPs. But African exports remain based on raw materials for which prices have been falling. Copper, for example, accounts for 70 per cent of Zambia's exports, and coffee for 73 per cent of Burundi's (BBC, 2004). Non-oil commodity prices have dropped by an average of 35 per cent since 1997 (Ismi, 2004. p.11), leading to declining terms of trade as export revenues fell relative to the price of imported goods. At the same time, under SAPs, Africa's external debt has actually increased by more than 500 per cent since 1980 (ibid., p.12). Alongside cuts in public expenditure, this results in Africa spending four times more in debt servicing than health. At the same time, HIV and AIDS has had a devastating impact with an estimated 28 million of the 40 million sufferers worldwide resident in sub-Saharan Africa (Potter *et al.*, 2004, p.209).

Reflect

➤ Is it still meaningful to talk of a 'Third World' in view of the increased economic divergence between developing regions and the end of the cold war?

268

Box 11.5

Contrasting experiences of development: Ghana and Malaysia

Both Ghana and Malaysia are former British colonies, which gained independence in 1957 and 1961 respectively. Both had fairly typical colonial economies, based on the export of raw materials, particularly cocoa and gold in Ghana and rubber, tin and timber in Malaysia. While poor, both were probably better placed than many other former colonies, with Malaysia enjoying higher incomes than many of its neighbours and Ghana regarded as one of the best prospects in sub-Saharan Africa in terms of the high value of exports, a well-developed infrastructure and a relatively skilled and educated workforce (Khan, 2002; Konadu-Agyemang, 2000).

Since 1960, however, the economic fortunes of the two countries have diverged dramatically, with Malaysia experiencing sustained economic growth while Ghana has suffered prolonged stagnation and decline (Doyle, 2005), before undergoing something of an upturn over the

last few years. Malaysia was ranked 61 in the UN's Human Development Index in 2003, with a GDP per capita (PPP US$) of US$9,512, compared with Ghana, ranked 138 with GDP per capita of US$2,238, although it is a few places above the bottom category of 'low human development' (UNDP, 2005, p.221). Ghana remains dependent on the export of raw materials, particularly cocoa and gold, while Malaysia has diversified successfully, producing a range of industrial goods, including cars, and advanced services.

How can this contrasting economic performance be explained? The role of the state is clearly of central importance. Malaysia has enjoyed four decades of political stability, with one politician, Dr Mahathir Mohamad, occupying the position of prime minister for an unbroken 20 years. This has offered a stable climate for investment and growth, with the government pursuing a consistent policy of rapid export-led

growth and poverty reduction (Khan, 2002). At the same time, it retained the flexibility to adapt the mechanisms through which this policy was implemented as global economic conditions changed. Ghana, by contrast, soon lapsed into political instability with the first of a series of military coups occurring in 1966 (Doyle, 2005). Corruption and economic mismanagement became rife, coupled with an overvalued currency and unfavourable terms of trade (Konadu-Agyemang, 2000). Malaysia retained control of its economic destiny, while high levels of debt and inflation, coupled with falling export revenues, meant that Ghana has been subject to a succession of World Bank and IMF-sponsored adjustment programmes. These have had the effect of reducing the access of the poor to social services and increasing foreign ownership and control of the economy, although Ghana's economic performance has certainly improved since the late 1990s (OECD, 2006a).

11.5 Resisting development: the growth of local social movements in developing countries

Orthodox forms of development based on the promotion of economic growth and open markets have often failed to reach those groups most in need of assistance. As we saw, neoliberal models of development based on the principles of deregulation, liberalization and privatization have often reduced the access of the poor to vital public services and increased

externally owned MNCs' control over local resources and services. As part of this project of development, Western values and expertise have been privileged over local knowledge and culture (Routledge, 2005, pp.211–12). The agency of local people in developing countries, viewed simply as passive subjects of development, is often ignored by mainstream approaches. This provides the context for assessing the activities of certain **local social movements** in resisting particular aspects of neoliberal globalization and the development projects undertaken by individual states. The activities of these groups have helped to fuel the growth of the **anti-globalization movement** (section 5.7), attracting the interest of researchers and activists in Northern countries.

According to the political geographer Paul Routledge, local social movements in developing countries are highly diverse, incorporating

> squatter movements, neighbourhood groups, human rights organizations, women's associations, indigenous rights groups, self-help movements amongst the poor and unemployed, youth groups, educational and health associations and artists' movements.
>
> (Routledge, 2005, p.214)

Such groups articulate an 'environmentalism of the poor', concerned with defending local livelihoods and maintaining communal access to local resources in face of their commodification or appropriation by the state and corporations and with freeing people from extreme poverty and domination by other groups (e.g. large landowners or private logging companies).

Local social movements in developing countries can be seen as expressions of conflict between different groups in society over the control and use of space. While local groups wish to retain control of local economic resources such as land, forests and water, utilizing them to meet their material needs on a day-to-day basis, states and private interests often want to exploit them for economic gain, threatening the basis of local livelihoods. Deforestation schemes, the construction of dams for hydroelectric power and irrigation and industrial development are often the focus for conflicts that can be seen as examples of 'resource wars' between the different parties. Many of the local resistance movements are place-based, asserting local values and identity in the face of knowledge and power of states and private corporations. One of the effects of policies of privatization and liberalization has been to make it easier for external corporations and interests to gain control over local resources, resulting in increased conflict with local groups. While most movements are locally based, their struggles have often assumed a wider dimension, involving imaginative use of the media and Internet to gain international support.

A number of examples of local social movements in developing countries can be highlighted. One of the most high-profile and successful in attracting international interest and support is the **Zapatista guerrilla movement** in the Chiapas region of Mexico, which rep-

resents the interests of the indigenous Mayan people. On New Year's Day 1994, a group of masked guerrillas emerged from the jungle, captured the local state capital and declared war on the Mexican state. The Zapatistas were protesting against poverty and the exploitation of local resources, problems that were enhanced by reforms associated with the North American Free Trade Agreement (NAFTA) between the US, Canada and Mexico, completed in 1994. This fuelled the growth of intensive agriculture for international markets, leading to the emergence of a small group of wealthy farmers and a large class of landless Indian labourers (ibid., p.215). The Zapatista movement employed a range of tactics designed to gain international media attention alongside their armed insurgency, engaging in a prolonged 'war of words' with the Mexican government through the pronouncements and communiqués of their mysterious spokesman, Subcomandante Marcos.

Another example is the **Movimento Sem Terra (MST)** in Brazil, a mass social movement of an estimated 1.5 million members (http://www.mst.org.br). The movement was founded in 1984 and is active in 23 out of Brazil's 27 states. It is made up mainly of landless labourers and peasants from rural areas, many of whom have been dispossessed and displaced by agricultural reforms, mechanization and land clearance. The context for the growth of the movement is the highly uneven distribution of land in Brazil, with most of the land owned by fewer than 50,000 people, while 4 million peasants share less than 3 per cent (ibid., pp.216–17). The glaring contradictions of the landholding system in Brazil are illustrated by the fact that over 42 per cent of privately owned land in Brazil is unused while large numbers of peasants and labourers have no access to land. The strategy of the 'Movimento' involves targeting large, unused private estates and illegally squatting and occupying the land. After areas are 'secured' in this way, the MST resettles large numbers of people on these sites, builds schools and houses and begins farming. Over 600,000 people have been resettled since 1991, leading to considerable violence as large landowners and their private armies have attacked and killed squatters (ibid., pp.216–17). The MST has organized several marches and congresses in the capital city of Brasilia to publicize its agenda for agricultural

Figure 11.11 An MST protest march.
Source: Luciney Martins.

reform (Figure 11.11). Since the late 1990s, the government has voiced some tentative support for land reforms, reflecting the success of the MST in mobilizing popular support for its cause.

A third example discussed by Routledge (ibid., p.214) is the protest against the **Narmada River Valley** project in central India, running through the states of Madhya Pradesh, Maharashtra and Gujarat (Figure 11.12). The river is regarded as sacred by the Hindu and tribal populations of India and provides water supplies for thousands of communities surrounding it. Originally initiated by the Indian state in 1961 and reformulated in the late 1980s, the project involves the construction of 30 large dams as well as 135 medium-sized and 3000 small dams, designed to provide water and energy for agriculture and industry (Routledge, 2003, p.244). With two of the major dams already constructed, much of the resistance was focused on the Sardar Sarovar reservoir, the largest individual scheme. This scheme alone was expected to flood 33,947 acres of forest land and to submerge 248 towns, threatening to destroy the local ecology and economy and undermine the spiritual significance of the river for the local population (Routledge, 2005, p.218). Resistance has been led by the Narmada Bachao Andolan (Save Narmada Movement), which is made up largely of

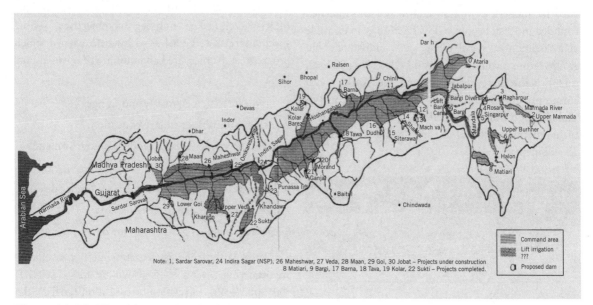

Note: 1, Sardar Sarovar, 24 Indira Sagar (NSP), 26 Maheshwar, 27 Veda, 28 Maan, 29 Goi, 30 Jobat – Projects under construction
8 Matiari, 9 Bargi, 17 Barna, 18 Tava, 19 Kolar, 22 Sukti – Projects completed.

Figure 11.12 The thirty major dams in the Narmada Valley, India.
Source: http://www.narmada.org/maps/nvdp.jpg

peasant farmers and indigenous people, demanding the abandonment of the scheme. Protestors have deployed a range of tactics, including demonstrations, blockages, public meetings and disrupting construction. A week-long 'Rally for the Valley' was held in July 1999, involving a fleet of fishing boats, their white sails festooned with slogans, protesting along the river. Despite such resistance and the generation of considerable international sympathy and support, however, construction of the dams continues.

These examples of local social movements from the developing world show how particular development projects can become the focus of substantial protest and resistance. In all three cases, resistance movements have emerged when local livelihoods have been undermined and local cultures threatened by economic reforms and large-scale development projects. They illustrate the continuing scope for 'bottom-up' action and initiative in an increasingly globalized economy, indicating that local people should be seen as active players in the development process rather than passive subjects of the state and international agencies. While most of these movements are place-based, mobilizing local culture and identity to animate and structure their campaigns, they have become increasingly active in appealing to international audiences and in forging links with similar movements in other countries, facilitated by modern communications and the Internet. Founded in 1998, People's Global Action is an example of a coalition of social movements, individuals and groups from around the world (Routledge, 2002, p.323). Networks such as this and the World Social Forum provide an emerging basis for a new model of **'globalization from below'**, envisaged and promoted as an alternative to the dominant neoliberal model of development.

> ### Reflect
>
> ➤ To what extent does the growth of local social movements resisting development indicate that conventional development policies are failing?

11.6 Current development issues and challenges: trade, debt and aid

11.6.1 Development strategies and goals

As we indicated in section 11.3.3, international development policy has experienced a shift of emphasis since the late 1990s, as the focus has moved from macroeconomic reform to poverty reduction, strengthening institutions and encouraging local participation (Mawdsley and Rigg, 2003). Whether the new anti-poverty agenda (Storey *et al.*, 2005) really represents a new post-Washington Consensus approach is questionable, since many of the central tenets of neoliberalism remain in place. As we saw, the World Bank and IMF continue to set strict conditions before support is offered, reflecting the underlying tendency to hold developing countries responsible for global poverty rather than focusing attention on broader aspects of the global environment such as trade regulations or financial regimes.

The establishment of the **millennium development goals (MDGs)** in 2000 provided a focus for the new anti-poverty agenda. Adopted by the UN in 2000, the MDGs identified a number of objectives, setting specific targets to be achieved by 2015 against which progress can be measured and monitored on an annual basis:

1. Eradicate extreme poverty and hunger.

2. Achieve universal primary education.

3. Promote gender equality and empower women.

4. Reduce child mortality.

6. Combat HIV and AIDS, malaria and other diseases.

7. Ensure environmental sustainability.

8. Develop a global partnership for development.

The latest analysis indicates that, while improvements are occurring, several key goals will be missed (UNDP, 2005, pp.39–45). In relation to education (goal 2), the universal enrolment target will be missed by at least a decade if current trends continue, with 46 countries in the developing world actually regressing. The gender equality goal (3) will be missed by an absolute figure of 6 million girls not in school, mostly in sub-Saharan Africa. The two-thirds reduction in child mortality (goal 4) will only be achieved in 2045 – 30 years too late (ibid., p.41). The target of halving the number of people without access to safe water sources will be missed by about 210 million people, mostly based in sub-Saharan Africa. More positively, the overall global target of halving extreme poverty will be met, largely because of strong growth in China and India. While current trends are not set in stone, meeting the goals will require a concerted effort from governments, development agencies and NGOs.

11.6.2 Trade and the politics of globalization

Trade has become one of the key battlegrounds of globalization in recent years, ever since protesters ended the WTO's meeting in Seattle in December 1999. Throughout the post-war period, GATT/WTO operated through successive 'rounds' of trade negotiations leading to agreements between member states to reduce tariffs (charges imposed on imported goods). Until the 1980s, GATT agreements concentrated on trade in manufactured goods in particular, with restrictions on agricultural goods and primary commodities remaining in place. This left many developing countries dissatisfied with a system that they felt was dominated by powerful northern states, requiring them to open up their markets to manufactured goods while they were confronted with barriers to the export of the primary goods in which they enjoyed a comparative advantage.

The most recent **Uruguay Round**, which ran from 1986 to 1994, was the most ambitious and wide-ranging of the GATT agreements. For the first time, it incorporated not only agriculture, textiles and clothing but also new agreements on services:

➤ The General Agreement on Trade in Services (GATS)

➤ Trade-Related Aspects of Intellectual Property Rights (TRIPs)

➤ Trade-Related Investment Measures (TRIMs).

The Uruguay Round led to an average tariff reduction of one-third by developed countries (Dicken, 2003, p.586). The agreement on services again seemed to reflect the interests of the North, representing an area in which it enjoyed a strong **competitive advantage** over the countries of the global South. In general, the Uruguay Round failed to resolve a number of tensions between developed countries and developing countries. One of the effects of the TRIPs agreement, for instance – pushed hard by Northern multinationals in sectors such as pharmaceuticals to give stronger protection to their patent rights – was to make essential lifesaving drugs more expensive for poor people in developing countries (Stiglitz, 2003).

The **Doha Round** of trade negotiations, launched in Qatar in late 2001, is presented as the 'development round' by the WTO, the EU and the US, who argue that a new trade agreement represents the best way to reduce global poverty. The implication is that the inequities stemming from the Uruguay Round would be addressed, with developed countries opening up their markets to agricultural goods and textiles while industrial tariffs would be slashed further. At the same time, the EU and the US aim to advance the liberalization of trade in services, introducing new rules for investment, competition, government procurement and trade facilitation (the so-called Singapore issues). If agreed, these new rules would open up the public sector in particular to trade, with countries prevented from favouring domestic suppliers of services over foreign companies.

Meanwhile, developing countries remained angry about continuing restrictions that developed countries imposed on trade in agriculture and textiles through **subsidies** favouring domestic producers (Box 11.7). The subsidies that the EU and the US offer to farmers result in the dumping of surplus products on world markets, driving down prices for farmers in developing countries (Oxfam, 2005, p.10). The $3 billion plus subsidies that the US offers to its 25,000 cotton farmers, for example, has undermined the livelihoods of more than 10 million farmers in West African countries like Benin, Chad and Mali (*The Economist*, 2003). Instead of launching a new round of trade talks, most developing countries wished to concentrate on implementation issues left over from the Uruguay Round, particularly

Box 11.7

Trade reform and development: the case of the EU sugar industry

One of the main barriers to economic development in the world's poorest countries is their lack of access to the agricultural markets of the global North. Within the current Doha Round, the agricultural protectionism of the EU and the US has become a major sticking point. The EU's Common Agricultural Policy (CAP) favours European producers through a range of subsidies and restrictions on imports. Sugar is one of the most protected sectors in European agriculture, with EU prices set over 300 per cent higher than the world market price in March 2002 (Gibb, 2006, p.6). A small proportion of sugar from third-world countries outside the EU is imported under special quotas offering preferential access.

The EU's trade relations with the developing world were governed by the Lomé Convention, which ran from 1975 to 2000. This offered the African, Caribbean and Pacific (ACP) group of states, made up of former European colonies, duty-free access to the European market for most of their exports. It effectively discriminated against other developing countries not part of the ACP group. As became increasingly apparent during the 1990s, such discrimination was not compatible with the WTO agenda of trade liberalization, prompting reform. In response, the EU has divided the existing ACP grouping into 'least developed countries' (LDCs) and 'developing countries', extending Lomé trade preferences to all the former while seeking to negotiate reciprocal free trade agreements with the latter. There are only nine developing countries that were not previously members of the ACP group (Gibb, 2006, p.4, for listings). Under its 'everything but arms' (EBA) system, introduced in March 2001, the EU introduced duty-free access for all exports from LDCs, (49 countries) with the exception of arms, rice, bananas and sugar.

The prospect of LDCs gaining free access to the European sugar market under a three-year transitional arrangement sparked vigorous lobbying by European producers to block the reform. As a result, the final version watered down the access offered to LDC sugar exporters, using a quota system to restrict access over an extended transitional period from 2001/2 to 2008/9. The burden of adjustment arising from this limited new access is placed squarely on the developing countries, who lose the preferences that allowed them to supply quota sugar to the EU. 'In short, poor countries are paying for the access privileges offered to the very poor' – 'an extraordinary outcome for an initiative designed to help the world's poorest countries' (ibid., pp.9, 14). In the case of sugar, then, the EU's reform of its trade regime offers only minimal access to one group of poor countries at the expense of another. In late 2005, EU agriculture ministers, under pressure from the WTO and other bodies, agreed a 36 per cent cut in sugar prices, sparking protest from sugar-producing LDCs affected by the price cut (BBC, 2005).

the failure of developed countries to reduce agricultural subsidies.

Against this backdrop, trade negotiations were always going to be difficult. Talks broke down at the Mexican resort of Cancun in September 2003. Basically, developing countries were not prepared to discuss EU-led proposals on new rules for investment and competition without large-scale reduction of agricultural subsidies and textile quotas, which the EU and the US were unwilling to accept. The emergence of organized coalitions of developing countries prepared to take on the EU and the US has been particularly significant. The so-called G21 group led by India, China and Brazil played a key role in advancing the interests of developing countries in Cancun, refusing to bow to

pressure. The process was 'saved' by a procedural agreement in July 2004, setting up a programme of further talks, but negotiations collapsed again in Geneva in July 2006.

The substantial difficulties experienced by the Doha Round of trade negotiations raises important questions of global economic regulation. A new assertiveness by developing countries in the face of continuing trade inequities has thrown the GATT/WTO model into doubt. While negotiations continue, there is little sign of any resolution on the underlying issues dividing developed and developing countries. The EU and the US remain unwilling to eradicate subsidies in agriculture and textiles, while continuing to push for further liberalization of trade in services and industrial goods.

Without a commitment to the former, developing countries will not agree to the latter. Without major concessions from the EU and the US, it seems unlikely that agreement will be reached. Paradoxically, however, failure is likely to damage developing countries more since improved access to Northern markets seems to represent a necessary, if not sufficient basis, for sustained economic development in the global South.

11.6.3 Debt relief

The **Heavily Indebted Poor Countries (HIPC) Initiative** was launched by the World Bank and IMF in 1996 in recognition of the need for a concerted effort to reduce the debts of the poorest countries in the world. It aimed to provide a permanent solution to the debt problem and to free up resources for investment in development. The Initiative creates a framework for all creditors to provide debt relief to the world's poorest and most indebted countries. To be eligible countries must be very poor (according to World Bank and IMF categories), face an unsustainable debt situation (defined as a debt to export ratio of above 150 per cent) and have sound macroeconomic policies consistent with poverty reduction and sustained growth.

The Initiative has identified 38 countries, 32 of them in sub-Saharan Africa, as potentially eligible for debt relief. To reach '**decision point**', allowing them to begin receiving debt relief, countries must have a stable economic policy framework in place and have prepared Poverty Strategy Reduction Papers. Reaching 'completion point', which unlocks the full amount of debt relief, depends on their success in maintaining economic stability, introducing economic and social reforms and implementing the PRSPs. The countries currently (January 2006) at the different stages are as follows:

➤ Completion point (18 countries); Benin, Bolivia, Burkina Faso, Ethiopia, Ghana, Guyana, Honduras, Madagascar, Mali, Mauritania, Mozambique, Nicaragua, Niger, Rwanda, Senegal, Tanzania, Uganda, Zambia.

➤ Decision point (10 countries): Burundi, Cameroon, Chad, Democratic Republic of Congo, the Gambia, Guinea, Guinea-Bissau, Malawi, São Tomé and Príncipe, Sierra Leone.

➤ Pre-decision point (10 countries): Central African Republic, Comoros, Republic of Congo, Ivory Coast, Laos, Liberia, Myanmar, Somalia, Sudan, Togo.

According to the World Bank, the external debt of the 28 approved countries has been cut by two-thirds, freeing up funds for poverty reduction. At the July 2005 G8 summit in Gleneagles, Scotland, world leaders agreed to write off the debts of the 18 countries that have reached completion point, amounting to some $40 billion. Significant progress has certainly been made through these initiatives, although campaigners have called for more countries to be included and for levels of relief to be increased. In a direct echo of complaints against SAPs, the issue of '**conditionality**' has also caused concern, with the World Bank and IMF requiring countries to implement policies of liberalization and privatization before they become eligible for relief.

11.6.4 Aid to developing countries

Aid is defined as official development assistance offered by developed countries to promote economic development and poverty reduction (Burnell, 2002, p.473). It has been an important aspect of development policy throughout the post-war period, although levels of assistance have often fallen short of what development activists and campaigners have called for. At the same time, aid has been used to further donors' strategic and political interests by, for example, tying it to the purchase of goods and services from that country. During the 1980s and 1990s, a particularly high proportion of aid from the UK was tied, reaching 74 per cent of bilateral aid in 1991 (Potter *et al.*, 2004, p.36), although all UK aid has been fully untied since April 2001.

In 1970, the UN urged developed countries to commit the equivalent of 0.7 per cent of their GNP to foreign aid by 1975, a demand reiterated by the Brandt Commission in 1980 and subsequent global summits in the 1990s (ibid., p.360). In reality, however, actual expenditure has fallen well short of this goal. Figure 11.13 shows that only five members of the Development Assistance Committee (DAC) of the OECD committed more than the target in 2004, with the majority

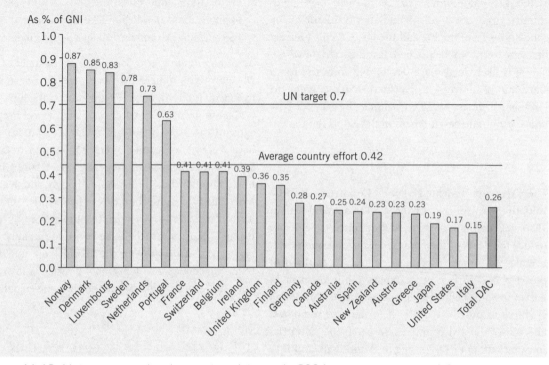

Figure 11.13 Net overseas development assistance in 2004 as a percentage of GNI.
Source: 'Net ODA in 2004 as a percentage of GNI, aid rising sharply, according to the latest OECD figures (13 December 2005)', © OECD, 2005.

remaining well below. World leaders agreed an additional $50 billion a year in aid to developing countries by 2010, with $25 billion going to Africa, while the EU countries pledged to reach a collective aid target of 0.56 per cent of GDP by 2010 and 0.7 per cent by 2015. Such commitments are certainly welcome, but it remains questionable whether these goals will actually be reached, in the light of past performance.

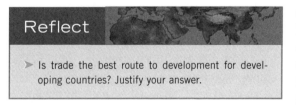

Reflect

➤ Is trade the best route to development for developing countries? Justify your answer.

11.7 Summary

The development of the former colonies in Africa, Asia and Latin America can be seen as a key political and economic project of the post-war era, launched by the

US and its allies in the late 1940s against a backdrop of cold war tensions. The so-called '**Third World**' was constructed as a realm of universal poverty and backwardness, with development requiring an injection of external knowledge and resources. The establishment of SAPs assumed particular importance in the context of the **debt crisis** in the 1980s and 1990s. Since the mid-1990s, the emphasis on economic reform has given way to a focus on poverty reduction, generating a new consensus among development agencies, governments and NGOs. In reality, however, the World Bank and IMF continue to dictate the terms of assistance, maintaining a regime of '**conditionality**' where developing countries have to meet their specific requirements before qualifying for assistance. A commitment to economic liberalization, openness and adjustment remains central, alongside the new focus on poverty reduction. The responsibility for alleviating poverty is allocated to poor countries themselves, while broader structural constraints such as the global rules underpinning trade and investment are largely ignored.

Despite some limited improvements over the last 30 years, global inequalities remain massive. Within the global South itself, processes of **uneven development** have resulted in a growing divergence between regions. Sustained growth in East Asia and improvements in South Asia contrast with economic decline in sub-Saharan Africa and stagnation in Latin America and the Middle East. According to the UNDP (2005, p.34), 'the war against poverty has witnessed advances on the eastern front, massive reversals in sub-Saharan Africa and stagnation across a broad front between these poles'. In recognition of the problem of global poverty, world leaders agreed the Millennium Development Goals (MDGs) in 2000, committing themselves to reaching a series of targets by 2015. While considerable progress has been made in relieving the debts of the poorest countries and increasing aid, the latest analyses suggest that many of these goals will be missed if current trends persist. A new trade deal to help developing countries is likely to remain elusive in the face of unresolved conflicts of interest between rich and poor countries. This returns us to the insight that uneven development is not simply a reflection of internal failures in poor countries but an expression of the relationships between core and periphery (Massey, 2001). As such, efforts to reduce poverty require action at the global level to create a more level playing field for trade and investment, in addition to investment within developing countries themselves.

Exercises

Select a developing country. Review its experience of development since the 1960s, using the websites listed below (in Useful websites) as starting points for your research. Assess the country's economic performance over time (referring to figures on growth, employment, income, investment, education, health, etc.).

1. What have been the key forces shaping development?

2. What development strategies has the country adopted?

3. Who has set the development agenda: the government, the World Bank/IMF, foreign MNCs, NGOs, domestic interests (e.g. landowners, industrialists, traders, workers, peasants)?

4. Have development strategies been informed by any of the theories discussed in section 11.3?

5. Are there any examples of local social movements resisting particular development projects?

6. How would you describe the country's development prospects at the present time?

Key reading

Desai, V. and Potter, R. (eds) (2001) *The Companion to Development Studies*, London: Arnold.
Contains a number of short overviews of various development themes and issues written by leading authorities. Extremely comprehensive, dealing with a wide array of topics including the meaning of development, the main theories, rural development, urbanization, industrialization, the environment, gender and population, health and education, violence and instability and key agents of development.

Mohan, G. (1996) 'SAPs and development in West Africa', *Geography*, 81: 364–8.
A brief and accessible review of the main impacts of structural adjustment programmes in West Africa. Provides a concise definition of SAPs and highlights their effects on trade and economic policy, poverty, health and welfare, employment and gender roles, resource extraction and the environment, and the state and democracy.

Potter, R.B., Binns, T., Elliott, J.A. and Smith, D. (2004) *Geographies of Development*, 2nd edn, Harlow: Pearson.
Probably the best contemporary textbook on development issues in geography. Provides a comprehensive and integrated treatment of development that covers the main theories and the historical legacy of colonialism; assesses the role of population, resources and key institutions; and examines spaces of development within developing countries. The second edition incorporates extended treatment of current issues such as globalization and debt.

Power, M. (2005) 'Worlds apart: global difference and inequality', in Daniels, P., Bradshaw, M., Shaw, D. and Sidaway, J. (eds) (2005) *Human Geography: Issues for the Twenty-first Century*, 2nd edn, Harlow: Pearson, pp.184–97.
An engaging discussion of development issues, emphasizing the scale of global differences and inequalities. Covers the key theories, institutions and history of development, highlighting its unequal impact on households and individuals within developing countries.

Routledge, P. (2005) 'Survival and resistance', in Cloke, P., Crang, P. and Goodwin, M. (eds) (2005) *Introducing Human Geographies*, 2nd edn, London: Arnold, pp.211–24.
A useful account of the growth of local social movements against particular projects. Covers the three examples outlined in section 11.5, stressing both the place-based origins of the movements and the increasing global links between them.

Willis, K. (2005) 'Theories of development', in ibid., pp.187–99.
A very clear summary of the main theories of development, covering the modernization school, dependency theory, neoliberalism and grass-roots development. Contains useful bullet point summaries of each approach.

Useful websites

http://home.developmentgateway.org/
A gateway to a wide range of development resources, including the main development agencies, aggregate data and statistics, and profiles of individual countries.

http://www.oxfam.org.uk/index.htm
The website of one of the leading British development NGOs, providing details of its history and current strategies and highlighting some of the key projects and campaigns it is undertaking. Contains information on how you can support Oxfam's activities through volunteering and fund-raising.

http://hdr.undp.org/
The UNDP site provides access to information about human development in the form of current statistics, the annual Human Development Report and details of how measures such as the Human Development Index are compiled. Contains interactive maps and tools illustrating human trends and outcomes.

http://www.worldbank.org/
The official World Bank site contains a wide range of information, reports and projects. Key resources include the annual World Development Report, World Development Indicators and the speeches of the president and other senior figures, giving an indication of current thinking. Data profiles of countries and country groupings are also available.

http://www.mstbrazil.org
The website of Brazil's Landless Workers Movement, the *Movimento Sem Terra* (MST). Contains a range of information about the movement's history, objectives and campaigns.

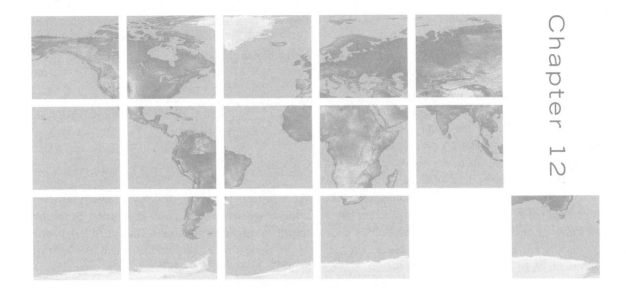

Chapter 12

Tourism, culture and economic development

Topics covered in this chapter

➤ The nature of tourism and the different forms it takes.

➤ The growth of tourism and the tourist industry in recent decades.

➤ Processes of tourist consumption, particularly new 'post-Fordist' forms of consumption.

➤ Tourism and economic development, focusing particularly on developing countries.

➤ The role of tourism and related cultural industries in facilitating processes of urban regeneration.

Chapter map

We begin by emphasizing the economic importance of tourism within the contemporary world economy and relating tourism to concepts of place, space and uneven development. In section 12.2, we define what tourism is and outline key characteristics of the global tourist industry. This is followed by a discussion of changing processes of tourist consumption, focusing particularly on the rise of new 'post-Fordist' forms of consumption such as heritage tourism, ecotourism and the visiting of theme spaces and mega-malls. Section 12.4 assesses the relationship between tourism and economic development in different parts of the world, concentrating on the experience of developing countries in particular. In section 12.5, we examine the links between tourism, culture and urban regeneration in developed countries. We summarise the main points covered in the chapter in section 12.6.

12.1 Introduction

Tourism is a key component of the service economy, broadly corresponding with the 'entertainment, hotels and leisure' category identified in section 8.2.1, in addition to long-distance passenger transport. It is commonly described as the world's largest industry, being expected to generate 10.3 per cent of total world GDP and 8.7 per cent of total world employment in 2006 (WTTC, undated). Like other service industries, tourism is based on a direct interaction between the buyer and seller of services and this interaction forms an important part of the tourist experience. Its main product is not a particular good or service, but an experience (I. Gordon and Goodall, 2000, p.291). What is particularly distinctive about tourism is the extended time period covered by this experience (the holiday period) and its inclusive nature (all the activities undertaken in this period) (ibid., p.291).

The British sociologist John Urry (2002) has developed the concept of the **tourist gaze**, arguing that a key part of the modern tourist experience is to encounter scenes and landscapes that are distinctive from those associated with everyday life. These other places are anticipated and viewed with a sense of excitement and curiosity. The ability to experience and consume unfamiliar and distant places is a product of unprecedented levels of personal mobility, reflecting successive advances in transport technology since the eighteenth century, as well as rising incomes and increased leisure time. For example, the number of trips abroad made by UK residents tripled from just over 20 million per year in 1984 to 64.2 million in 2004 (ONS, 2005b, p.19). Contemporary tourism is heavily implicated within the broader phenomenon of **globalization**, generating large-scale flows of people between places. In a similar fashion to other industries, tourism is characterized by **uneven development**, with the overwhelming majority of activity focused on the three core regions of Europe, North America and East Asia. At the same time, while individual tourists travel across geographical space, those who provide goods and services to them are fixed in particular locations. The geographical fixity of the providers means that they are dependent on attracting a sufficient number of visitors to the place in which

they are based in the face of intense competition from other places. As a result, **place promotion** and marketing become important.

12.2 The development of tourism

12.2.1 Defining tourism

While tourism is difficult to define precisely, the term brings together tourists (the customers) with the tourist industry (the providers). The tourist is someone who travels outside their normal home area to visit another destination in which they do not normally reside. This can be either in another part of the home country (domestic tourism) or abroad (international tourism). Different kinds of travel can be distinguished according to the purpose of the visit and the length of the stay. The main purposes of travel by tourists are (Hunter and Green, 1995, p.2):

➤ pleasure (e.g. sightseeing, sport, relaxation, dining out);

➤ visiting friends or relatives;

➤ work-related business (e.g. conferences and conventions);

➤ personal business (e.g. education study trips, medical appointments).

While we tend to think of tourism in terms of just 'going on holiday', it important to appreciate the importance of business tourism, which generates significant revenue for providers such as hotels and restaurants (Hunter, 2005). Industry bodies such as the World Tourism Organization define tourists as visitors who stay at least one night but less than one year in the host area.

12.2.2 The growth of tourism

Prior to the Industrial Revolution, human mobility was overwhelmingly local in nature. Longer-distance travel was small-scale and undertaken for particular pur-

Figure 12.1 Thomas Cook advertisement for a tour of Scotland, late nineteenth century.
Source: © Imaging Scotland: Tradition, Repesentation and Promotion in Scottish Tourism since 1750, Gold, J.R. and Gold, M., 1995, Ashgate, p.105.

poses, particularly religion (visiting sacred sites) or education. It was confined to a wealthy and privileged elite who followed recognized routes between renowned sites as part of their general education and development, acting as a marker of social status (Urry, 2002, p.16). The European 'Grand Tour' of the seventeenth and eighteenth centuries is an example of this pattern of travel. During the nineteenth century, tourism became a mass industry, driven by social changes and transport improvements, spreading to the growing middle classes by the 1850s and the working classes by the 1880s and 1890s. The **mass tourist industry** of the late nineteenth century found particular expression in the growth of working-class resorts like Blackpool in England, clearly distinguished from more prestigious middle-class resorts offering 'relaxation and restoration', often in areas of natural beauty (Cater, 1995, pp.185–6). Companies such as

Thomas Cook played a crucial role in developing mass tourism from the late nineteenth century onwards, offering a wide range of tours by rail and ship (Figure 12.1).

Over the last century, the tourist industry has grown steadily, experiencing particularly rapid expansion from the 1950s. International tourist arrivals, for example, grew from a relatively low base of 25.3 million in 1950 to an all-time high of 808 million in 2005 (Cater, 1995, p.190; World Tourist Organization, undated). In general terms, the main factors behind the growth of tourism since the 1950s are:

➤ increased leisure time and rising disposable incomes in developed countries;

➤ the growth of commercial jet aircraft;

➤ the availability of relatively cheap oil;

➤ the entry of multinational corporations into the industry, particularly in areas such as hotels and travel agencies, offering convenience and value for money.

International tourism, however, is estimated to account for only around 10 per cent of total activity, with the remainder being made up of domestic tourism. Expenditure by domestic tourists in the UK in 2004 was estimated to be over £59 billion, compared with around £13 billion by overseas tourists (Visit Britain, undated). At the same time, international tourism has grown more rapidly, from 11 million visits to the UK in 1976 and 15.5 million in 1987 to 27.8 million in 2004 (Urry, 2002, p.6; Visit Britain, undated). The growth of domestic tourism has been slower, with 151 million trips made in the UK in 2003 compared with 147.8 million in 1995 (Visit Britain, 2004).

12.2.3 Destination and source countries

The development of tourism has been highly uneven in geographical terms, reflecting broader patterns of inequality. At the global level, the three core regions of Europe, the Americas (with the US and Canada making up the bulk of these totals) and Asia and the Pacific play a dominant role, accounting for 91 per cent of arrivals and 93 per cent of departures, with Africa and the

Table 12.1 International tourism arrival and origin regions, 2004

	Arrivals (millions)	%	Origins (millions)	%
World	763.3	100	763.3	100
Europe	416.4	54.5	431.3	56.5
Asia and the Pacific	152.5	20.0	151.2	19.8
Americas	125.8	16.5	127.7	16.7
Africa	33.2	4.4	18.2	2.4
Middle East	35.4	4.6	22.0	2.9
Not specified			12.9	1.7

Source: World Tourism Organization, (undated).

Middle East relatively insignificant by comparison (Table 12.1). It is worth noting that 81 per cent of tourist travel occurs within the same region. As one might expect, Asia and the Pacific has grown most rapidly since 1990 as both a destination for and source of tourists. The US was the most important tourist destination in terms of expenditure in 2004, followed by Spain, France, Italy, Germany, the UK and China (Figure 12.2). Germany is the top spender, followed by the US, the UK, Japan and Italy (World Tourist Organization, undated). The pattern of regional inequality indicated by these figures demonstrates that tourism reflects broader patterns of **uneven development** within the world economy.

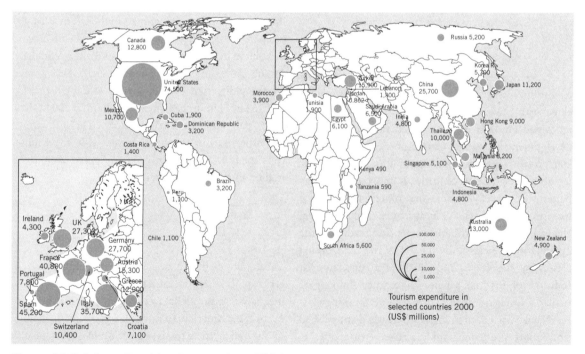

Figure 12.2 International tourism receipts, 2004.
Sources: Adapted from Desforges, 2005, with 2004 data from World Tourism Organization, undated.

12.2.4 Changing forms of tourism

The mass tourist sector dominated the wider industry from the nineteenth century until at least the 1970s. In the 1960s and 1970s, the growth of new Mediterranean destinations eclipsed the role of the traditional seaside resorts, which fell into decline. Since the 1980s, however, mass tourism has been supplemented by the growth of other forms of travel. These are designed to offer the tourist a more distinctive and individualized experience, compared with a standard package holiday, serving specialized niche markets (Shaw and Williams, 2004, p.115). As a result, the tourism market is characterized by increased diversity and flexibility, although mass tourism remains important. Short-break holidays, including weekends, have become more important over the past couple of decades, accounting for 64 per cent of holiday trips by all UK residents in 2000 (ibid., p.118).

Other forms of tourism, alongside the traditional package holiday, include:

➤ Business tourism, including conferences, exhibitions, meetings, conventions and trade fairs. It is a growing sector, with expenditure by overseas residents visiting the UK for business purposes increasing from 1.1 billion in 1984 to 3.7 billion in 2004 (ONS, 2005b, p.18).

➤ Heritage tourism has also expanded rapidly, with over 1,000 new museums registered in Britain in the mid-1970s, together with over 210,000 listed buildings, while the membership of the National Trust for England, Wales and Northern Ireland grew from 278,000 in 1971 to almost 2 million by 2000 (Shaw and Williams, 2004, p.121).

➤ Ecotourism, defined as 'responsible travel to natural areas that conserves the environment and improves the well-being of local people' (International Ecotourism Society, undated). Activities undertaken by ecotourists include birdwatching, viewing the local landscape and scenery, trekking and backpacking. It tends to attract independent travellers who value the conservation of local environments and cultures. The World Tourist Organization estimated in 1998 that ecotourism and all nature-related tourism accounted for approximately 20 per cent of all international travel (International Ecotourism Society, 2000).

➤ Adventure and sporting holidays, focused on activities such as mountaineering, skiing, surfing and white-water rafting.

➤ Visiting **theme parks** and large mega-malls where tourism and retail overlap. Examples include Disneyland in the US, Disneyland Paris, Alton Towers and the Mall of America in Minneapolis. A large number of theme parks and attractions were opened in the 1980s and 1990s across a range of countries. In the UK, for instance, there were 800 visitor attractions in 1960, 2,300 in 1983 and 6,500 by 2000, attracting 395 million visits in 1998 (Department of Culture, Media and Sport, undated; Urry, 2002, p.6).

Reflect

➤ Compare and contrast conventional mass tourism with the other forms of tourism listed above.

12.3 Consuming places: tourist experiences

12.3.1 Tourist consumption and social change

Tourism is a key form of consumption in contemporary societies, meaning that it is bound up in wider debates about the changing nature of consumption. Influenced by theories of postmodernism, a number of commentators have emphasized the increased importance of consumption to economic life. The growth of more segmented and individualized consumption practices, where the purchase of goods and services can be regarded as an expression of a person's lifestyle and identity, is a major trend (Miller, 1998b; Slater, 2003). One of the key themes of recent debates is that the boundaries between tourism and other forms of consumption such as shopping, leisure, entertainment, education and sports are becoming increasingly blurred (Urry, 2002). Such '**de-differentiation**' is a central aspect of postmodern culture. What remains distinctive about tourist consumption is that it is based on the **commodification** of place where commodification refers to the process by which particular objects, relationships and ideas are turned into goods and services that are bought by individual consumers through the market (Desforges, 2005, p.521).

In very broad terms, two main modes of tourist consumption can be identified. First, there is the Fordist mode of mass consumption catering for large numbers of tourists. Such consumption is highly standardized and rigid, underpinned by the economies of scale developed by the large tour companies, hotel chains and airlines. It held sway from the late nineteenth century to the 1970s, epitomized by the traditional seaside resorts of England and, later, the Mediterranean resorts and *costas*, based on the provision of standardized package holidays. Mass tourism is characterized by a collective **tourist gaze** based on the presence of large numbers of people, which creates an atmosphere and sense of excitement (Urry, 2002, pp.43–4). It is a fairly passive form of consumption, limited to key sites such as the hotel, beach and bar, together with attendance at certain packaged events (Shaw and Williams, 2004, p.154)

By contrast, **post-Fordist differentiated consumption** involves the creation of more specialized niche markets, suited to meeting fluctuating patterns of tourist demand (Table 12.2) (ibid., p.115). Flexibility is the key theme, expressed in terms of less structured and more independent forms of tourism, in contrast to the standardized package holiday. This is sometimes described as the **new tourism**. The rise of short breaks and the taking of multiple holidays of different types over a year have been key trends. Such flexibility has been facilitated by two important technological developments: credit cards, which release unearned income to allow more travel, and Internet booking, which has made the selection, booking and payment of holidays much easier (ibid., p.118). Mass tourism remains important, of course, with the number of holiday packages taken by UK residents increasing by 22 per cent between 1986 and 1997 (ibid., p.127).

New forms of consumption are linked to broader social changes and particularly the rise of new middle-class groupings (Urry, 2002, p.79). The role of the so-called **service classes** – defined as groups that do not own capital or land but are employed in servicing capital, generally in professional and managerial posts in activities such as financial services, advertising, research

Table 12.2 The shift to post-Fordist consumption in tourism

Post-Fordist Consumption	Tourist examples
Consumers increasingly dominant and producers have to be more consumer-oriented	Rejection of certain forms of mass tourism (holiday camps and cheaper packaged holidays) and increased diversity of preferences
Greater volatility of consumer preferences	Fewer repeat visits and the proliferation of alternative sights and attractions
Increased market segmentation	The multiplication of types of holiday and visitor attractions based on lifestyle research
The growth of a consumers' movement	Much more information about alternative holidays and attractions through the media
The development of many new products, each of which has a shorter life	The rapid turnover of tourist sites and experiences because of rapid changes in fashion
Increased preferences expressed for non-mass forms of production/consumption	The growth of 'green tourists' and of forms of refreshment and accommodation that are individually tailored to the consumer (such as country house hotels)
Consumption as less and less functional and increasingly aestheticized	The 'de-differentiation' of tourism from leisure, culture, retailing, education, sports, holidays

Source: Urry, 1995, p.151.

and development and management consultancy – is particularly important in terms of changing forms of tourist consumption. The status and power of these service-class groupings are derived from the accumulation of what the French sociologist Pierre Bourdieu termed 'cultural capital' rather than being grounded in traditional notions of economic capital (ibid., p.80).

Such groups provide the key market for postmodern culture, accumulating cultural capital through individualized types of consumption such as shopping, visiting museums and art galleries and travel. New forms of tourism are of central importance here, reflecting a search for unique and 'authentic' experiences. The **tourist gaze** in question here is romantic rather than collective, emphasizing solitude, privacy and the personal significance of the object of the gaze, in contrast to the collective ethos of mass tourism (ibid., p.43). These groups have been referred to as 'post-tourists', defined by a high level of self-awareness and knowledge, manifest in an awareness of the choice between a diverse range of potential destinations and of the futility of searching for a single, authentic tourist experience (ibid., pp.90–91).

12.3.2 New forms of tourist consumption

Heritage tourism

As we indicated earlier, heritage tourism has grown rapidly in developed countries such as the UK since the 1970s. As well as the continued appeal of rural heritage, represented by stately homes and historic monuments, with over 12 million people visiting those owned by the National Trust in England, Wales and Northern Ireland (National Trust website: http://www.nationaltrust. org.uk), there has been an unprecedented rediscovery of industrial heritage. The creation of the Millennium Commission and the Heritage Lottery Fund in the mid-1990s gave a major boost to heritage development, providing a large source of funding for community groups seeking support (Shaw and Williams, 2004, pp.262–3). For example, the Commission provided funding for some 25 major visitor attractions between 1994 and 2002, distributed fairly widely across British cities, including the now defunct Millennium Dome in Greenwich, East London (ibid., p.263).

The rediscovery of industrial heritage has coincided with the disappearance of heavy industry in northern England, central Scotland and Wales. Urry (2002, p.97) identifies two key aspects of the relationships between the rediscovery of the industrial past and the process of **deindustrialization**. First, the latter has generated a sense of loss for past technologies and ways of life. Second, local authorities in the areas most affected by deindustrialization have been particularly active in promoting economic redevelopment and regeneration, with heritage-based tourism providing one of the very few alternative sources of employment, in the face of intense competition for visitors and investment. As Robins (1991, p.580) argues, even in 'the most disadvantaged places, heritage … can be mobilized to gain competitive advantage in the race between places' (quoted in Waitt, 2000, p.840). As a result, some unlikely places such as Bradford, Liverpool and the Rhondda Valley of South Wales have become centres of heritage tourism, presenting packaged versions of their industrial history for tourist consumption (Urry, 2002, p.105). Similar developments have taken place in other industrialized countries, with Lowenthal remarking of the US that 'the trappings of history now festoon the whole country' (ibid., p.106).

The growth of heritage can be seen as an expression of **postmodern** culture as the distinction between different time periods is relaxed, and the past is reconstructed and represented according to the needs of the present. Accordingly, history (like place) becomes a commodity for tourist consumption (Shaw and Williams, 2004, p.121). The Rocks area of Sydney is one prominent example of a heritage development where the past is reinterpreted and reconstructed in particular ways for tourists (Box 12.1). The demand for heritage tourism has been stimulated by rising incomes, increased education and the rise of the new middle classes (Richards, 1996). For instance, in the UK, the percentage of the service class visiting museums and heritage centres in any one year is about three times that of manual workers (Urry, 2002, p.96). At the same time, there is much wider support for conserving historical buildings and sites, with nine out of ten people supporting the use of public funds to preserve heritage (ibid., p.96).

Box 12.1

Consuming heritage: The Rocks development in Sydney

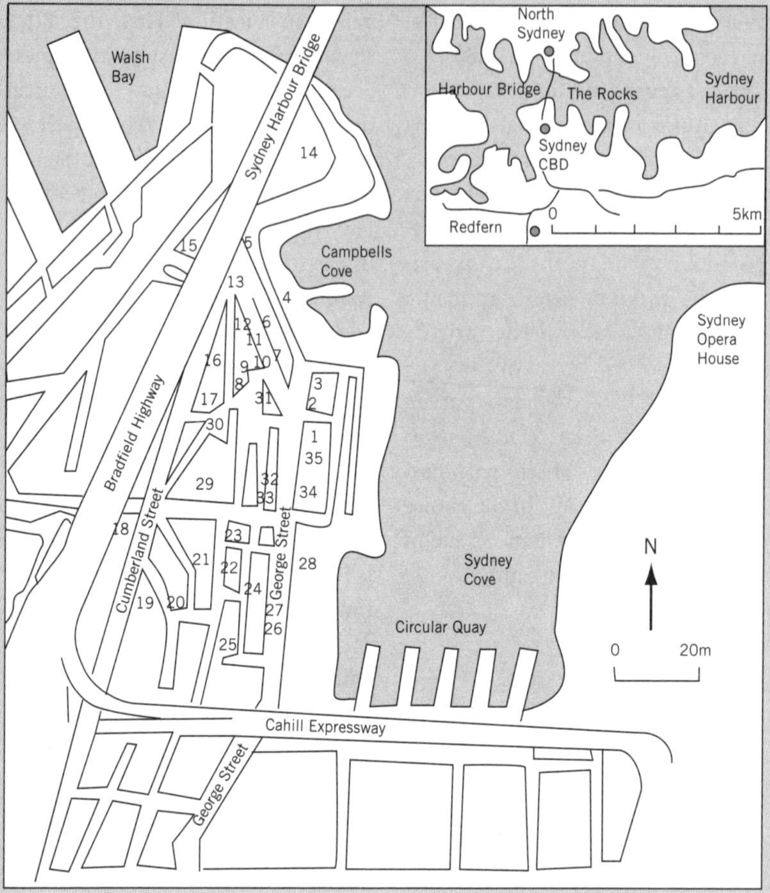

Figure 12.3 Map of The Rocks, Sydney.
Source: Waitt, 2000, p.842.

Tourist attractions

1 Coroner's Court (1907)
2 Mariner's Church (1856–59)
3 Australian Steam Navigation Co. Building (1884)
4 Campbell's Storehouse (1839–61)
5 Electric Power Station (1902)
6 Metcalfe Stores (1912–16)
7 Bushell Place restored terraces (1880–81)
8 Atherdon Street restored terraces (1880–81)
9 Westpac Banking Museum
10 Union Bond Store (1841)
11 Regency town house (1848)
12 Sergeant Major's Row (1866)
13 Mercantile Hotel (1914–15)
14 Dawes Point Battery (the first fortified position in Australia)
15 Antique cast-iron pissoir
16 Plaque to Eber Bunker, an American whaler who captained the first whaling expedition from Sydney
17 Foundation Park
18 The cobbles of Argie Street. The Argyle Cut excavation was begun by convict labour in 1843 and completed with free labour in 1859
19 The Australian Hotel (1914)
20 Susannah Place (1844). The terrace is conserved for display as a house museum
21 The Stafford Apartments, converted terraces built around 1886
22 The Habour Rocks Hotel, a converted warehouse and cottages dating from the 1890s.
23 Renold Cottage (1830), the home of the first Irish blacksmith, William Reynolds
24 Nurses Walk, dedicated to Australia's first nurses, who were also convicts
25 Shopfronts between Globe and Argyle Streets dating from 1838
26 Former bank, dating from 1886, built of sandstone, in an example of Gothic
27 Former Police Station (1882)
28 The Museum of Contemporary Art
29 Restored Argyle Stores (1826–81)
30 The Argyle Terraces (1875–77)
31 First Impression. A tribute sculpture in sandstone to the convicts, soldiers and free settlers of Australia
32 The Coach House (1853–54)
33 Uwin's Stores (1843–46). A network of narrow alleys and courtyards are found behind these buildings
34 Cadman's Cottages (1816). The oldest surviving building in Sydney
35 Sailors' Home (1864–1926)

Box 12.1 (continued)

The redevelopment of redundant waterfront districts, formerly based upon port and warehousing functions, into heritage precincts designed to attract tourists has been a prominent feature of urban regeneration policies since the 1970s (section 12.5). A good example is The Rocks area of Sydney, which has been subject to major redevelopment along such lines under the auspices of the Sydney Cove Redevelopment Authority (SCRA), established in 1968 (Waitt, 2000). The area is located on a peninsula in Sydney's Central Business District, but cut off from the rest of the CBD by two major roads, the Harbour Bridge Approach Road and the Cahill Expressway (Figure 12.3) (ibid., p.842). It represented the first area of European colonization from 1788, becoming the hub of much of Sydney's port trade.

The redevelopment of The Rocks has transformed a run-down area of abandoned buildings and facilities into a major tourist attraction (Figure 12.3). The redeveloped area is made up of heritage sites, ambience, artefacts and buildings along with entertainment, dining, museums, galleries, shops and accommodation (Figure 12.4). The officially sanctioned and commercial history presented by the SCRA emphasizes the area's role as the 'cradle' and 'birthplace' of the Australian nation, containing various Australian 'firsts' including the first church, village, fortified position and Sydney's oldest homes (ibid., p.836). It focuses on Australian national identity and origins, emphasizing its European roots. The version of history presented by the SCRA is obviously selective and partial, determined by its judgement of what is commercially attractive and marketable to tourists, ignoring other aspects of the history of areas such as social inequality, conflict and racial discrimination.

The redevelopment of The Rocks has been highly successful in commercial terms, attracting over 1.2 million overseas visitors in 1996, making it one of Australia's premier attractions after the Sydney Opera House (1.7 million) (ibid., p.839). Waitt found that the desire to obtain historical experiences was the most important reason for visiting the area, with most tourists perceiving the history presented in the redeveloped area as authentic. The built environment and overall setting were viewed as important in portraying the area as authentic, including features such as stone steps, a cobbled street and terraced houses. There was no meaningful relation between level of perceived authenticity and income, suggesting that this example of heritage tourism has broad appeal beyond the new middle classes. What this research indicates is that official versions of history presented by the heritage industry tend to be accepted as authentic and meaningful by tourists, with criticism of such histories as biased and selective having a limited wider impact.

Figure 12.4 The Rocks, Sydney
Source: D. Mackinnon.

Ecotourism and adventure holidays

Ecotourism and adventure holidays represent another growth sector of the tourist market. Such activities are characterized by a considerable degree of 'de-differentiation' in that tourism is intertwined with the pursuit in question, for example birdwatching, backpacking, surfing or bungee jumping (Shaw and Williams, 2004, p.122). An earlier ecotourist market profile found that 35–54 was the most common age band, with an experienced ecotourist prepared to spend more than general tourists. The two key motivations for travelling were to enjoy scenery/nature and to access new experiences/places (International Ecotourism Society, 2000). Most studies have found that most ecotourists tend to be well educated and have above-average incomes, fitting the characteristics of the new middle classes (Shaw and Williams, 2004, p.124). A study of Canadian ecotourists found that they were characterized by high levels of education and income, with high levels of environmental awareness (ibid., p.155). Key activities included learning about the natural environment, birdwatching and photographing wildlife. An example of ecotourist activity centred on a particular location is whale- and dolphin-watching in Kaikoura, New Zealand, a booming ecotourist destination that attracts 1.5 million tourists a year (Figure 12.5) (Cloke and Perkins, 2005, p.908).

Themed spaces

The growth of themed spaces such as theme parks and mega-malls is part of the wider '**de-differentiation**' of tourist consumption under postmodernism, overlapping with retail and local leisure and entertainment. Examples of theme parks include Disney World and Disneyland Paris, providing models for their growth and global spread over the past couple of decades. Attendance at US theme parks, for example, increased by 24 per cent between 1980 and 1990 (Shaw and Williams, 2004, p.126). Theme parks and malls provide a safe, clean and family-oriented environment for consumers to spend their leisure time, merging consumption and play, providing an exciting experience but one that is highly standardized and controlled (ibid., p.126). Recent years have also seen the growth of themed environments at a range of scales, including restaurants, cafes and marketplaces, themed in relation to speciality shopping, leisure and entertainment (ibid., p.256). Themed landscapes have also emerged on a broader scale, packaged and promoted by the tourist

Figure 12.5 Kaikoura, New Zealand.
Source: D. Mackinnon.

industry, and often associated with television, films, music and literature (ibid., p.258). Examples include the establishment of book towns, such as Wigtown in Scotland and Hay-on-Wye in England, literary trails, such as Catherine Cookson Country in the north of England, and efforts to associate particular places in New Zealand with locations in the film of J.R.R. Tolkien's *Lord of the Rings* (ibid., p.258).

The development of themed spaces has prompted discussion of the '**McDisneyification**' of tourism, drawing on Ritzer's theory of 'McDonaldization' that argued that society in general was becoming increasingly predictable, subject to manipulation by controlling technologies (ibid., p.124). In the sphere of tourism, the 'McDisneyification' thesis contends that tourists want their holidays to be predictable, efficient and controlled. This is related to the post-Fordist form of consumption through the notion of mass customization, where the market is broken down into segments and flexible packages offered to groups of consumers. This argument is contentious, however, with critics viewing it as over-generalized, presenting 'McDisneyification' as a monolithic process, ignoring individual choices and the diversity of tourist experiences (ibid., p.125).

Reflect

➤ In what ways are the new forms of tourist consumption discussed above bound up with the growth of postmodern culture?

12.4 Tourism and economic development

12.4.1 Tourism and uneven regional development

Tourism has traditionally been seen as an activity based primarily in peripheral and outlying regions. As such, it could be regarded as a counterbalance to many other industries in which firms tend to cluster together in core regions. In the UK, for example, mass tourism was traditionally concentrated in poorer regions, although not exclusively so, with important coastal resorts in the South-east (Agarwal *et al.*, 2000, p.246). Subsequently, of course, mass tourism has been developed in regions

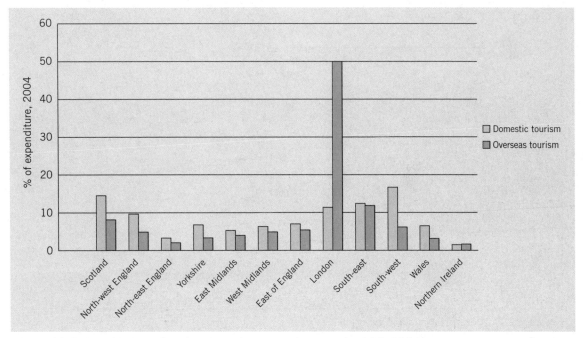

Figure 12.6 Distribution of regional tourist expenditure in the UK, 2004.
Source: ONS, 2005, *Travel Trends 2004*. Reproduced under the terms of the Click-Use Licence.

Box 12.2

The uneven development of tourism in Portugal.

Modern mass tourism in Portugal took off from the mid-1960s, with the number of international visitors rising from less than 100,000 in 1950 to an estimated 26 million in 1998 (Williams, 2002, p.86). Tourism was estimated to generate 17.9 per cent of total exports and 1.7 per cent of total employment in 2006, demonstrating its importance in macroeconomic terms (WTTC, 2006a, p.6). In common with other Mediterranean destinations, Portuguese tourism is dominated by a small number of source countries, mainly in northern Europe, particularly the UK and Germany, which generated 53.3 per cent of visitors in 1997 (Williams, 2002, p.87). Tourism has been a key force shaping regional development patterns in Portugal since the 1960s, reproducing and accentuating the traditional divide between the highly developed coastal regions and a backward interior. In particular, it has transformed the Algarve region in the south from an underdeveloped backwater to a leading centre of mass tourism (Figure 12.7).

The Algarve accounted for 42 per cent of total tourist overnight volume and 48 per cent of international

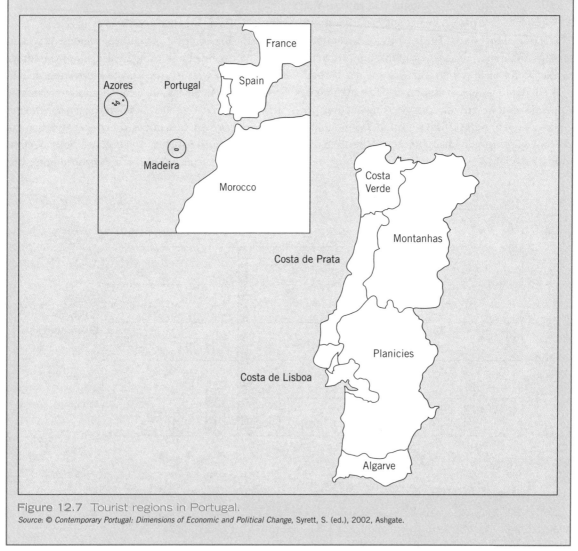

Figure 12.7 Tourist regions in Portugal.
Source: © Contemporary Portugal: Dimensions of Economic and Political Change, Syrett, S. (ed.), 2002, Ashgate.

Box 12.2 (continued)

tourism in Portugal in 2002 (WTTC, 2003, p.14). According to Williams (2002, p. 91), this represents 'a level of regional polarisation which is almost unsurpassed in Europe', although the level of concentration has fallen slightly over the last decade. The Algarve is heavily dependent on tourism, which is esti-

mated to generate 66 per of regional GDP and 60 per of total employment when its indirect effects are included (WTTC, 2003, p.6). Tourism is also important in Lisbon and the island of Madeira and, on a smaller scale, in the Costa Verde and Costa de Prata regions (see Figure 12.7). The Algarve accounted for 40 per cent of

tourist expenditure and the Lisbon area 33 per cent in 1997, above its share of visits. This reflects the prevalence of mass tourist facilities owned by foreign companies in the Algarve and Lisbon's capacity to attract more lucrative business trips and cultural tourism.

---

such as the Mediterranean, and many developing countries have invested in tourism facilities and infrastructure. A wide range of localities in the developed world have also sought to encourage tourism, often based on their history and heritage.

In the Mediterranean, tourist development is largely concentrated in **coastal strips** such as Spain's Costa del Sol or Portugal's Algarve and islands such as Majorca and Crete. As such, it has played an important role in shaping patterns of uneven regional development within these countries (Box 12.2). In developed countries in north-west Europe and North America, however, the geography of tourist development is more complex. Tourism seems to have become more dispersed across different areas since the 1960s as the traditional resorts and destinations have experienced decline and other regions that promoted tourism and provided facilities have prospered. In the case of the UK, Figure 12.6 shows that domestic tourism is relatively evenly distributed across the regions, with traditional destinations such as Scotland, south-west England and north-west England attracting the most visitors, along with London and the South-east. Overseas tourism, however, is very strongly concentrated in London, which accounts for 49.8 per cent of total overseas tourist expenditure in the UK (Figure 12.6). Other world cities of major historical interest, such as New York, Paris and Rome, also tend to attract large numbers of tourists. Tourism has become an important part of the dynamic service-based economies of such regions, serving to widen the gap between them and other non-core regions.

12.4.2 The local impact of tourism

Like all other economic activities, tourism has both positive and negative impacts on the local areas in which it is located. Positive impacts include the provision of employment, the generation of investment, the creation of new facilities and attractions, the improvement of infrastructure and a greater awareness of the region as a tourist destination (Hall and Page, 2002, p.132). Unlike some other economic sectors such as most retail or construction, which depend on local markets, tourism offers scope for the expansion and development of local economies through the injection of additional expenditure by visitors from outside the locality. The overall impact of such expenditure is measured in terms of its **multiplier effects** (Box 6.2), referring to the amount of additional income that is generated within the economy through the initial injection of tourist expenditure as tourist-oriented businesses buy goods and services from other firms and as their employees spend their wages. The multiplier effects of tourism seem relatively high compared with other forms of expenditure that might occur locally, with studies indicating that around half of tourism expenditure will remain in the locality (Urry, 2002, p.106), although much depends on the structure and ownership of the industry. As such, tourism has been embraced by many local authorities and agencies as a key element of their development strategies.

At the same time, tourism also has negative local impacts, including overdevelopment of an area so that its original attraction is eroded, outflows of profit to externally owned companies, environmental degradation (pollution, depletion of resources, visual intrusion on the landscape) and a vulnerability to potential future downturns in the tourist market, which is notoriously fickle from year to year and place to place. The main destinations and attractions all have an underlying **carrying capacity**, referring to the maximum number of tourists that a destination or site can support without damaging the environment (Hunter and Green, 1995, pp.66–8). Once this point is reached, tourist enjoyment and satisfaction will be compromised by overdevelopment and environmental degradation. Increasing numbers of visitors have forced many sites and attractions to adopt management plans and measures that seek to balance development and environmental protection. In Yellowstone National Park in Wyoming, for example, the impact of 2.5 million visitors a year has created problems of environmental conservation and management for the authorities, requiring the introduction of conservation measures such as banning cars, restricting the number of visitors and vehicles allowed in each day and closing camping grounds (Dove, 2003).

12.4.3 The evolution of tourist destinations

The most influential theory of how tourist areas evolve over time is Butler's **tourist-area cycle model**. This essentially takes the idea of product life cycles, influential in economics and management, and applies it to tourist resorts. The basic idea is that the life of a resort goes through a number of stages from initial exploration to ultimate decline or rejuvenation (Hunter and Green, 1995, pp.64-6) (Figure 12.8). The exploration phase is one of small-scale development based on a limited number of visitors seeking new experiences and sites. This is followed by the involvement stage, characterized by slow growth as more local residents and businesses provide facilities for tourism. This feeds into rapid growth during the development phase as more businesses invest in the area in the face of rising tourist demand and as marketing and development activities expand. Local control tends to give way to the growing influence of outside corporations and interests at this stage.

The consolidation phase is marked by the increasing

maturity of the industry, with a substantial portion of the local economy now dependent on tourism. The number of visitors continues to grow, but at a much slower rate than the previous stage. The Algarve region (Box 12.2) reached this stage in the early 1990s when the growth in tourist numbers began to slow. Stagnation refers to a phase in which the capacity of the destination has been reached or exceeded, resulting in congestion, environmental degradation and growing costs. Other, newer destinations become more attractive and tourist numbers begin to drop. While the Algarve has not yet reached this stage, concern has been expressed about the overdevelopment of tourist facilities detracting from the appeal of the region. The final stage is open to a number of possible scenarios, depending on the strategies adopted by tourist providers and the responses of the market. The best-case scenario is that of rejuvenation when redevelopment plans, involving, for instance, the establishment of new attractions and facilities and marketing initiatives, are successful and visitor numbers start to rise again. Butler's model provides a useful general framework for understanding the development of tourist resorts, seeming to fit the experience of many of the classic seaside resorts in Britain that have been in decline for decades. At the same time, however, it is

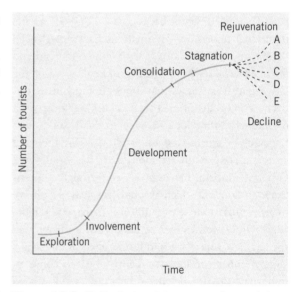

Figure 12.8 Butler's tourist-area cycle model.
Source: Reprinted by permission of Sage Publications Ltd from Shaw, G. and Williams, A.M., *Tourism and Tourism Spaces*, p.192. Copyright (© Gareth Shaw and Allan M. Williams, 2004). Modified from Butler 1980 by Shaw and Williams.

Box 12.3

Tourist development strategies in South Africa

In South Africa tourism has been 'viewed as an essential sector for national reconstruction and development' since the early 1990s (Visser and Rogerson, 2004, p.201). While considerably wealthier than most other developing countries, the apartheid era left South Africa with a horrendous legacy of social inequality, poverty and racial discord. At the same time, nearly 500,000 jobs in labour-intensive sectors such as mining and resource-based manufacturing were lost in the second half of the 1990s (Binns and Nel, 2002, p.236). In response, the government has adopted economic growth and diversification policies within which tourism has played a key role, attempting to capitalize on the country's rich natural and cultural resources.

Following the end of apartheid and the reintegration of South Africa into the international community, tourism grew rapidly. The number of international arrivals, for instance, increased from 3.1 million in 1993 to 5.7 million in 1998 (WTTC, 2002, p.16). South Africa also has a well-established domestic tourist industry, unlike many other developing countries, accounting for 33.5 million

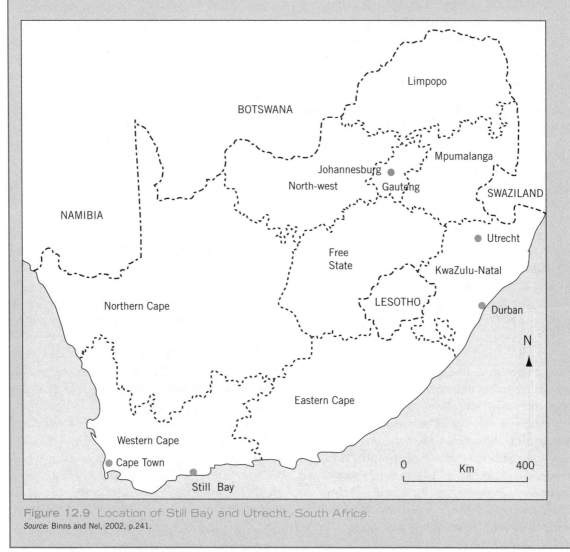

Figure 12.9 Location of Still Bay and Utrecht, South Africa.
Source: Binns and Nel, 2002, p.241.

Box 12.3 (continued)

overnight trips in 2000–1 (p.18). In 2006, the tourist industry was estimated to account for 8.2 of total GDP, 7.5 per cent of total employment and 13.9 per cent of total exports, leaving considerable scope for further growth (WTTC, 2006b, p.6). The provinces of Gauteng and the Western Cape attracted the most foreign air arrivals in 2000 by a considerable margin, followed by KwaZulu-Natal (WTTC, 2002, pp.17–18).

At the local level, a range of local authorities have adopted tourist-led development strategies, emphasizing the importance of community participation and local ownership. Many such strategies have an explicitly pro-poor focus, aiming to alleviate or eliminate poverty through the generation of jobs and income (Binns and Nel, 2002, p.237). Binns and Nel examine two local initiatives in particular, located in Still Bay in the Western Cape and Utrecht in

KwaZulu-Natal (Figure 12.9). The former is located in one of the country's premier tourist regions, with the town itself characterized by a natural beauty and tranquillity. This had been underexploited until the early 1990s, when the tourism-led development strategy was established by community leaders and local entrepreneurs in response to the collapse of the historically dominant fishing industry (ibid., p.240). Subsequent development succeeded in transforming the area, with increasing numbers of tourists attracting private investment and creating employment. Tourist-led development in Utrecht was prompted by a similar realization of the need to diversify the local economy away from coal mining, which fell into steep decline from the early 1990s. Although the results are not as dramatic as Still Bay, with job creation rather limited, the community has been refocused around the tourist-based strategy,

attracting financial support from the government.

Both these local examples demonstrate the potential for tourist-based local development in South Africa, with visionary local leaders and entrepreneurs playing a key role in drawing different agencies and interests together and obtaining financial support (ibid., p.245). At the same time, however, little real integration of the poor into development initiatives has taken place. There is also a real danger of a large number of uncoordinated local initiatives leading to market saturation through the duplication of facilities such as game parks and craft centres. Successful initiatives in favoured locations such as the main urban centres and tourist regions are able to generate a market-led momentum, but few of those developed in poorer, outlying townships tend to be economically viable, relying heavily on government subsidies.

important to appreciate that the model is a descriptive generalization, and not all resorts will necessarily go through all stages of the cycle in a linear sequence.

12.4.4 Tourism and developing countries

Many developing countries have adopted tourism as a key element within their economic development strategies since the 1980s (Box 12.3). In particular, it provides them with foreign exchange and a means of **economic diversification** away from overdependence on a small number of primary commodities. The aim is to attract international visitors to tourist resorts and attractions, capitalizing on the spread of cheap air travel. In the Caribbean island of Barbados, for example, the number of visitors grew from 15,000 in 1955 to 369,915 in 1980, compared with a national

population of 248,983 (Potter, 1983). The development of tourist facilities such as luxury resorts and wildlife parks has become fairly widespread, providing a so-called 'pleasure periphery' for rich Western tourists. At the same time, however, it is important to appreciate that domestic and regional tourism has also grown in a number of developing countries, particularly those, such as China and Thailand, that have experienced rapid economic growth since the mid-1980s (Shaw and Williams, 2004, pp.160–61).

While tourism has created employment and generated foreign currency to help pay for imports, the wider economic benefits (section 12.4.2) have not materialized for many developing countries. A key factor here is the external ownership and control of key parts of the industry. The main transport companies, hotel chains and tour operators operate globally, but tend to be based in the developed countries where most of the world's tourists are resident. Many visitors to developing countries will have their trips arranged and coordinated by tour companies based in their home countries, travel in Western airlines and stay in transnational hotel chains. Foreign ownership is particularly prevalent in the upmarket sections of the tourist industry, with 74 per cent of Class I hotel bed spaces in Barbados, for instance, owned by non-nationals (Potter *et al.*, 2004, p.171). The major hotel chains strive to provide a consistent and standardized service across the world, based on the norms of their home countries, with the Hilton intending each of its establishments to act as a 'little America' (quoted in Cater, 1995, p.210).

This concentration of foreign ownership limits the benefits of tourist development for host countries because it means a large share of the profits are repatriated abroad, along with the wages of managerial staff, while centralized purchasing policies mean that many inputs are supplied from abroad (ibid., p.201). These are referred to as economic **leakages**, in that many of the benefits of development are 'leaked' abroad due to the external ownership and control of the hotel industry (Figure 12.10). At the same time, a large proportion of tourist-related jobs are low paid, with the wages of Kenyan hotel staff, for example, lower than those on offer in many other sectors of the economy (Potter *et al.*, 2004, p.325). Tourism employment, in both developed and developing countries alike, is also often highly seasonal in nature, with many staff employed only during the peak season and facilities sometimes closed during the 'off season' (Cater, 1995, p.203).

The wider cultural impacts of tourism on developing countries have also been the subject of much debate (Cater, 1995; Potter *et al.*, 2004, pp.171–5). On the positive side, it is often claimed that tourism is itself a form of cultural exchange, allowing visitors to learn about the cultures and histories of other societies. In reality, however, the activities of the tour companies, hotel chains and promotional agencies ensure that the whole 'tourist experience is packaged so that it becomes a bubble wherein local traditions and practices ... are tailored to meet the needs of tourists' (Cater, 1995, p.209). The huge income gaps between tourists and the local population are starkly apparent, symbolized by the Western clothes, jewellery and photographic equip-

ment of the former, who frequent hotels, bars and restaurants that few locals could afford to visit (ibid., p.207). In many traditional societies, particularly Muslim ones, the behaviour of Western tourists may be offensive to the cultural and religious beliefs of the local population, generating tension and conflict. Despite such cultural concerns, economic imperatives such as the need to earn foreign exchange and provide employment for growing populations mean that the governments of many developing countries will continue to promote tourism as a key element of their development strategies.

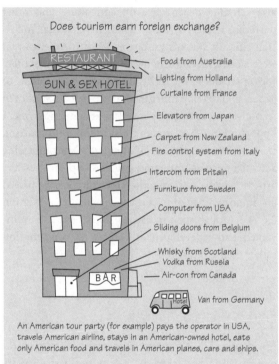

An American tour party (for example) pays the operator in USA, travels American airline, stays in an American-owned hotel, eats only American food and travels in American planes, cars and ships.

Who profits most?

Figure 12.10 Leakage in the tourist industry.
Source: Carter, 1995, p.202.

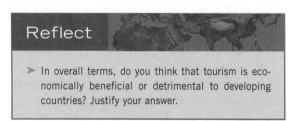

Reflect

➤ In overall terms, do you think that tourism is economically beneficial or detrimental to developing countries? Justify your answer.

12.5 Tourism, culture and economic regeneration

Tourism and culture have come to play a key role in local economic strategies in recent years (Kong, 2000). In this section, we are primarily concerned with local economic development in developed countries, having discussed the impact of tourism in developing countries in the previous section. We also focus on urban areas, which provide the most striking examples of **culture-led regeneration**, although tourism has played an important role in diversifying many rural economies away from a dependence on agriculture and other declining primary sector activities. In particular, industrial cities in Western Europe and North America that were faced with catastrophic declines in their manufacturing sectors in the 1970s and 1980s have sought to attract and develop tourism and associated activities in response, generating new investment and employment opportunities. Overlapping considerably with tourism, the cultural or creative industries – encompassing activities such as museums, galleries and theatres, the arts, the media, design and architecture – have played an important role in urban regeneration (section 8.6.2). Through such initiatives, cities have been reinvented and redeveloped as centres of consumption rather than production.

12.5.1 Urban entrepreneurialism

A widespread shift from urban managerialism to entrepreneurialism has taken place over the past 25 years (Harvey, 1989b). **Urban entrepreneurialism** is defined by the new focus on economic development and regeneration that emerged from the late 1970s and early 1980s, in response to problems of economic restructuring and deindustrialization. The managerialism of the post-war decades, by contrast, was primarily concerned with managing the delivery of welfare services such as housing, transport, education and welfare to local residents. The entrepreneurial approach saw cities focus on the need to generate growth and employment, seeking to attract new investment and funds from outside and generate new

business and income from within. The construction of facilities such as retail malls, conference centres, hotels, theatres and concert halls has been of central importance, alongside the attraction of cultural events and festivals. Cities have sought to promote themselves as vibrant and dynamic places within an increasingly global market, reinventing themselves as centres of consumption rather than production. More specifically, three key features of urban entrepreneurialism can be identified.

12.5.2 Public–private partnership

First, urban entrepreneurialism is based on a public–private partnership between local authorities and business. In many US cities during the 1970s and 1980s (e.g. Pittsburgh, Detroit and Boston), local government and business interests such as developers, construction companies and realtors (estate agents) came together to promote and redevelop the city following **deindustrialization**, aiming to attract new investment and employment (Cox and Mair, 1988). These groups were brought together by their shared interest in the economic fortunes of the locality, which provided the ultimate source of tax revenues and profits. They have been described as urban '**growth machines**' and coalitions (Molotch, 1976). Through entrepreneurial strategies, traditional local 'boosterism' (whereby local leaders promote and 'talk up' the area as a place to live and work) is integrated with the use of local government powers to try to attract external funding and investment (Harvey, 1989b, p.7) In Western Europe, regeneration has been driven and coordinated by the public sector, particularly local authorities, with business leaders and associations tending to play a secondary role (Ward, 1996).

12.5.3 Place marketing

Second, place marketing and promotion is a central element of **urban entrepreneurialism**. Local authorities and business leaders in a range of cities have developed specific projects and initiatives to sell the city to potential investors, tourists and residents, often employing techniques borrowed from the marketing and advertising industries. The aim is to reinvent the city within

Figure 12.11 'Glasgow's Miles Better'.
Source: © National Museums Scotland. http://www.scran.ac.uk.

a post-industrial economy, giving it a positive image as a dynamic, attractive place offering modern facilities and attractions. Cities are promoted as both places to work and places to play, although these strands have tended to overlap and merge in recent years, reflecting, again, the postmodern trend of '**de-differentiation**' (Urry, 2002, pp.84–7).

Key themes emphasized by promotional agencies to attract business include the qualities of the workforce (skilled and hard working), the benefits of a city's location (proximity to markets, borders, ports, communications, time zones), a pro-business climate of low taxes and assistance programmes, a high quality of life and local cultural and education facilities (Short, 1999, p.48). Specific slogans have come to play a key role within such promotional campaigns, summarizing the attraction of a city in a single short and memorable phrase. Slogans adopted by US cities in the 1990s include 'More Services, More Choices' (Baltimore), 'At the Heart of Everything' (Chicago) and 'The Business City that Never Sleeps' (New York) (ibid., p.49). In the UK, probably the best-known promotion slogan is

'Glasgow's Miles Better' (Figure 12.11), launched as part of a broader redevelopment strategy by the city council in 1983 (Paddison, 1993).

12.5.4 The creation of a new urban landscape

The third key element of urban entrepreneurialism is the creation of a new urban landscape, linked to the manipulation of image and culture. Old industrial spaces within cities have been transformed through new investment and development, creating new attractions and facilities for investors, residents and tourists. Particular types of redevelopment include the construction of shopping malls, cultural centres, heritage parks, conference centres and science parks. Through such investments, downtown areas and city centres have been transformed, while out-of-town developments such as the Metro Centre in Gateshead and Meadowhall in Sheffield have also played a crucial role in the regeneration of the wider city region.

Figure 12.12 Gateshead Quays Regeneration area.
Source: D. Mackinnon.

Waterfront districts, formerly functioning as ports and industrial areas, have often become the sites for flagship developments and renewal projects. The model for such developments is Baltimore's Harborplace, with other notable examples including Boston's Fanueil Hall Marketplace, South Street Seaport in New York and Bilbao's Abandoibarra district (Box 12.4) as well as Sydney's Rocks area (Box 12.1). Such areas are typically redeveloped as leisure and entertainment districts, focused on tourism in particular, coupled with some upmarket housing and flagship office complexes. The most high-profile waterfront redevelopment in the UK is London Docklands, where a derelict industrial area was transformed into the prestigious district of office complexes and housing containing Canary Wharf. Other examples include the Albert Dock in Liverpool, Birmingham Canal, Salford Quays in Manchester and Newcastle's Quayside. The latter contains a range of restaurants, bars and hotels on the Newcastle side of the River Tyne and a new arts and cultural quarter on the Gateshead side, linked by the award-winning Millennium Bridge (Figure 12.12).

12.5.5 The evolution and effects of entrepreneurial strategies

Tourism and culture have played a central role in entrepreneurial strategies, offering a source of new investment and employment – along with other, overlapping sectors like retail and leisure – to replace the old manufacturing industries. Museums, galleries, concert halls and theatres have provided a focus for investment and facilities to attract local residents and tourists alike. Bilbao provides a good example of culture-led regeneration (Box 12.4). In the UK, cities such as Glasgow, Liverpool and Newcastle have used culture and the arts as mechanisms of regeneration (Figure 12.12). Glasgow was designated as European City of Culture 1990, prompting further investment in cultural facilities and programmes, while Liverpool beat off competition from Newcastle to gain this award for 2008.

Entrepreneurial strategies have generated new investment and jobs in sectors like tourism, retail, culture and leisure. Glasgow, for example, has become the fourth largest tourist destination in the UK (and the most important retail centre outside London), gaining a reputation as a vibrant and fashionable city. New

Box 12.4

The regeneration of Bilbao

Located in the Basque region of northern Spain, Bilbao's economy was based on the heavy industries of iron and steel, coal and shipbuilding from the nineteenth century. Like other industrial cities, Bilbao experienced a catastrophic rundown of its industrial base, losing almost half of its manufacturing jobs between 1975 and 1996, with the share of indus-trial employment dropping from 46 to 23 per cent. Unemployment rose from 2.3 per cent in 1975 to 26 per cent in 1986, remaining around this level for the best part of a decade before falling in the late 1990s (Rodríguez et al., 2001, p.163).

After a period of considerable inertia, the city council and Basque regional government developed an entrepreneurial strategy in response to this local economic crisis in the early 1990s. The aim was to revitalize the city and create a new image of it as a dynamic and vibrant place. The main focus of the regeneration effort was the Abandoibarra district located in the city centre waterfront area (Figure 12.13) (ibid., p.170). The master plan for Bilbao

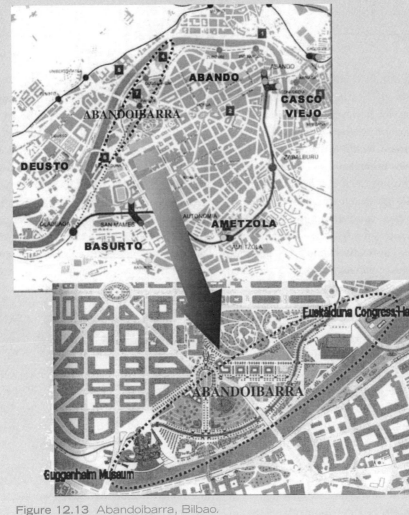

Figure 12.13 Abandoibarra, Bilbao.
Source: Rodríguez et al., 2001, p.169.

Box 12.4 (continued)

proposed transforming Abandoibarra into a new business centre based on advanced services, high-income housing, retail and leisure facilities and cultural attractions. It was promoted as the flagship scheme that would drive the regeneration of the wider city region. The new facilities included an international conference and concert hall and the Guggenheim Bilbao Museum, based on a partnership between local and regional government and the Guggenheim Foundation, which became the

central anchor and symbol of the whole development.

The former opened in 1999 and the latter in 1997, requiring substantial amounts of funding from the public sector. The development was highly successful in terms of securing the physical renovation of the area and changing the image of the city. The Guggenheim Museum, designed by Frank Gehry, has been widely praised as an architectural landmark and symbol of urban regeneration, attracting 1.4 million visitors in its

first year of operation (ibid., p.174). Yet there is little evidence that the museum has played a broader generating role in attracting other investment and employment to the area. This is a common problem with property-led regeneration that is often successful in creating new flagship buildings and facilities for incomers and tourists, but fails to deliver real benefits for existing residents and communities.

investment has successfully transformed many derelict and run-down districts, creating a new urban landscape. Rather ironically, however, while development agencies have been concerned to promote their places as distinctive and unique within a highly competitive environment, the new urban landscapes that they have created through projects such as shopping centres, hotels and business parks tend to look rather similar. In this sense, inter-urban competition for investment, public funds and visitors has fostered a certain sameness, often erasing distinctive industrial histories. Competitive pressures within an increasingly saturated market mean that some cities are likely to struggle to attract enough visitors and business to make new facilities and attractions viable, while others will surge ahead.

The benefits of regeneration have often failed to reach low-income groups in run-down inner-city neighbourhoods or peripheral public housing schemes. Many of the jobs tend to be low paid and part-time, compared with the relatively well-paid employment in manufacturing industries that has been lost. They have often been taken up by women and younger people rather than older workers made redundant from industry. Regeneration schemes have been criticized as having a rather limited and superficial impact, with prestigious new developments serving to mask the continuing realities of poverty, unemployment and poor housing facing some established

residents (Imrie and Thomas, 1999). The images and slogans adopted by local authorities and promotion agencies have been criticized as false and misleading, aimed at external capital and potential tourists, distorting and sanitizing the real industrial histories of cities like Glasgow, Milwaukee and Chicago (see Boyle and Hughes, 1994). The influx of new residents and businesses has changed the social mix of redeveloped areas, sometimes leading to the marginalization of existing residents. Links with local community groups have often been limited, with residents' groups coming together to oppose developers' plans in some cases (Imrie and Thomas, 1999).

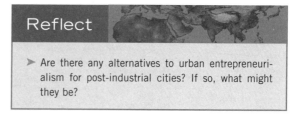

Reflect

➤ Are there any alternatives to urban entrepreneurialism for post-industrial cities? If so, what might they be?

12.6 Summary

The starting point for this chapter was the economic importance of tourism, often described as the world's biggest industry. Global tourist patterns remain highly unequal, however, with the three core regions of Europe, North America and Asia and the Pacific both generating and receiving the overwhelming majority of the world's tourists (Table 12.1). Since the 1970s, the dominance of mass tourism has been challenged by the rise of new forms of tourist consumption, including heritage tourism, ecotourism, adventure holidays and visits to theme parks and mega-malls. In contrast to the standard package holiday offered by the 'Fordist' mass producers, these 'post-Fordist' products serve specialized market niches and segments. They are underpinned by a romantic version of the **tourist gaze** associated with privacy, solitude and a personally meaningful relationship with the objects of the gaze, in contrast to the collective gaze that structures mass tourism (Urry, 2002). The growth of 'post-Fordist' forms of tourist consumption is related to the rise of new middle-class groups who derive much of their social status from the accumulation of cultural capital. More independent forms of travel that involve learning about other cultures, environments and histories are a key means of generating such cultural capital. As part of this phenomenon, the boundaries between tourism and other spheres of consumption, such as shopping, entertainment, education and sport, are becoming increasingly blurred, a process known as 'de-differentiation'.

Based on the anticipated economic benefits, many governments and local authorities have promoted tourism as a key part of their economic development strategies over the last couple of decades. In many developing countries, however, the potential benefits of tourist development have been limited by the external ownership and control of key sectors of the industry, resulting in substantial 'leakage' of profits and revenue to external interests. In developed countries, tourism and associated cultural industries have been promoted as key vehicles of economic diversification and regeneration. In formerly industrial cities in particular, the authorities have followed entrepreneurial strategies of place promotion and large-scale redevelopment, seeking to reinvent their cities as spaces of consumption within a post-industrial economy. Centrally located waterfront districts, such as Baltimore's Harborplace, London Docklands, Sydney's Rocks area and the Abandoibarra district of Bilbao, have been selected for flagship development schemes. In many cases, the effects have been dramatic as the construction of new cultural attractions, office blocks, apartments, shops, bars and restaurants has transformed the urban landscape. More broadly, however, the benefits of such regeneration have often failed to trickle down to lower-income communities in neighbouring areas, contributing to increased levels of social polarization and inequality.

Exercises

With reference to a specific area within a post-industrial city of your choice, critically assess the costs and benefits of tourist-led regeneration.

1. What have been the main elements of the regeneration strategy?
2. What organizations and agencies have been involved?
3. What new facilities and attractions have been established?
4. What campaigns and slogans have been deployed to promote the area?
5. To what extent can the regeneration scheme be seen as an example of urban entrepreneurialism?
6. How many jobs have been created?
7. What social groups are employed?
8. How does the quantity and quality of jobs compare with the old industrial economy?
9. Have any groups been marginalized?
10. Who has benefited from the redevelopment of the area?

Key reading

Cater, E. (1995) 'Consuming spaces; global tourism' in Allen, J. and Hamnett, C. (eds) *A Shrinking World? Global Unevenness and Inequality*, Oxford: Oxford University Press, pp.182–231.
A stimulating and wide-ranging discussion of the growth and impacts of tourism and travel, although rather dated now. Emphasizes how tourism is implicated in the 'shrinking' of space associated with processes of time–space compression. The unevenness of tourism on a global scale and the limited benefits of tourist development for developing countries are key themes.

Desforges, L. (2005) 'Travel and tourism' in Cloke, P. Crang, P. and Goodwin, M. (eds) (2005) *Introducing Human Geographies*, 2nd edn, London: Arnold, pp.517–26.
A brief and lively discussion of the relationship between tourism and place, in terms of both the cultural and economic impacts and how places are promoted and packaged for tourist audiences. Suggests that the popular critique of tourism for commercializing and 'spoiling' previously untouched places is simplistic and misleading.

Harvey, D. (1989b) 'The transition from managerialism to entrepreneurialism: the transformation of urban governance in late capitalism', *Geografiska Annaler* 71 B: 3–17.
A very important paper that has informed a whole generation of research on urban entrepreneurialism. After identifying the main elements of entrepreneurial policies, Harvey explains the widespread adoption of such policies in terms of the impact of deindustrialization, the pressures of inter-urban competition and the shift from Keynesian to neoliberal forms of economic regulation

Shaw, G. and Williams A.M. (2004) *Tourism and Tourism Spaces*, London: Sage.
Probably the best recent textbook on the geographies of tourism. It adopts an integrated approach that brings together the processes of production and consumption with a consideration of particular types of tourist spaces. The commodification of the tourist experience is a central theme.

Urry, J. (2002) *The Tourist Gaze*, 2nd edn, London: Sage.
A highly influential theoretical assessment of tourism in the context of wider economic, social and cultural processes. In this book, Urry develops the notion of the tourist gaze and examines its evolution in relation to different forms of tourism.

Useful websites

http://www.world-tourism.org/
The official site of the World Tourist Organization, a specialized agency of the United Nations. Contains useful and up-to-date data on global tourism.

http://www.wttc.org/
The website of the World Travel and Tourism Council, established in 1990 and made up of representatives of the leading private sector companies. Again, contains useful summary statistics and more detailed information on the economic impact of tourism in particular countries.

http://www.staruk.org.uk//default.asp
The official site of the UK Research Liaison Group, made up of representatives of the different national tourist boards and central government. Contains a range of data on trends in UK domestic and inbound international tourism.

http://www.ecotourism.org
The website of the International Ecotourism Society, offering a range of information on ecotourism.

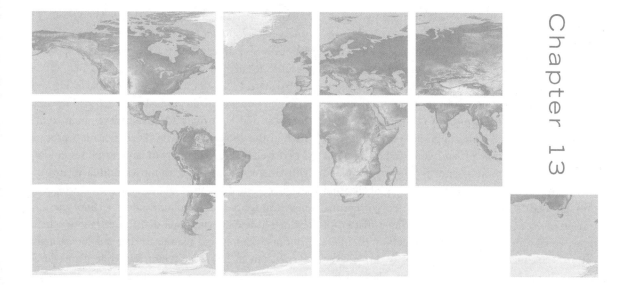

Chapter 13

Conclusion

In this book, we have examined the changing geography of economic activity within the contemporary world economy, focusing particularly on the location of different types of activity, the economic relationships between different regions and the economies of particular localities. Our approach to these issues is based on a revised political economy approach informed by cultural and institutionalist insights, making it more open to the importance of context, difference and identity. From this perspective, we view the economic geography of the world as shaped by the interaction between general processes of capitalist development and pre-existing local conditions and practices. This broad framework, developed in Part 1 of the book, has underpinned and informed our detailed examination of contemporary topics in economic geography in Part 2. In this short concluding chapter, we summarize the key themes of the book, answering the questions outlined on p.2, and offer some reflections on contemporary economic development policy at the regional and global scales.

13.1 Summary of key themes

The three interrelated themes running though the book are globalization and connections across space, uneven development and place. **Globalization** should be viewed as an ongoing process, not a final outcome or 'end state', that is shaped by the actions of a wide range of individuals, organizations, firms and governments (Dicken *et al.*, 1997). The economy has become more globally integrated in terms of significantly increased flows of goods, services, money, information and people. The process of globalization has been facilitated by the development of a new set of 'space-shrinking' transport and communications technologies since the 1960s, resulting in **'time–space compression'** as the costs of moving materials, capital and information across space have been dramatically reduced (Harvey, 1989a). MNCs, financial markets and international economic organizations like the World

303

Bank and the IMF can be viewed as key agents of globalization. States should be viewed as active facilitators, rather than passive 'victims', through the adoption of neoliberal policies including the abolition of exchange and capital controls, the reduction of trade barriers and the implementation of privatization programmes. At the same time, the project of neoliberal globalization has encountered strong opposition in recent years from a loose coalition of 'anti-globalization' groups, sparking periodic protests and the beginnings of an alternative model of 'bottom-up' globalization through bodies such as the World Social Forum. Instead of economic integration creating a 'flatter' or more equal world through the reduction of international inequalities, globalization is an uneven process, creating prosperity in some places while others become increasingly marginalized. On a global scale, this is symbolized by the dramatic growth of East Asia, especially China, and the stagnation of sub-Saharan Africa since 1980.

Uneven development is a basic characteristic of capitalism as a mode of production, reflecting the tendency for capital and labour to move to the areas where they can secure the highest returns (profits and wages respectively). In this way, growth becomes concentrated in core regions though a process of 'cumulative causation' (Myrdal, 1957) (Box 4.2). Surrounding regions get left behind, becoming subordinate peripheries supplying resources and labour to the core. Patterns of uneven development are not static, however, with capital and labour moving between locations in search of higher profits and wages. The process of economic growth in a particular region eventually tends to undermine its own foundations, leading to overheating as the prices of labour, land and housing spiral inexorably and underinvestment in infrastructure creates congestion. Lower costs in other underdeveloped regions attract capital and labour. This 'see-sawing' of capital between locations creates shifting patterns of uneven development at different geographical scales (N. Smith, 1984). This process of uneven regional growth and decline is apparent from the decline of so-called 'rustbelt' regions (northern England, the American Midwest) and the rise of 'sunbelt' ones (the south and the west of the US, southern Germany).

An important contemporary manifestation of uneven development processes is the **spatial division of labour**, where different parts of the production process become located in different types of regions. On a global scale, MNCs moved routine manufacturing and assembly operation in industries such as textiles, footwear and electronics to certain developing countries in the 1960s and 1970s. This relocation process was driven by the availability of large surpluses of low-cost labour in developing countries, and facilitated by the increasing division of labour in manufacturing and new transport and communication technologies. It created a **new international division of labour** where routine assembly and manufacturing was increasingly carried out in low-wage countries, while higher-level functions such as strategic management and research and development remained based in developed countries. More recently, the offshoring of IT-enabled services to developing countries such as India has prompted talk of a '**second global shift**', creating a new international division of labour in services (Bryson, 2006). A key question is whether 'offshoring' will extend to higher-value services such as legal research and financial analysis.

Patterns of uneven development are structured by the conflicting processes of spatial concentration and dispersal. The question of which of these processes is predominant will vary according to a range of factors, including the type of economic activity in question, the level of available technology, the development of the (technical) division of labour and the size and location of the market being served. ICTs are a key force in reshaping the economic landscape, facilitating the relocation of both manufacturing and services on a global scale. Their influence does not herald the 'end of geography', however, as shown by the concentration of corporate headquarters and advanced financial and business services in world cities. This brings us to the crucial observation that high-value economic activities show a marked tendency towards spatial concentration in large urban areas, while lower-value activities are far more susceptible to dispersal to lower-wage locations. We regard this as a key underlying 'rule' or 'law' of economic geography. Various qualifications can be made – for instance, high-value financial services are supported by a range of low-status services from cleaning to catering in clusters such as the City of London and Wall Street – and various exceptions found, of course, but these do not diminish the force of the basic point.

The role of **place** represents the third key theme of the book, reflecting, in part, the effects of wider processes of economic development in creating distinctive forms of production in particular places. These general processes interact with pre-existing local conditions and practices (e.g. resource bases, employment patterns, skills, income levels, cultural values, institutional arrangements, political orientations) to create new geographies of production and consumption. Rather than being a production of geographical isolation, the distinctiveness or uniqueness of places is actually reproduced through this process of interaction (Johnston, 1984). From this perspective, place can be regarded as a key meeting point or node constructed out of wider social relations and connections (Massey, 1994, pp.154–5). Such relations and connections span the spheres of production, consumption and circulation, with transport and communication networks and financial flows playing important roles in connecting places together. The volume of these spatial flows and connections has increased markedly under globalization, inspiring Massey's effort to develop 'a global sense of place'.

The assumption that globalization is erasing the distinctiveness of place, making distant localities appear increasingly similar, is only true at a very superficial level. In terms of consumption, for instance, the spread of global brands such as McDonald's and Coca-Cola is often cited, together with the profusion of large retail malls and centres. Yet the associated idea of a single global consumer culture is crude and simplistic, masking a more complex process whereby different local cultures have become increasingly mixed and entangled (P. Crang, 2005). Local sites of consumption remain highly significant, not least in terms of the growth of 'post-Fordist' tourist consumption in which place itself is a central object of consumption. In the sphere of production also, place remains important in the global economy, although many of the specialized industrial regions of the past have disappeared in the face of wider processes of economic restructuring. Particular places continue to be associated with dis-

tinctive forms of economic activity. Numerous examples have been cited and discussed through the text, from the densely clustered financial districts of Wall Street and the City of London, the high-tech centre of Silicon Valley and the burgeoning service industries of Bangalore to the deindustrialized regions of Wales and north-east England and the impoverished villages of southern Africa. As this suggests, some places have prospered under globalization while others have been left behind. The effects that the processes of uneven economic development will have in different places is not something that can be easily predicted or modelled in advance, since they are the product of the actions of a large number of individuals and institutions in different locations. This means that the shaping of the economic landscape is characterized by a basic openness and unpredictability, a quality that lends economic geography much of its interest as an academic subject.

13.2 Globalization, neoliberalism and regional development

Alongside its role in describing and explaining the location and distribution of economic phenomena, there is also a need for economic geography to engage with economic development policy across different geographical scales (section 1.3.2). Traditionally, local and regional development policies have been the main objects of attention, but questions of international and national economic regulation have become increasingly pertinent in recent years, not least in terms of their influence on local and regional strategies. As we have emphasized, neoliberalism has become the dominant economic policy framework since the late 1970s, providing the overarching framework for local and regional development. This is based on inter-regional competition, focusing on supply-side measures that aim to stimulate enterprise and innovation with regions (section 6.5.4). This supply-side approach presents regional development as a 'race to the top', where all regions can be winners if they focus on enhancing their competitiveness in terms of skills, innovation and new-firm formation (Bristow, 2005). In reality, however, the unrestrained mobility of capital that neoliberal reforms have encouraged has allowed MNCs to play regions off against one another, on the basis of labour costs and the financial assistance offered by government agencies, sparking fears of a 'race to the bottom' as wage rates and living standards are progressively undermined. Any consideration of an alternative model of regional development needs to consider basic questions about what kind of regional development for whom, raising normative or ethical issues of equity and justice (Pike *et al.*, 2006). In our view, a more progressive policy needs to focus on the problem of uneven development, rather than regional growth per se, aiming to reduce regional disparities.

The prevailing supply-side approach focuses on the high-value sectors of the formal economy, influenced by theories of the knowledge economy and business clusters. Research and development, innovation and the commercialization of academic research tend to be heavily emphasized, focusing support on industries such as biotechnology, electronics and computing, information technology and certain financial and business services, together with tourism and the cultural industries. Beneath these high-value sectors, however, lie a range of other more mundane activities, some based on essentially non-capitalist motives and practices, as symbolized by the 'iceberg' model of diverse economies (section 1.3.1) (Gibson-Graham, undated). These tend to be overlooked by orthodox regional development strategies, despite their importance in providing employment, particularly in disadvantaged areas. Beyond the heady rhetoric of globalization, many regional economic transactions remain tied to meeting local demand for services and materials (Krugman, 1996, p.9). This means that there is considerable scope for the introduction of demand-led approaches that seek to boost the overall demand for employment in the economy, stimulating local labour markets through classic Keynesian multiplier effects (Box 6.2). At the same time, there is a need to go beyond traditional top-down Keynesian initiatives to encourage bottom-up development and participation, involving support for community development and emphasis on the potential of the so-called social economy to generate employment and growth. A focus on the inter-regional dimension is also a vital component of any approach that seeks to address regional disparities, remembering the basic insight that these are a product of the relationships between regions rather than the inadequacies of poorer regions per se (e.g. a lack of enterprise or innovation) (Massey, 2001). In particular, there is a need for governments to adopt a more interventionist approach that seeks to direct investment into less-favoured regions while actively controlling growth in prosperous core regions such as the south-east of England. In the absence of such measures, there is little prospect of reducing or even containing regional disparities.

At the global level, neoliberalism underpinned the **Washington Consensus** of the 1990s (section 6.5.2), based on its enthusiastic adoption by the US government, the IMF and the World Bank. Governments across the world have sought to liberalize and deregulate their economies, adopting policies of low inflation, open competition and privatization. Such policies have proved controversial, often becoming associated with cuts in public expenditure and increased levels of inequality. In the context of developing countries particularly, IMF- and World Bank-inspired structural adjustment programmes have foisted often inappropriate policies of liberalization and deregulation on heavily indebted economies, with harsh social consequences in terms of, for instance, the reduction of health and education expenditure and the introduction of user charges for basic services.

Local social movements protesting against particular development projects and initiatives have become increasingly interconnected in recent years, forming the basis of the so-called 'anti-globalization' or Global Justice Movement (section 5.7). In reality, it is the neoliberal project of liberalization and deregulation that is the main target of protest for this movement, although some groups term themselves anti-capitalist, aiming to overthrow the established economic system. While protest groups have been criticized in the mainstream media for not formulating a coherent alternative to 'corporate globalization', despite the establishment of bodies such as the World Social Forum, the significance of their demonstration that an alternative approach is even possible should not be underestimated. This is because of the situation in the 1990s when neoliberalism went virtually unchallenged at the level of global policy and ideology – notwithstanding various local protests – in the wake of the collapse of communism, celebrated by US conservatives as the 'end of history', that marked the final triumph of liberal capitalism over socialism (Fukuyama, 1989). In reality, however, the shape and direction of the economy is actively contested between different ideas and social movements rather than being structurally determined by abstract economic forces. This is something that has become increasingly apparent with the rise of the Global Justice Movement since the late 1990s. Partly as a result of opposition to their policies, the World Bank and the IMF have adjusted their approach, embracing the so-called 'post-Washington' poverty consensus (Maxwell, 2005), although the underlying economic reform agenda (e.g. reducing the role of the state through privatization, opening up to trade and investment) remains largely in place.

A central element in any effective alternative economic strategy involves the regulation and reform of global institutions, including the World Bank and the IMF, and their rules, eliminating the conditionality measures that allow them to force the adoption of inappropriate policies on developing countries. **Conditionality** should be replaced by recognition of the benefits of countries opening up their economies gradually, according to local social priorities. This would allow them to protect 'infant industries' and maintain capital and exchange controls, in much the same way as developed and East Asian countries did during the nineteenth and late twentieth centuries respectively. While the decision by the G8 leaders to write off the debts of 18 countries represents a welcome step in the right direction, there is a need to extend this to more countries and to increase the level of relief on offer. The provision of aid by developed countries remains very low, and real movement towards the target of 0.7 per cent of GDP by 2007 is crucial. Unfortunately, however, progress here has been rather disappointing, with most countries actually counting the debt relief agreed in 2005 as aid, rather than com-

mitting additional resources as they originally agreed (Oxfam, 2006). The collapse of the ill-fated **Doha Round** of trade negotiations raises real questions about the future of the WTO and the model of multilateral trade negotiations that it embodies. Supporters of the WTO commonly assert that poorer countries are likely to lose the most from a movement away from the WTO model towards bilateral deals between two or more countries. The experience of southern European countries such as Spain and Portugal, however, shows that less-favoured economies can gain significantly from regional trade agreements when an open trade regime is accompanied by a substantial redistribution of resources from richer to poorer states. Indeed, this brings us to the crucial underlying point: any serious effort to tackle international inequalities and introduce a more equitable form of globalization depends on rich Northern countries committing more resources to economic development. As such, the question is primarily one of political will, requiring the governments of the G8 countries in particular to look beyond their own short-term economic interests.

A final consideration to note at this point is the relationship between global capitalism and the environment. Without resorting to 'Club of Rome'-style pessimism about the 'limits to growth' (Meadows et al., 1972), there is certainly a sense in which the apparent scarcity of key resources like oil and water and mounting evidence of global warming (Klare, 2002) question some of the taken-for-granted assumptions that underpin the global economy. Not least among these are cheap energy and low transport costs, facilitating large-scale movements of goods, services, raw materials and people over long distances. While green arguments for the relocalization of the economy, implying a reversal of globalization, can seem unrealistic, the era of 'peak oil' may lead to a partial rebalancing of the economy, together with growing consumer demands for greater local authenticity and sustainability (Hudson, 2005, pp.205–11). Rather ironically, this could see economic geography return to some of its traditional concerns, emphasizing the importance of transport costs and the need for proximity to raw material sites and markets.

Glossary

Absorptive capacity. This refers to a firm's ability to recognize, assimilate and exploit knowledge. It seems to rely upon the existence of a common corporate culture and language: that is, everybody shares the same broad outlook and sense of the company's overall purpose and objectives.

Accumulation function. This refers to the state's activities in supporting and promoting economic development within its territory, ensuring that business can accumulate capital for investment and growth.

Agglomeration economies. A set of economic advantages for individual firms that are derived from the concentration of other firms in the same location – such as the availability of skilled labour, proximity to a large number of customers, access to specialist suppliers of services, and the provision of an advanced infrastructure.

Aid. Official development assistance offered by developed countries to promote economic development and poverty reduction in developing countries.

Anti-Globalization Movement. The movement that has arisen to oppose the inequalities brought about by free market and **neoliberal** economic policies; also known as the **Global Justice Movement.** The movement is significant in bringing together diverse social groups (for example, farmers, trade unionists, environmentalists) from both the global North and South.

ATTAC (Association pour la Taxation des Transactions Financieres pour l'aide aux Citoyens). A social movement originally created in France by political journalists, but now an international network of around 30,000 activists in over 30 countries campaigning for a global tax on international currency speculation to be used to fund development in the global South. The tax, referred to as the Tobin Tax, after US economist James Tobin, is one of the key demands of the **Global Justice Movement.**

Ba. A Japanese term used to emphasize the importance of context in shaping innovation and learning processes. While not necessarily corresponding to a particular physical space, *ba* does refer to a specific time and space defined by the interaction between individuals involved in the process of knowledge creation.

Back office functions. Routine clerical and administrative tasks such as the maintenance of office records, payroll and billing, bank cheques and insurance claims.

Backwash effects. The adverse effects of growth in a core region on a surrounding region in terms of capital and labour being 'sucked out' of the latter region into the former which offers higher profits and wages. In this situation, the virtuous circle of growth in the core is matched by a vicious circle of decline in the periphery, expressed in classic symptoms of underdevelopment such as a lack of capital and depopulation.

Basic needs. An approach to development prominent in the 1960s and 1970s that focused attention on the everyday lives of the poor, reflecting concerns that these were being neglected by orthodox modernization policies. The concern for identifying and meeting such needs in relation to food, shelter, employment, education, health, etc., is shared by the recent work on **grass-roots development.**

Blended shore strategies. An approach adopted by companies whereby they combine elements of 'off-shoring', 'near-shoring' and 'onshoring' across their overall business operations.

Branch plant economy. A term coined to describe the way particular regions become dominated by 'branch plants', which undertake basic production activities but are controlled from outside. Branch plant economies emerged through economic restructuring processes in the 1960s and 1970s whereby processes of corporate growth at national and international level produced a new spatial division of labour taking over from, or displacing, local ownership.

Bretton Woods. The global financial agreement (named after the town of the same name in New England), brokered by the leading capitalist nations, intended to create a stable global economy in the aftermath of the Second World War. The agreement, based upon keeping exchange rates fixed against the dollar, facilitated a growth in international trade and an expansion in the global economy and lasted until the early 1970s.

Business process outsourcing. The relocation of standardized operations such as **call centres** and 'back office' **functions**.

Business services. Services offered to other businesses. These are usually high-value activities employing skilled labour. Examples include accountancy, legal services, research and development, management consultancy, advertising and marketing.

Call centres. Offices that provide services to customers remotely over the telephone, focusing on marketing, routine customer enquiries and more sophisticated technical support. Recently rebranded as 'contact centres' or 'customer service facilities', they have grown very rapidly since the early 1990s.

Capital accumulation. The process of investment and reinvestment in production to generate higher profits. Accumulation lies at the heart of the capitalist system, representing the basic driving force for economic growth and innovation.

Capital switching. The process by which capital is moved between sectors of the economy and regions in response to changing investment opportunities. In geographical terms, capital is often transferred from regions dominated by declining sectors to 'new industrial spaces' in distant regions offering more attractive conditions for investment.

Capitalism. A particular mode of production, dominant since the eighteenth century, based on the private ownership of the means of production – factories, equipment, etc. – and the associated need for the majority of people to sell their labour power to capitalists in exchange for a wage.

Carrying capacity. The maximum number of visitors that a destination or site can support without damaging the environment. Once this point is reached, tourist enjoyment and satisfaction will be compromised by over-development and environmental degradation.

Central place theory. Developed by Walter Christaller in the 1930s, this is perhaps the best known of the **German locational models**. Based on the assumption of economic rationality and the existence of certain geographical conditions such as uniform population distribution across an area, central place theory offers an account of the size and distribution of settlements within an urban system. The need for shop owners to select central locations produces a hexagonal network of central places.

Centralization of finance. The process by which financial markets have become larger and more integrated over time with local and regional banks giving way to national and, increasingly, global systems.

Circuit of capital. The basic process of producing commodities for profit under capitalism, involving the combination of the means of production – factories, machines, materials, etc. – and labour power to produce a commodity for sale, generating a profit above the initial money outlay. Part of this profit is reinvested back into the production process, which recreates the circuit anew and forms the basis of capital accumulation. In addition to the primary circuit of capital in production, secondary and tertiary circuits are also commonly identified, referring to investment in the built environment and education, health and welfare, respectively.

Clusters. A more popular term for agglomeration involving the concentration of economic activities in a particular location. The term has become associated with the work of Michael Porter, a Harvard business economist, in recent years. Porter defines clusters as 'geographical concentrations of interconnected companies, specialized suppliers, service providers, firms in related industries, and associated institutions (for example universities, standards agencies and trade associations) that compete but also co-operate'.

Codified knowledge. Also termed 'explicit knowledge', this refers to formal, systematic knowledge that can be conveyed in written form through, for example, programmes or operating manuals. *See also* **Tacit knowledge**.

Cold War. The political and ideological conflict between Western capitalist countries led by the USA and a communist block led by the Soviet Union, lasting from the mid 1940s to the late 1980s.

Colonialism. A political and economic system based on territorial empires, involving the direct political control of colonies by the colonial powers.

Commercial geography. A key component of the traditional approach to geography, prominent from the 1880s to the 1930s. Closely linked to European imperialism, it provided knowledge about colonial territories in Africa, Asia and Latin America, identifying and mapping key resources, crops, ports and trade routes, and relating these to climate and settlement patterns.

Commodification. The process by which particular objects, relationships and ideas are turned into goods and services that are bought by individual consumers through the market. The commodification of place through tourism and other activities is of particular interest to geographers.

Commodity chains. The network of connections involved in the production, circulation and consumption of a commodity, covering the different stages from the supply of materials to final consumption. They typically incorporate a range of actors, for example, farmers, subcontractors, manufacturers, transport operators, distributors, retailers and consumers. Commodity chains have a distinct geography, linking together different stages of economic activity carried out in different places.

Commodity fetishism. The tendency for the geographies of production and distribution that actually generate the goods on sale in Western shops to be obscured by the emphasis consumers attach to the physical appearance and price of goods.

Communities of practice. A term derived from organizational studies, which emphasizes the close informal contacts and ties built between individuals working together in the same field. Increasingly, it is argued, such communities of practice can be developed and maintained through networks that operate across distance, using technologies such as email, conference calls and video conferencing. As such, the notion of communities of practice is often opposed to the traditional emphasis on geographical communities.

Comparative advantage. A key principle of international trade theory, classically expressed by David Ricardo in 1817. It states that a country should specialise in exporting goods that it can produce more cheaply than other goods and import goods that are more expensive to produce domestically. Through specialization both countries gain by focusing on the goods in which they have a comparative or relative cost advantage and importing those in which a country has a comparative disadvantage.

Competence or resource-based theory of the firm. Derived from the work of the economist Edith Penrose, the competence perspective views firms as bundles of assets and competencies that have been built up over time. Competencies can be seen as particular sets of skills, practices and forms of knowledge.

Competence trust. The confidence that other firms – for example, suppliers or customers – will meet their commitments, based on expectations of performance derived from reputation and/or past experience.

Competition state. The idea that a new type of state is emerging in developed countries, the key task of which is the promotion of innovation and economic competitiveness rather than the provision of social welfare services.

Competitive advantage. The dynamic advantages derived from the active creation of technology, human skills, economies of scale, etc., by firms. This can be contrasted with the rather static and naturalistic notion of **comparative advantage**, which relates efficiency to pre-existing endowments of the main factors of production. Competitive advantage helps to account for patterns of trade and regional specialization at a more detailed level.

Competitiveness. A key concept underpinning economic development policy since the early 1990s. It refers to the underlying strength of the economy in terms of its capacity to compete with other countries, based on the assumption that nations and regions compete for global market share in a similar fashion to firms. Key aspects of competitiveness include levels of innovation, enterprise and workforce skills and government is charged with fostering these capacities.

Conditionality. The requirement for developing countries to meet strict conditions set by the World Bank and

International Monetary Fund in return for providing grants and loans. In particular, the Bank and IMF have required countries to undertake a set of economic reforms, generally known as **Structural Adjustment Programmes**, based on the principles of the **Washington Consensus**.

Consumer culture. A key feature of contemporary society, which is increasingly organized around the logic of individual choice in the marketplace with shopping and consumption an increasingly central part of people's lives and identities. Modern consumer culture developed through the targeted of middle-class women in particular during the nineteenth century with the establishment of large department stores in cities playing a key role.

Consumption. The processes involved in the sale, purchase and use of commodities by individuals.

Creative destruction. A term coined by the Austrian economist Joseph Schumpeter to capture the dynamism of capitalism in terms of innovation and the development of new technologies. As new products and technologies are developed and adopted (creation), they often render existing industries obsolete, unable to compete on the basis of quality or price (destruction).

Culture-led regeneration. A set of policies for the economic redevelopment of depressed areas, which emphasizes the importance of culture for growth through attracting investment and jobs. Culture-led regeneration typically includes the repackaging of the history and heritage of an area for tourism, museums, galleries, theatres and arts festivals.

Cultural 'turn'. An important development in human geography and the social sciences from the late 1980s, involving a shift of focus from economic to cultural questions.

Cultural industries. The cultural or creative industries include: performance, fine art, and literature; their reproduction: books, journal magazines, newspapers, film, radio, television, recordings on disk or tape; and activities that link together art forms such as advertising. Also considered are the production, distribution and display processes of printing and broadcasting, as well as museums, libraries, theatres, nightclubs, and galleries. Analyses often tend to begin with distinct products such as film, music recordings or books, emphasizing the links between the interrelated processes of production, processing, distribution and consumption. The cultural industries have been characterised as innovative, flexible and creative, operating at the intersection of the local and the global scales, often by packaging and selling local distinctiveness through global distribution networks.

Cumulative causation. A model of the process of uneven regional development derived from the work of the Swedish economist Gunnar Myrdal. This explains the spatial concentration of industry in terms of a spiral of self-reinforcing advantages that build up in an area and the adverse effect this has on other regions, creating a core–periphery pattern.

Debt crisis. The problem of large-scale indebtedness facing many developing countries since the 1980s, undermining development efforts and, at times, threatening the viability of the world financial system. The origins of the debt crisis lie in the interactions between three sets of factors: the borrowing of large sums by developing countries from Northern banks and institutions in the 1970s; rising interest rates in the late 1970s and early 1980s; and reduced commodity prices, and thereby export earnings, since the early 1980s.

Decision point. A key threshold that countries must reach before they can begin receiving debt relief from the World Bank and IMF, requiring them to have a stable economic policy framework in place and to have prepared 'Poverty Strategy Reduction Papers'.

Dedifferentiation. The blurring of boundaries between tourism and other forms of consumption regarded as a central aspect of postmodern culture, such as shopping, leisure, entertainment, education and sports.

Deindustrialization. A decline in the importance of manufacturing industry as a sector of the economy, expressed in terms of its share of employment or output. Generally associated with the growth of service industries and the closure of older heavy industries such as coal, steel and shipbuilding, deindustrialization has been a common experience across developed countries since the 1960s.

Dependency theories. The most prominent set of the **structuralist theories**, particularly associated with the radical economist Andre Gunder Frank. According to Frank, the metropolitan core exploits its 'satellites', extracting profits (surplus) for investment elsewhere. **Colonialism**

was a key force here, creating unequal economic relations that were then perpetuated by the more informal imperialism characteristic of the post-war period.

Deregulation. The reduction of the rules and laws under which business operates, a key component of **neoliberalism**.

Derivatives. New financial instruments, defined as contracts that specify rights/obligations based on (and hence derived from) the performance of some other currency, commodity or service used to manage risk and volatility in global markets. Key forms of derivatives include futures, swaps and options.

Deskilling of labour. The removal of more rewarding aspects of work such as design, planning and variation, often associated with an increased division of labour. The sub-division and fragmentation of the labour process increases employers' control of production, reducing workers to small cogs in the system and making them vulnerable to replacement by machines, which often perform standardized tasks more efficiently. This is what Marx meant when he spoke of the alienation of labour under capitalism, with industrial workers playing an increasingly minute role within tightly controlled factories, becoming divorced from the products of their labour.

Development. In the context of economic policy, the term conveys a sense of positive change over time, making a particular country or region more prosperous and advanced. As an economic and social policy, development has been directed at those 'underdeveloped areas' of the world that require economic growth and modernization.

Developmental state. A particular type of state that is heavily involved in the promotion of economic development. The term is derived from the experience of East Asian countries such as Japan, South Korea and Singapore where the activities of the state were crucial in fostering export-led expansion. Weiss's definition of the developmental state emphasizes three key criteria: the aim of increasing production and closing the economic gap with the industrialized countries; the establishment of a strong government department to co-ordinate and promote industrial development; and close cooperative ties with business.

Dialectics. A way of thinking that sees change as driven by the tensions between opposing forces, usually in the form of thesis–antithesis–synthesis. Capitalist society in particular is characterised by continual change and flux, driven by the search for profits in the face of competition. Dialectics is originally derived from the eighteenth century German idealist philosopher Hegel, before being adapted by Marx and subsequent Marxists like David Harvey.

Diamond model. Michael Porter's representation of how business clusters actually operate, enhancing competitiveness and productivity by fostering the interaction between four sets of factors: demand conditions; supporting and related industries; factor conditions; and firm strategy, structure and rivalry.

Digital divide. Differences in access to information and communication technologies (ICTs) between social groups and countries.

Digitization. The general shift of data storage and transmission from analogue to digital formats, reflecting the merging of computer services and telecommunications following the microelectronics revolution of the 1960s and 1970s. When coupled to the spread of the Internet, digitization allows information to be collected, packaged and distributed far more rapidly and efficiently than before.

Discourses. A key term derived from poststructuralist philosophy that refers to networks of linked concepts, statements and practices that produce distinct bodies of knowledge. Crucially, meaning is generated through particular discourses that, instead of simply reflecting an underlying reality, actively create it.

Disintermediation of production networks. A process, usually associated with the spread of new technologies like the Internet, which removes the need for intermediaries between the buyers and sellers of goods.

Diverse economies. A term that signifies an emphasis on the different forms of economic activity and organization that co-exist with **capitalism** such as domestic work, cooperatives and gift-buying.

Division of labour. A key principle of industrial society, which has technical, social and geographical dimensions. The technical division of labour can be defined as the process of dividing production into a large number of highly specialized parts, so that each worker concentrates on a single task rather than trying to cover several. Under industrialization, as Adam Smith argued, an increased division of labour results in huge rises in productivity.

Doha Round of trade negotiations. The latest round of discussion launched in Qatar in late 2001 under the auspices of the World Trade Organization (WTO) (which replaced GATT in 1995).

Dual city hypothesis. The theory, associated with the work of the urban sociologists Manual Castells and Saskia Sassen, that leading world cities are characterized by a growing divide between rich and poor as a result of changes in the economic base. Relatively well-paid manufacturing jobs have been lost as factories shut down or moved out of the city in search of lower costs. The new occupational structure is characterised by high-status, high-paid jobs in business and financial services and a range of routine, low-paid jobs such as cleaners, porters, waiters and bartenders, which exist to serve the lifestyle needs of the rich.

Dual labour markets. The idea that labour markets are divided into workers of different status and job quality. Core workers are usually those in permanent, and often skilled, occupations that are relatively well paid and recompensed (for example, with pensions and social benefits). Peripheral workers, on the other hand, occupy lower paid and less stable forms of work, being typically unskilled.

E-commerce. This can be defined as 'trade that actually takes place over the Internet through a consumer visiting a seller's website and making a transaction there'. E-commerce is divided into business-to-business (b2b) and business-to-consumer (b2c) transactions.

Economic diversification. The process by which a regional or national economy changes in orientation, dependent on accessing new markets. It is associated with the creation of a broader base of economic activities and reducing dependence on an existing number of sectors, particularly if the former are in decline.

Economic geography. A major branch of human geography that addresses questions about the location and distribution of economic activity, the role of uneven geographical development and processes of local and regional economic development.

Economic leakage. The process whereby income and employment are lost to a local area, accruing to individuals and organizations based elsewhere. Such leakage is often associated with the external ownership and control of key industries.

Economics. An important neighbouring discipline to economic geography that views the economy as governed by market forces which basically operate in the same fashion everywhere, irrespective of time and space. In contrast to the diversity and open-endedness of economic geography, economics is a formal theoretical discipline based on modelling and quantification.

Economies of scale. The tendency for firms' costs for each unit of output to fall when production is carried out on a large scale, reflecting greater efficiency. Industrialization led to huge economies of scale through the establishment of large factories employing sophisticated machinery and an elaborate **division of labour**.

Embeddedness. A key concept – derived from economic sociology – adopted by economic geographers in recent years. Originally referring to the notion that economic action is grounded in social relations, geographers have 'spatialized' the concept to emphasize how particular forms of economic activity are rooted in particular places.

Employment. The act of selling your labour to work for an employer in the formal economy, usually paid in the form of a weekly wage or a monthly salary. Paid employment as the norm is a form of work distinctive to capitalist society.

Equity finance. Funds raised by investors buying a stake or share in a **firm**. For large, publicly quoted firms, this occurs through the issuing of shares on the stock market.

Ethnography. A research methodology that involves in-depth, close interaction and observation by the researcher, with research subjects, usually employing qualitative techniques. Ethnographic approaches have become increasingly popular in economic geography with the institutional and cultural turns and emphasis upon understanding the economy as an embedded process.

Eurocentric/Eurocentrism. A preoccupation with European or Western experiences and practices, reflecting an assumption that Western values and methods are always superior to those found in developing countries. This has been widely criticized as arrogant and condescending, particularly by post-colonial writers.

Explicit knowledge *see* **Codified knowledge**

Export platform. Production enclaves in parts of the less developed world set up by **multinational corporations** (or their suppliers) from the developed world to supply markets in advanced economies. These are typically disconnected from the local or national economy in which they are situated and often have low levels of regulation and labour standards, prompting claims of exploitation, especially of female workers.

Export processing zones. Similar to export platforms, but offering added tax and investment incentives to foreign firms to encourage relocation. Typically these will include up to 100 per cent rebates on local taxation, the provision of all infrastructure, and the relaxation of the usual rules governing foreign ownership. Some of the most notorious of these zones have been set up in Mexico along the border with the United States and have led to the establishment of branch plants known as 'maquiladoras' (literally 'assemblers').

Export-orientated industrialization (EOT). Often regarded as the opposite strategy to **import-substitution industrialization**, EOI involves countries producing goods and services for selling in external markets. It is compatible with traditional notions of free trade and comparative advantage in contrast to ISI, which involves high levels of protection and state intervention.

Factors of production. The different elements that are bought together to produce particular goods and services: capital, labour, land and knowledge.

Fair or ethical trade. The emergence of an alternative system of international trade in commodities and products, such as clothes, that links concerned consumers (primarily in richer countries) with farmers and producers in the global South. Such alternative trade networks aim at giving producers a fair and stable price that allows decent living standards and is not subject to global market fluctuations or cost pressures from dominant multinationals. Fair trade is a trading partnership, based on dialogue, transparency and respect, which seeks greater equity in international trade.

Fictitious commodity. A material or product that appears to be a commodity but cannot be regarded as a 'proper' one. Land, labour and money were originally identified as 'fictitious commodities' by the economic sociologist Karl Polanyi in his analysis of nineteenth-century industrialization. Such commodities are not directly produced for sale on the market and often require state intervention to balance supply and demand.

Financial exclusion. Defined as those processes that serve to prevent certain social groups and individuals from gaining access to the financial system. It is based on income, compounding the difficulties facing disadvantaged individuals and groups.

Financial globalization. The massive increase in global financial flows that has occurred since the 1970s; facilitated by **neoliberal** policies, information and communication technologies and the development of new financial and monetary products.

Financial services. Service based on the exchange and processing of money, including commercial and investment banking, and trading in financial products and insurance of all types. The commercial and residential real estate industry is often bracketed alongside financial services.

Firm, the. A legal entity involved in the production of goods and services, owned by individual capitalists or, more commonly, a range of shareholders. The standard organizational form of capital.

Fiscal crisis of the state. A situation in which the state lacks the resources to meet its financial obligations. The term was coined to describe the financial difficulties facing many national and local governments in the developed world in the 1970s as the post-war era of economic growth gave way to recession. This meant that tax receipts fell while welfare expenditure continued to grow fuelled by rising unemployment. New York City became technically bankrupt in 1975, for example; while the UK was forced to seek an emergency loan from the IMF in 1976.

Fisher-Clark thesis. A theory, originally developed in the 1930s, which argued that rising levels of productivity through the development of new technologies in one sector mean that workers move into next sector. Thus, increased productivity in manufacturing in the post-war period was releasing labour that became employed in services. A direct parallel was drawn between this and increases in agricultural productivity in the eighteenth and nineteenth centuries which enabled workers to move into the manufacturing sector.

Flexible labour markets. An employment regime in which wages, conditions and worker attitudes become more responsive to the pressures of competition and the needs of business, requiring workers to accept varying pay rates and hours of work while learning new skills and undertaking new tasks. The creation of flexible labour markets has been a key goal of neoliberal policy since the 1980s.

Flexible production. A new form of production regarded in the 1980s and early 1990s as the successor to **Fordism**, based on the rise of new sunrise industries, such as advanced electronics, computers, financial and business services and biotechnology. In the sphere of production, flexibility was associated with the widespread use of information and communications technologies that enabled processes and equipment to be continually reprogrammed and reset. In the sphere of consumption, it was rooted in a new emphasis on niche markets and customization compared to the standardized mass markets of the post-war period.

Flexible specialization. Closely related to flexible production, this is a form of economic organization based on the flexible use of information and communication technologies in production and the growth of customized and niche markets. It was identified by industrial sociologists and economic geographers as an important new form of production to rival mass production in the 1980s, based on the rise of craft-based industrial districts such as those of the 'Third Italy'.

Fordism. A regime of accumulation, dominant from the 1940s to the 1970s, based on a crucial link between mass production and mass consumption, provided by rising wages for workers and increased productivity in the workplace. This regime of accumulation was supported by a Fordist–Keynesian mode of regulation where the state adopted highly interventionist policies of demand management, full employment, welfare provision, trade union recognition and national collective bargaining. Fordism is named after the American car manufacturer, Henry Ford, who pioneered the introduction of mass production techniques. As such, the term is also used in a narrower sense to refer to an **intensification of the**

labour process, based on a highly elaborate division of labour and the introduction of moving assembly lines. *See also* **Neo-Fordism**.

Foreign direct investment. Overseas investment in production or service provision.

Geographical expansion. The spatial expansion of the economy outwards form original core areas to encompass new territories, an inherent feature of capitalism since the sixteenth century, reflected in the development of an integrated world economy from an initial core in Western Europe. Geographical expansion has been driven by profit-seeking and competition between firms, generating a search for new markets, new sources of raw materials and new supplies of labour. It gained new momentum as the industrial revolution took off, culminating in the 'age of empire' between 1875 and 1914. The shift towards globalization since the 1970s can be seen as the latest chapter in this ongoing story.

German location theory. A body of spatial economic theory from the nineteenth and early twentieth centuries – based on the work of German theorists such as Von Thunen, Weber, Christaller and Losch – which developed models of the economic landscape derived from the assumptions of **neoclassical economic theory**. *See also* **Central place theory**.

Global consumer culture. A phrase emphasizing the creation of a single global market, centred upon brands like McDonalds, Coca-Cola and Nike. For many commentators, this is erasing the distinctiveness of local places and cultures, heralding the 'end of geography'. As a number of studies have shown, however, this cultural homogenization argument is highly simplistic.

Global days of action. Mass protests against **neoliberalism** that converge on cities on particular dates to coincide with the meeting of global elites (for example, finance ministers of the G8, World Bank or IMF meetings). In recent years, protests in cities such as Genoa and Prague have drawn hundreds of thousands of protesters and resulted in violent clashes with police.

Global Frame Agreements. Global-wide agreements between trade union federations and **multinational corporations** that agree to certain basic rights and conditions for all workers employed by the corporation. These usually include the rights to organize in trade unions, decent wages, freedom from discrimination on the grounds of gender or race, and the end to child labour. GFAs are not legally binding but are voluntary agreements and therefore do not have the status of collective bargaining agreements signed at the national level.

Global Governance Institutions. Quasi-government organizations (for example, World Bank, IMF, WTO) originally established after the Second World War to help manage the global economy but increasingly playing a role in policy formation, monitoring and regulating the actions of national governments.

Global Justice Movement *see* **Anti-Globalization Movement**

Global pipelines. A term that refers to the construction of channels of communication with selected partners outside a cluster or region. Successful establishment of global pipelines requires firms to develop a shared organizational context that enables them to learn and solve problems together.

Global production networks. An approach in economic geography that integrates an analysis of commodity chains with regional development by considering how key actors at within global business networks interact with regional and national actors, in a process termed 'strategic coupling', to produce the space economy.

Global sense of place. An attempt to rethink place in an era of globalization, associated by the geographer Doreen Massey. This approach rejects the idea of 'place' as isolated and bounded, viewing it as a meeting place, a kind of node where wider social relations and connections come together.

Globalization. A process of economic integration on a global scale, creating increasingly close connections between people and firms located in different places. Manifested in terms of increased flows of goods, services, money, information and people across national and continental borders.

Globalization from below. A new model of 'bottom-up' globalization that emerged out of the increasing linkages between the local social movements, and facilitated by modern communications and the Internet. Networks such as People's Global Action and the World Social Forum exemplify this, providing an alternative to the dominant neoliberal model of 'globalization from above'.

'Goodwill' trust. A 'deep' form of trust associated with collaborative ventures where firms are satisfied that the commitment of partners goes beyond explicit contracts and agreements.

Governance. A term used to refer to the growth of new ways of governing societies in place of the traditional notion of 'government'; it incorporates special purpose agencies, business interests and voluntary organizations alongside government bodies. Geographers are particularly interested in the emergence of new forms of local governance since the 1980s.

Grass-roots development. An approach to development that is directly concerned with the practical problems and needs of poor people in developing countries. The ethos is one of 'helping people to help themselves'; it involves small-scale projects that directly benefit individuals, families and households, supporting local services and livelihoods.

Green revolution. The intensification and mass commercialisation of agriculture in the less developed world; a process ongoing since the 1970s. One of its effects has been the release of agricultural labour onto urban labour markets as a low-wage resource.

Gross Domestic Product (GDP). The indicator commonly used to measure a country or region's level of income or wealth, based upon a calculation of all the output produced by an economy and its value.

Heavily Indebted Poor Countries (HIPC) Initiative. An initiative launched by the World Bank and IMF in to reduce the debts of the poorest countries in the world. To be eligible countries must be very poor (according to World Bank and IMF categories), face an unsustainable debt situation (defined as a debt to export ratio of above 150 per cent) and have sound macroeconomic policies consistent with poverty reduction and sustained growth.

Host regions. Destination regions for foreign direct investment (FDI).

Household or domestic labour. Work undertaken within households as opposed to in workplaces and usually associated with the performance of tasks associated with social reproduction (that is, looking after children or performing work for the family). Under capitalism, such work is usually unpaid and predominantly performed by women.

Human Development Index (HDI). A widely known composite measure of development established by the United Nations Development Programme (UNDP), and published annually since 1990 in its *Human Development Report*. The HDI measures the overall achievement of a country in three basic dimensions of human development – longevity, knowledge and a decent standard of living.

Immaterial nature of service outputs. A defining characteristic of services is that their outputs do not take a physical of material form, but involves the performance of particular tasks, the arrangement of events and activities, and the offering of advice. This is generally defined against the tangible nature of manufactured goods.

Import-substitution industrialization (ISI). Occurs when a country attempts to produce for itself goods that were formerly imported. Newly created '**infant industries**' are protected from outside competition through the erection of high tariff barriers, allowing the country's economy to become diversified and dependence on foreign technology and capital to be reduced. *See also* **Export-orientated industrialization**.

Income elastic demand. The tendency for demand for a good or services to rise with increased incomes. Demand for service and high-value manufactured goods is generally income elastic compared to many agricultural goods and raw materials, where demand tends to remain static.

Incremental innovation. Relatively small improvements in the design and operation of particular products and services, as emphasized by recent interactive models. Innovation, in this sense, can be seen as a continuous process of technical improvement or learning, akin to the '*kaizen*' concept derived from Japanese management practice.

Industrial districts. Specialized industrial areas based on networks of small firms and craftsmen. Associated with regions like the Sheffield cutlery district (from the nineteenth century until the 1980s); and a number of scholars draw attention to the revival of industrial districts in central and north-eastern Italy.

Industrial society. Used usually to define the period of mass industrialization in western (that is, North American and Western European) society, from around 1750 through to the 1960s, when society went through major social and economic changes, not least of which was the transition from a rural and feudal way of life to an urban and capitalist one. Increasingly used as a term to distinguish it from the current post-industrial (post-1970) society.

Industrial working class. The class of workers that emerged with the development of modern industry from 1750 onwards who, as Marx noted, were distinguished by the fact that their sole means of 'earning a living' was employment or waged labour. As an increasingly large and concentrated group of workers, with considerable collective strength through their trade unions, the industrial working class were viewed as being at the forefront of struggles for social change in the first half of the twentieth century. With the **deindustrialization** of many societies in the global North, this class is perceived as being in terminal decline and increasingly marginal to mainstream **capitalism**, although this is a very western-centric perspective. A new industrial working class is growing in many parts of the global South (particularly in China) with the growth of manufacturing industry.

Infant industries. New industries at an early state of development that are regarded as strategically important by the state, which seeks to foster their development. This involves protection from outside competition until such industries are strong enough to compete in global markets.

Informal sector. The unofficial part of the economy, existing outside of the officially recognized, money-based and state-regulated economy. It is estimated to account for between 25 and 40 per cent of output in developing countries, consisting mainly of low-value services such as street hawking, shoe shining, car washing, etc.

Innovation. The creation of new products and services or the modification of existing ones to gain a competitive advantage in the market. The commercial exploitation of ideas is crucial, distinguishing innovation from invention.

Institutional foundations of markets. The wider social rules, norms and practices upon which the workings of markets depend, allowing transactions to take place and contracts to be respected.

Institutions. Broad social and organizational conventions, practices and rules that shape economic life. Key institutions include firms, markets, the monetary system, the state and trade unions.

Intellectual property rights. The ownership rights, often known as copyright, held by a firm or individual over a particular product such as a piece of music, a book, or a drug.

Intensification of the labour process. An increase in the volume and speed of work within a factory or office, often associated with the introduction of new technology and organizational techniques. **Fordism** involved an intensification of the labour process, based on the introduction of mechanized assembly lines.

Interactive approach to innovation. A more recent perspective that views innovation as a circular process based on cooperation and collaboration between manufacturers or services providers, users (customers), suppliers, research institutes, development agencies, etc.

International division of labour. A pattern of geographical development that involves different countries specializing in different types of economic activity. The classic 'old' international division of labour of the nineteenth century, associated with colonialism, involved the developed countries of Europe and North America producing manufactured goods while the underdeveloped world specialized in the production of raw materials and foodstuffs. This can be contrasted with the **'new' international division of labour** of the 1970s and 1980s.

International policy transfer. The transfer of policies between states, often involving initiatives and experiments that have been introduced in one country being adopted in another.

International strategic alliances. These refer to the growing number of collaborations between firms who are competitors in the same markets. They are particularly pronounced in sectors that require high levels of research and product development and therefore have high start-up costs (for example, computers and biotechnology). Their growth highlights the vulnerability of **multinational corporations** to changes in global markets and the desire to minimize uncertainty.

Just-in-time (JIT) production. A form of production organization prevalent in Japanese companies whereby firms maintain minimal stocks of supplies and components by maintaining a close relationship with suppliers. JIT systems were viewed by many commentators as being more flexible and efficient than Fordist production methods, providing Japanese car manufacturers

with an important source of competitive advantage over US and European rivals during the late 1970s and early 1980s.

Keynesian economic policies. A set of economic policies, widely adopted by developed countries between the 1940s and the 1970s, derived from the economic theories of John Maynard Keynes. Keynesianism involved the state taking an active role in managing the aggregate level of demand for goods and services in the economy. This involved stimulating demand in downturns by increased government expenditure – supporting public works and employment schemes, for example, road building projects – or reducing taxes. Then, in periods of economic growth, the state would dampen down demand by increasing taxes or reducing expenditure, thus preventing inflation.

Keynesian Welfare State *see* **Welfare state**

Knowledge process outsourcing. The relocation of high-value services such as medical research and financial analysis.

Knowledge-based economy. The idea that a new type of economy has developed since the early 1990s in which knowledge has become the key resource and learning the key process for firms and individuals.

Kondratiev Cycles. Named after the Soviet economist Kondratieff who first identified such cycles in the 1920s, the term refers to long waves of economic development, based on distinctive systems of technology. Five Kondratiev cycles are usually distinguished since the late eighteenth century. Each cycle consists of two distinct phases: one of growth (A) and one of stagnation (B).

Labour geography. A growing branch of economic geography that is concerned with how workers and their trade unions help to influence the changing geography of the economy. A central concern of labour geography is to provide a balance to perspectives that are dominated by business and state actors.

Learning regions. A concept of the region as a nexus of innovation and learning processes, rooted in the local concentration of **tacit knowledge** and formal and informal ties and relationships between firms.

Legitimation function of the state. The range of activities undertaken by the state to maintain social order, ensuring that the capitalist system and the associated social order is regarded as legitimate and 'natural' by the majority of citizens.

Liberalization. The opening up of protected sectors of the economy to competition, a key component of **neoliberalism**.

Linear model of innovation. The traditional understanding of innovation as a series of well-defined stages running from the research laboratory to the production line, marketing department and retail outlet. The linear approach is focused on large corporations, emphasizing formal research and development involving scientists and engineers operating separately from other divisions of the company.

Living Wage movement. Local campaigns waged at the local (usually city) level by trade unions and other social groups to create decent minimum wages and conditions for low-paid workers, typically in services sectors such as cleaning, security, etc. Campaigns are distinguished by being area-based rather than targeting single employers, tactics thought to be more effective in the dispersed workforce of the service sector. Living wage campaigns are also significant because they often involve women and ethnic minority groups, traditionally ignored by the established trade union movement.

Local 'buzz'. An intangible dimension of the agglomeration process derived from the concentration of a large number of firms in a location. The term conveys the vibrancy and excitement of daily life within a cluster generated through a proliferation of activities and events. It operates through different modes of communication such as chatting, gossiping, brainstorming, in-depth discussions and problem solving.

Local labour markets. A term used to signify the importance of the local scale in the operation of most forms of employment, whereby the majority of the population live and work within relatively small spatially delineated areas. The term 'travel-to-work-area' has been developed to statistically define local labour markets, identifying geographical boundaries that correspond to the limits within which the majority of a population of an area live and work. A major contribution of economic geography is to highlight the way that labour markets vary widely in their processes and practices, both at the local and national level.

Local social movements. Community-based groups that have been established to protest against extreme poverty and to defend local livelihoods in face of their commodification or appropriation by the state and corporations in developing countries.

Localization economies. A particular type of **agglomeration economy** stemming from the concentration of firms from the *same* industry in a particular area.

Macro-regions. Large world regions incorporating a number of nation-states such as North America or Europe which account for significant levels of trade within their borders and increasingly associated with the emergence of new supra-national bodies (EU, NAFTA) designed to enhance economic integration and growth.

Marxism. A set of social and economic theories derived from the writings of Karl Marx. Marxism adopts a materialist view of society, stressing the importance of underlying social relations and forces over ideas. The economy is structured by a capitalist **mode of production** defined by the antagonistic social relations between the capitalist and working classes. The exploitation of workers forms the basis of capitalist profit, but this contradictory relationship will also ensure of capitalism is ultimately overthrown by socialism.

Mass consumption. A form of consumption based on the purchase of standard consumer durables such as automobiles, fridges and washing machines by large numbers of households. This was a key dimension of post-war **Fordism**, linked to mass production by rising wages for workers.

Mass tourist industry. Large-scale holidays and package tours that tend to be highly standardized and rigid in nature, underpinned by the economies of scale developed by the large tour companies, hotel chains and airlines. Mass tourism held sway from the late nineteenth century to the 1970s and is epitomized by the traditional seaside resorts of England and, later, the Mediterranean resorts and *costas*, based on the provision of standardized package holidays.

McDisneyification. The idea, derived from Ritzer's theory of 'McDonaldization', that tourism is becoming increasingly predictable and standardized, subject to manipulation by controlling technologies and large corporations. The 'McDisneyification' thesis contends that tourists want their holidays to be predictable, efficient and controlled.

Millennium Development Goals. A set of objectives for development agreed by the United Nations in 2000, providing a focus for the new anti-poverty agenda. The Millennium Development Goals list specific targets to be achieved by 2015 against which progress can be measured and monitored on an annual basis. These include the eradication of extreme poverty and hunger, the achievement of universal primary education, and the promotion of gender equality.

Modernization School. The dominant approach to development in the 1950s and 1960s. The basic idea is that developing countries undergo a linear process of transformation (modernization), analogous to the changes experienced by developed countries in the nineteenth century following the industrial revolution. Economic growth is paramount, generating increased income and employment opportunities, which, it was assumed, would 'trickle down' to the poorest groups in society. This thinking was famously expressed in Rostow's **stages of economic growth** model.

Modes of production. Economic and social systems by which societies are organized, determining how resources are utilized, work is organized and wealth is distributed. Economic historians have identified a number of modes of production, principally subsistence, slavery, feudalism, capital and socialism.

Modes of regulation. A concept developed by the French regulationist school of political economy that emphasizes the role wider processes of social regulation play in stabilizing and sustaining capitalism. These wider processes of regulation find expression in specific institutional arrangements that mediate and manage the underlying contradictions of the capitalist system, enabling renewed growth to occur. Modes of regulation are focused on five key aspects of **capitalism** in particular: labour and the wage relation, forms of competition and business organization, the monetary system, the state and the international regime. The postwar period of economic growth was based upon a Fordist–Keynesian mode of regulation. *See* **Fordism**.

Movimento Sem Terra (MST). Based in Brazil, this is a mass social movement with an estimated 1.5 million members. Founded in 1984, it is made up of mainly landless labourers and peasants from rural areas, many of whom have been dispossessed and displaced by agricultural reforms, mechanization and land clearance. The Movimento's strategy involves targeting large unused private estates and illegally squatting and occupying the land.

Multi-National Corporations (MCNs). Companies which conduct operations in a number of countries, allowing them to access different markets and to take advantage of geographical differences in conditions of production, such as the skills and costs of labour. MNCs have been key agents of **globalization** since the 1970s.

Multiplier effects. The effects of additional expenditure in the economy in creating jobs and generating income as businesses buy goods and services from other firms and their employees spend their wages. The attraction of tourists is a good source of such multiplier effects within a local or regional economy.

Narmada River Valley project. A large-scale hydro-electric project in central India, running through the states of Madhya Pradesh, Maharashtra and Gujarat. It involves the construction of 30 large dams as well as 135 medium-sized and 3,000 small dams, designed to provide water and energy for agriculture and industry. The scheme will result in the inundation of large areas of land; the river is also regarded as sacred by the Hindu and tribal populations of India. This has sparked protest by the Narmada Bachao Andolan ('Save Narmada Movement'), which is made up largely of peasant farmers and indigenous people.

Nation, the. A group or community of people who feel themselves to be distinctive, on the basis of a shared historical experience and cultural identity, which may be expressed in terms of ethnicity, language or religion.

National collective bargaining. A system in which representatives of employers, trade unions and government got together to agree pay rates and awards at a national level, often on an annual basis. This form of bargaining was a key feature of **Fordism** between the 1940s and 1970s with labour gaining higher wages in exchange for increased productivity.

Near-shoring. The relocation of services to (usually neighbouring) countries of a broadly similar level of development (for example, UK to Ireland, USA to Canada). *See also* **Offshoring**; **Onshoring**.

Neil Smith's theory of uneven development. This important contribution to Marxist geography explains uneven development in terms of the movement of capital between locations. Capital is attracted to areas that offer high profits for investors, with their resultant development leaving other areas behind. Over time, however, development leads to rising costs in core areas, prompting capital to move to other, less-developed regions where costs are lower. It is this 'see-sawing' of capital between regions, driven by the need to maintain profit levels, that creates patterns of uneven development.

Neoclassical economic theory. Mainstream economic theory on economics based on the assumption that people and firms act in a rational and self-interested manner, continually weighing up alternatives on basis of cost and benefits. The market is viewed as essentially self-regulating, tending towards a state of equilibrium or balance through the role of the price mechanism in mediating between the forces of demand and supply.

Neoconservativism. A term used to refer to the approach of the US Republican administration under George W. Bush since 2000 in particular. Neoconservatism emphasizes the need for social solidarity, order and pre-emptive state action in the face of an unpredictable range of external threats and dangers such as global terrorism and 'rogue states'. The need for a strong state to ensure security is prioritized over the neoliberal emphasis on promoting competition and enterprise.

Neo-Fordism. A phase of reinforced Fordism in the 1960s and 1970s as mass production technologies became increasingly routine and standardized.

Neoliberal approach/neoliberalism. A political and economic philosophy and approach to economic policy that seeks to reduce state intervention and embrace the free market, stressing the virtues of enterprise, competition and individual self-reliance.

New economic gography. A label applied, somewhat loosely, to describe the new culturally and institutionally informed research that has grown since the early 1990s, characterised by a focus on the links between economic action and social and cultural practices in different places. Confusingly, the same term is also used by spatial economists to describe their research, a body of work that is better defined as the '**new geographical economics**'.

New economy. A rather vaguely defined term, based on the spread of the advanced information and communication technologies, particularly the Internet.

New geographical economics. An approach to economic geography developed by the economist Paul Krugman and others, involving the application of mathematical modelling techniques to analyse issues of industrial location. The new geographical economics addresses questions such as why and under what conditions do industries concentrate? It applies the methods of mainstream economics, devising models based on a number of simplifying assumptions.

New industrial spaces. Areas distinct from the old industrial cores that became centres of flexible production from the 1970s onwards. These areas offered attractive environments and a high quality of life for managerial and professional workers. Three different kinds of 'new industrial spaces' have been identified in Europe and North America: craft-based industrial districts such as central and north-eastern Italy, centres of high-technology industries such as Silicon Valley in California, and clusters of advanced financial and producer services in world cities.

New international division of labour. A term that refers to the process by which MNCs based in Western countries have shifted low-status assembly and processing operations to developing countries where costs are much lower. It is a form of the **spatial division of labour**, operating at the global scale, facilitated by the increasing division of labour in large **multinational corporations**, new transport and communications technologies, and the creation of a pool of available labour in developing countries.

New labour internationalism. A term used increasingly by academics to reflect the need for trade unions to 'scale up' their organizing activities to the global level to compete with **multinational corporations**. Unlike previous efforts to organize internationally, it is argued that a new labour internationalism must be firmly rooted in grass-roots struggles rather than a top-down effort, controlled by union elites.

New model of local and regional development. The prevailing approach to economic development since the 1980s has stressed the need to facilitate growth and enhance the competitiveness of the regional economy. In contrast to traditional regional policy, this new model is more 'bottom up' in nature, focusing on the need to develop local skills and stimulate enterprise with regions.

New regionalism. A collective label used to describe a body of research in economic geography that stresses the renewed importance of 'the region' as a scale of economic organization under late capitalism.

New tourism. A more flexible and specialized form of tourism, which is often contrasted with standardized mass tourism. The rise of short breaks and the taking of multiple holidays of different types over a year are key trends.

New trade theory. An approach developed by the economist Paul Krugman and others since the 1970s that recognizes that comparative advantage does not simply reflect pre-existing endowments of the **factors of production**. Rather, it is actively created by firms through the development of technology, skills, economies of scale, etc.

Newly industrialized countries (NICs). Formerly underdeveloped countries that experienced rapid industrialization and economic development, as well as significant foreign direct investment, between 1960 and 1990, thus reaching a level of development that make them broadly comparable to industrialized countries in Europe, North America and Japan. Hong Kong, Singapore, South Korea and Taiwan are commonly identified as the four Asian NICs, with Brazil, Argentina and Mexico also bracketed under this label. The next tier of growth economies in Asia such as China, Malaysia and Thailand are also sometimes included, but they have not yet reached the requisite level of development.

Non-government organizations (NGOs). Organizations, often of a voluntary or charitable nature, which make up the so-called 'third sector' belonging to neither the private nor public sectors.

Offshoring. The relocation of economic activities from developed countries to low-wage economies. A characteristic of manufacturing since the 1960s, creating **the new international division of labour**, it has become a feature of service operations in recent years, popularised by move of some call centres to developing countries like India. *See also* **Near-shoring; Onshoring.**

Onshoring. The relocation of services from one region to another within the same country, usually from a core area to a peripheral location, where costs (particularly wages), are lower. *See also* **Near-shoring; Offshoring.**

'Open' political economy approach. A revised political economy framework informed by the culturally and institutionally informed approaches to economic geography that have been developed in recent years. This revised approach is not limited by its adherence to fixed categories, evolving in line with capitalism as its object of analysis.

Outsourcing. The tendency for firms, particularly large corporations, to buy in expertise, services and components from other specialist firms rather than providing such materials internally.

Over-production. The tendency for the volume of output to grow more rapidly than market demand. This is viewed by many Marxist and institutionalist economists as being inherent to a capitalism system based on decentralized decision-making.

Partnership. The growing tendency for different organizations to work together in order to address particular problems and to coordinate services, involving the development of common objectives and the sharing of resources.

Path dependency. A key evolutionary idea adopted by economic geographers, referring to how past decisions and experiences shape the economic landscape, particularly in terms of structuring and informing economic actors' responses to wider processes of economic change.

Place. A particular area, usually occupied, to which a group of people have become attached, endowing it with meaning and significance. Often associated with notions of family, home and community.

Place promotion. The promotion or marketing of particular places to attract investment or visitors, often employing techniques borrowed from the marketing and advertising industries. Place promotion draws on a long tradition of local 'boosterism' emphasizing the positive qualities of a place, in terms of its workforce, facilities, quality of life, attractive environment, etc.

Places of consumption. The particular sites at which goods and services are bought and consumed, including the department store, the mall, the street, the market and the home, as well as a host of more inconspicuous sites of consumption (for example, charity shops and car boot sales).

Political economy. A broad perspective on economic life which analyses the economy within its wider social and political context, focusing on production and the distribution of wealth between different sections of the population as well as the exchange of commodities through the market.

Positivism. A philosophy of science that states that the goal of science is to generate explanatory laws that explain and predict events and patterns in the real world.

Post-Fordist consumption. This is defined by flexibility with markets becoming fragmented into distinct segments and niches since the 1970s. Accordingly, consumption is defined individually rather than collectively with choice and identity becoming increasingly important as individual consumers regard the purchase and consumption of commodities as expressions of their lifestyles and aspirations.

Post-industrial society. A concept that emphasizes that services have become the dominant sector of the economy in developed countries, taking over from manufacturing. The theory of post-industrial society was developed by the American sociologist Daniel Bell in the late 1960s and early 1970s, highlighting the increased importance of white collar employment, the role of knowledge and information as the key resources in the economy, and the liberation of individual workers from routine manual labour.

Postmodernism. Defined as a movement in philosophy, the arts and social sciences characterised by scepticism towards the grand claims and grand theory of the modern era, and their privileged vantage point, stressing in its place openness to a range of voices in social enquiry, artistic experimentation and political empowerment.

Poverty Reduction Strategies (PSRs). The development of strategies by the governments of developing countries to address poverty, based on consultation with the World Bank and IMF, NGOs and local communities.

Power. The ability or capacity to take decision that involve or affect other people. While neglected by mainstream economists, alternative approaches such as political economy tend to stress the importance of power in

shaping the operation of the economy, particularly in terms of the (social) relations between different economic actors.

Privatization. The policy of transferring state owned enterprises into private ownership), a key component of **neoliberal** reform programmes since the early 1980s.

Producer and consumer services. A common distinction made between different types of services, referring to the sources of demand for their products, stemming from either other businesses (producers) or individual consumers.

Qualitative state. A conception of the state as a dynamic process rather than a fixed 'thing' or object. The term, coined by the Australian economic geographer Philip O'Neill, reflects a general shift of emphasis from a concern with quantitative aspects of state intervention to an interest in its qualitative characteristics.

Radical geography. A new kind of geography, developed in the USA in the late 1960s in response to the perceived irrelevance of much mainstream geography in the face of pressing social problems such as inner city poverty, racism and inequality. Initially finding expression in a reformist liberalism, radical geography embraced Marxism in the 1970s.

Radical innovation. An emphasis on major innovations, such as the development of new technologies and industries, ignored by most traditional approaches.

Redlining. This is defined as cases where goods and services are made unavailable, or made available only on less than favourable terms, to people because of where they live. Typically, low-income inner city areas are regarded as high risk by financial institutions and are 'redlined' or marked out on a map. As a result, people living within these areas are either denied access to credit and insurance services or, more commonly, charged higher premiums.

Regime of accumulation. Another key regulationist term, referring to a relatively stable form of economic organization that structures particular periods of capitalist development, creating a balance between production and consumption. Regimes of accumulation are supported by particular **modes of regulation**.

Regional division of labour. A particular form of the **spatial division of labour** that operates at the macro-regional scale. It generally involves the location of high-value activities in economically advanced 'core' areas and low-values ones in less advanced peripheries. A good example is East and, particularly, South East Asia where Hong Kong and Singapore have focused on high-level managerial and professional activities while production is carried out in neighbouring low-wage countries such as China, Indonesia, Thailand and the Philippines.

Regional geography. Another key element of the traditional approach, structuring human geography as a whole between the 1920s and 1950s. It was defined as a project of 'real differentiation' which describes and interprets the variable character of the earth's surface, expressed through the identification of distinct regions.

Regional inequalities. Differences in income and wealth between regions. These can be seen as products of reflecting processes of uneven development.

Regional policy. A set of programmes and measures established by governments to promote regional growth and development. As a key instrument of 'spatial Keynesianism', Conventional regional policy involved governments inducing companies to locate factories and offices in depressed regions by offering grants and financial incentives. At the same time, development in core regions such as South East England and Paris was restricted. Classical regional policy reached its peak in the 1960s and 1970s, helping to reduce the income gap between rich and poor regions in Europe.

Regional sectoral specialization. A pattern of industrial location during the nineteenth century where particular regions become specialized in certain sectors of industry. Characteristically, all the main stages of production from resource extraction to final manufacture were carried out within the same region.

Reintermediation of production networks. This is the opposite of disintermediation, reflecting how new technologies reshape the structure of markets, creating opportunities for new firms to mediate between other businesses and customers, sometimes at the expense of established firms who are unable to react quickly enough.

Relative immobility of labour. The tendency for labour to remain tied to particular places, reflecting its dependence on family and community for reproduction. This is often contrasted with the geographical mobility of capital.

Reproduction of labour. The daily processes of feeding, clothing, sheltering and socializing which support and sustain labour; processes that rely on family, friends and the local community, and occurring outside the market.

Resource wars. Conflicts over the use of local economic resources such as land, forests and water between communities, which rely on them to meet their material needs on a day-to-day basis, and states or private interests who often want to exploit them for economic gain, threatening the basis of local livelihoods.

Resurgence of local and regional levels of government. The increased prominence of local and regional organizations in economic development policy, particularly since the early 1980s.

Scale. The different geographical levels of human activity: local, regional, national, supra-national and global.

Scientific management or 'Taylorism'. An approach to industrial organization, associated with the Fordist mass production system, named after its founder and principal advocate, F.W. Taylor. Taylorism involved the reorganization of work according to rational principles designed to maximize productivity, based on an increased division of labour, enhanced coordination and control by management and the close monitoring and analysis of work performance and organization.

SECI approach. A theory of innovation developed by Japanese management theorists, drawing on the practices employed by Japanese corporations. Four stages are identified: socialization, involving the articulation and exchange of tacit knowledge; externalization, based on the transformation of this tacit knowledge into codified form; the combination of different bodies of codified knowledge into more complex and integrated systems; and internalization, by which firms embody codified knowledge in the skills of workers and the routines and work practices of the firm, turning it back into tacit knowledge. The process is viewed as a spiral, based on continuous interaction between tacit and codified knowledge.

Second 'global shift'. The movement of service activities to developing countries like India, where labour costs are much lower than in developed countries. This represents a new stage in the evolution of the international division of labour, following the original global shift based on the relocation of manufacturing activities in the 1970s and 1980s.

Service classes. Groups which do not own capital or land but are employed in servicing capital, generally in professional and managerial posts in activities such as financial services, advertising, research and development and management consultancy.

'Shrinking' of space. A reduction in the effects of geographical distance through the development of new transport and communications technologies which effectively reduce the time and costs of moving goods, services, capital and information.

Social division of labour. The vast array of specialized jobs that people perform in society.

Social identity. A term referring to the way that individuals and groups are identified and characterised within society usually in contradistinction to other individuals and groups (for example, by class, gender, race, age).

Social movement unionism. A form of trade unionism that campaigns on broader social issues beyond the workplace, often in coalition with other parts of civil society (for example, environmental, human rights, gender issues).

Social networks. The informal social ties between individuals working in different firms, providing a channel for the sharing of information and ideas. The role of such networks has often been cited in accounts of the rise of new industrial spaces like Silicon Valley.

Social partnership. An approach to employment relations that recognizes the legitimate role of trade unions and attempts to develop consensus between unions and employers on economic decision-making. Often referred to as a component of a Fordist model of economic management.

Social polarization. A term referring to growing inequalities in economic and social wellbeing, particularly prevalent in contemporary society through the effects of neoliberalism and a shift to a more market orientated society. Some of the greatest levels of social polarization are found in the global South where the top 5 per cent of society often own over 50 per cent of a nation's wealth.

Social relations. The sets of relationships between different groups of economic actors. The relations between

employers and workers have attracted particular attention, but other relations include those between producer and consumers, manufactures and suppliers, supervisors and ordinary workers, and government agencies and firms.

Space. An area of the Earth's surface, for example, that between two particular points or locations. Often thought of in terms of the distance and time it takes to travel or communicate between two points.

Spatial agglomeration. The tendency for industries to cluster in particular places, underpinned by the operation of **agglomeration economies.**

Spatial analysis. An approach to economic geography that became influential in the 1960s and 1970s as part of the so-called 'quantitative revolution' in geography. Spatial analysis in economic geography involves the use of statistical and mathematical methods to analyse problems of industrial location, distance and movement.

Spatial dispersal. The opposite process to agglomeration where industries or firms move out of existing centres of production into new regions.

Spatial division of labour. A concept developed by Doreen Massey (1984) to explain how an increasing division of labour within large corporations produced new spatial patterns. Companies were locating the higher-order functions in cities and regions where there are large pools of highly educated and well-qualified workers; while lower-order functions, such as assembly, were increasingly being located in those regions and places where costs (especially wage rates) are lowest. *See also* **New international division of labour.**

Spatial fix. The establishment of relatively stable geographical arrangements that facilitate the expansion of the capitalist economy for a certain period of time. Examples include imperialism during the nineteenth century, **Fordism** in the post-war period and **globalization** since the 1980s which has involved the **deindustrialization** of many established centres of production in the 'rustbelts' of North America and Western Europe and the growth of new industry in 'sunbelt' regions and the newly industrializing countries of East Asia.

Spatial Keynesianism. The application of Keynesian economic principles to spatial issues based on the redistribution of resources from rich and poor regions in order to close the gap in income between them.

Specialized industrial regions. Regions that became dependent on a particular set of industries, providing the basis of the local economy. Regional specialization was a product of industrialization in the nineteenth and early twentieth centuries, leading to a profusion of mining villages, steel towns and shipbuilding districts in Europe and North America. Many of these regions have experienced severe **deindustrialization** since the 1960s, although some have diversified successfully in recent years.

Spread effects. A contrasting set of effects that allow surrounding regions to benefit from increased growth in the core region. One important mechanism here is increased demand in the core region for food, consumer goods and other products, creating opportunities for firms in the peripheral regions to supply this growing market. At the same time, rising costs of land, labour and capital in the core region, together with associated problems like congestion, can push investment out into surrounding regions.

Stages of economic growth. An influential model of economic development produced by the US economist Walt Rostow in the late 1950s, which identified distinct stages of growth. The process of 'take off' is crucial in bringing about the transition from traditional society to the final stage of 'high mass consumption'.

Stagflation. A term coined to describe the unprecedented combination of economic stagnation and rising inflation that occurred in the 1970s. Generally, economists had regarded high unemployment and inflation as incompatible, as the former was thought to reflect stagnation or recession and the latter was regarded as an expression of overly rapid growth or 'over-heating'.

State, the. A set of public institutions that exercise authority over a particular territory, including the government, parliament, civil service, judiciary, police, security services and local authorities.

Structural adjustment programmes (SAPs). An economic reform package developed by the IMF and World Bank in the 1980s and 1990s as part of the **Washington Consensus**, SAPs have been adopted by a large number of developing countries in exchange for financial assistance. They encompass a range of measures requiring countries to open up to trade and investment and to reduce public expenditure. *See also* **Conditionality**.

Structuralist theories. A set of theories of development that explained global inequalities in terms of the structure of the world economy, particularly the relationship between developed countries and developing countries. They emphasized the mode of incorporation of developing countries into the world economy, viewing this as a key source of exploitation. The structuralist approach was particularly associated with a group of theorists (such as Raul Prebisch) and activists based in Latin America in the 1950s and 1960s. *See also* **Dependency theories**.

Structured coherence. A term introduced by David Harvey to describe the social, economic and political relations that develop in association with particular forms of production in specific places. It has been particularly associated with the main centres of heavy industry in developed countries during the late nineteenth and twentieth centuries, which became known as working-class regions with strong socialist and trade union traditions.

Subsidies. Financial assistance or support provided to companies by governments, usually favouring domestic producers over foreign competitors. As such, these are seen as protectionist measures that distort trade, allowing the countries whose producers are being discriminated against to complain to the WTO. The agricultural subsidies provided by the EU and US to their farmers are particularly controversial, driving down prices for farmers in developing countries.

Subsistence agriculture. The production of food and crops predominantly for use by the household itself rather than for sale as a commodity. Subsistence agriculture still characterises much of the global South.

Sunk costs. The costs of investment not directly recoverable if a firm were to pull out of a particular location. A neglected topic in economic geography but potentially important in restricting the mobility of firms.

Supply-side of the economy. This can be is defined in terms of the quality of the main **factors of production** such as labour (training, skills), capital (emphasizing enterprise and innovation) and land in terms of sites and infrastructure for investors. Improving these supply-side factors is the central focus of the new model of economic development described above, seen as vital to the **competitiveness** of the regional economy.

Supranational tier of government. This term refers to the increased prominence of supranational agencies and organizations operating above the level of the national state. Examples include the European Union, World Trade Organization, International Monetary Fund, World Bank and United Nations.

Tacit knowledge. In contrast to **codified knowledge**, this refers to direct experience and expertise, which is not communicable through written documents. It is a form of practical 'know-how' embodied in the skills and work practices of individuals or organizations.

Taylorism *see* **Scientific management**

Third way. A term used to refer to the policies of centre-left leaders like Blair, Clinton and Schröder in the late 1990s, which aimed to find a new path between the conflicting extremes of free market **capitalism** and state socialism. Marrying the efficiencies of markets to a revived social democratic notion of social justice is the key notion. The term has been widely criticized as vague and meaningless.

Third World. A collective label for the 'underdeveloped areas' in Africa, Asia and Latin America widely used between the 1960s and 1990s. The term is a product of the Cold War, with the 'Third World' distinguished from the 'First World' of Western democracies and the 'Second World' of communist states in the USSR and Eastern Europe.

Time-Space compression. A term that refers to the effects of information and communication technologies in reducing the time and costs of transmitting information and money across space. This reduces the 'friction of distance', which geographers have traditionally stressed.

Tourist gaze. A concept developed by the British sociologist John Urry, who argued that a key part of the modern tourist experience is to encounter scenes and landscapes which are distinctive from those associated with everyday life. These other places are anticipated and viewed with a sense of excitement and curiosity.

Tourist-area cycle model. An influential model of how tourist areas evolve over time, adapting the idea of product life cycles, derived from economics and management. The basic idea is that the life of a resort goes through a number of stages from initial exploration to rapid development, consolidation, stagnation and ultimate decline or rejuvenation.

Trade unions. Collective organization representing workers who grew in strength in the late nineteenth and early twentieth centuries, and in many countries were associated with the formation of parliamentary Labour Parties.

Traditional approach to economic geography. This was factual and descriptive in nature, focusing on the compilation of information about economic conditions and resources in particular regions. It was the dominant approach from the late nineteenth century to the 1950s.

Transnational capitalist elite. Senior managers, directors and decision-makers in the financial and business communities, who are connected into transnational corporate networks, and whose actions wield considerable influence and power in the global economy.

Transnationality Index (TNI). An index compiled by the United Nations to measure the extent to which an MNC's activities are internationalized, based upon foreign employment, sales and assets.

Uneven development. The tendency for some countries and regions to be more economically prosperous and advanced than others. Uneven development is an inherent feature of the capitalist economy, reflecting the tendency for growth and investment to become concentrated in particular locations that offer profitable opportunities for investment. Over time, patterns of uneven development are periodically restructured as capital moves or 'seesaws' between locations in the search for profit.

Untraded interdependencies. A term introduced by the economic geographer Michael Storper to capture the informal linkages and relationships that tie firms together within a region. Untraded interdependencies are comprised of an intangible set of skills, attitudes, habits and understandings that become associated with specialized forms of production.

Urban 'growth machines'. Coalitions of local government and business interests such as developers, construction companies and realtors which came together to promote and redevelop cities following deindustrialization in the 1970s and 1980s, aiming to attract new investment and employment. These groups are brought together by their shared interest in the economic fortunes of the locality that provides the ultimate source of tax revenues and profits. The term was initially developed in the context of US cities such as Pittsburgh, Detroit and Boston, but has since been applied to cities in Western Europe.

Urban entrepreneurialism. A new focus on economic development and regeneration which became a key part of urban policy from the early 1980s. It is often contrasted with the managerialism of the post-war decades, which was primarily concerned with managing the delivery of welfare services to local residents. The entrepreneurial approach saw cities focus on the need to generate growth and employment, seeking to attract new investment and funds from outside and generate new business and income from within.

Urbanization economies. A second type of **agglomeration economy**, derived from the concentration of firms in *different* industries in large urban areas.

Uruguay Round of trade negotiations. The last of eight rounds of trade negotiations held between the late 1940s and the mid 1990s under the umbrella of the General Agreement on Trade and Tariffs (GATT). Running from 1986 to 1994, the Uruguay Round was the most ambitious and wide-ranging of the GATT agreements, incorporating not only agriculture, textiles and clothing but also services for the first time.

Venture capital. A form of private equity finance provided by outside investors to new or growing firms that are generally not quoted on the stock exchange. Such investment tends to be high risk, focusing on firms with a high growth potential and attracting investors who aim to make high returns by selling their stake at a later date.

Voluntary work. Forms of unpaid work carried out willingly, typically work and the work of many carers (particularly family members), looking after older people, children and the disabled. The retreat of the state from welfare provision in many countries since the 1980s has meant that unpaid care work has become increasingly critical both to supporting the economy.

Washington Consensus. A set of economic policies, based upon **neoliberal** economic principles, adopted and implemented by the US Treasury and World Bank and IMF, all based in Washington, DC. Key elements include reducing public expenditure, eoconomic liberalization, **priviatization** and the promotion of exports and foreign direct investment. The role of the IMF and World Bank in imposing such policies on highly indebted poor countries has generated substantial controversy. *See also* **Conditionality.**

Welfare state. A particular type of state constructed between the 1930s and 1970s in developed countries, based on the establishment of elaborate welfare systems to spread the benefits of growth to all sections of the population and to offer social protection to its citizens against the vagaries of the market, including unemployment and ill health.

Work. The basic physical tasks needed to reproduce daily life. These would include hunting or finding food, finding shelter, making clothes, looking after and raising children, etc.

Workfare. A system introduced in the US that requires people to work in exchange for welfare benefits and payments.

World cities. These are defined by their coordination and control of the world economy, not size. Three main economic functions of world cities are commonly identified as: to operate as command centres for large **multinational corporations**; to host major concentrations of business services; to function as financial centres. London, New York and Tokyo are usually identified as the leading world cities, followed by larger numbers of major and secondary centres.

Zapatista guerrilla movement. A prominent example of a local social movements in a developing country, based in the Chiapas region of Mexico which represents the interests of the indigenous Mayan people. The Zapatistas have protested against poverty and the exploitation of local resources, employing a range of tactics designed to gain intentional media attention alongside an armed insurgency.

References

Adler, G. and Webster, E. (1999) 'The labour movement: radical reform and the transition to democracy in South Africa', in R. Munck and P. Waterman (eds) *Labour Worldwide in the Era of Globalization*, Basingstoke: Macmillan, pp.133–57.

Agarwal, S., Ball, G., Shaw, G. and Williams, A.M. (2000) 'The geography of tourism production: uneven disciplinary development?', *Tourism Geographies* 2: 241–63.

Aglietta, M. (1979) *A Theory of Capitalist Regulation: The US Experience*, London: New Left Books.

Agnew, J. and Grant, R. (1997) 'Falling out of the world economy? Theorising Africa in world trade', in Lee and Wills (1997) pp.219–28.

Albert M. (1993) *Capitalism against Capitalism*, London: Whurr.

Allen, J. (1988) 'Towards a post-industrial economy?', in Allen, J. and Massey, D. (eds) *The Economy in Question*, London: Sage, pp. 184–228.

Allen, J. (2000) 'Power/economic knowledge: symbolic and spatial formations', in Bryson, J., Daniels, P., Henry, N. and Pollard, J. (eds) *Knowledge, Space, Economy*, London: Routledge, pp.15–33.

Allen, J. (2003) *Lost Geographies of Power*, Oxford: Blackwell.

Allen, J. and du Gay, P. (1994) 'Industry and the rest: the economic identity of services', *Work, Employment and Society* 8: 255–71.

Allen, J. and Henry, N. (1997) 'Ulrich Beck's "Risk Society" at work: labour and employment in the contract services industries', *Transactions of the Institute of British Geographers* NS 22: 180–96.

Allen J. and Thompson, G. (1997) 'Think global, then think again: economic globalization in context', *Area* 29: 213–27.

Amin, A. (1999) 'An institutionalist perspective on regional economic development', *International Journal of Urban and Regional Research* 23: 365–78.

Amin, A. (2000) 'Industrial districts', in Sheppard and Barnes (2000), pp.149–68.

Amin, A. and Cohendet, P. (1999) 'Learning and adaptation in decentralised business networks', *Environment and Planning D: Society and Space* 17: 87–104.

Amin, A. and Thrift, N. (1994) 'Living in the global', in Amin, A. and Thrift, N. (eds) *Globalization, Institutions and Regional Development in Europe*, 1–22. Oxford: Oxford University Press, pp. 1–22.

Amin, A. and Thrift, N. (2004) 'Introduction', in Amin, A. and Thrift, N. (eds) *The Blackwell Cultural Economy Reader*, Oxford: Blackwell, pp.x–xxx.

Amin, A., Massey, D. and Thrift, N. (2003) *Decentring the Nation: A Radical Approach to Regional Inequality*, London: Catalyst.

Amsden, A. (1989) *Asia's Next Giant: South Korea and Late Industrialization*, New York: Oxford University Press.

Ancien, D. (2005) 'Local and regional development policy in France: of changing conditions and forms, and enduring state centrality', *Space and Polity* 9: 217–36.

Antipode (2001) 'Debating economic geography: more than responses to Amin and Thrift' *Antipode* 33 (2): 47–227.

Armstrong, P., Glyn, A. and Harrison, J. (1991) *Capitalism Since 1945*, Oxford: Blackwell.

Arnott, R. and Wrigley, N. (2001) 'Editorial', *Journal of Economic Geography* 1, pp.1–4.

Asheim, B.T. (1996) 'Industrial districts as "learning regions": a condition for prosperity', *European Planning Studies* 4: 379–400.

Atkins, P. and Bowler, I. (2001) *Food in Society: Economy, Culture, Geography*, London: Arnold.

Atkinson, J. (1985) *Flexibility, Uncertainty and Manpower Management*, Brighton: Institute of Manpower Studies, Report No. 89.

Bair, J. and Gereffi, G. (2001) 'Local clusters in global chains: the causes and consequences of export dynamism in Torreon's blue jeans industry', *World Development* 29: 1885–903.

Baran, P. and Sweezy, P. (1966) *Monopoly Capital*, New York: Monthly Review Press.

Barnes, T.J. (1997) 'Introduction: theories of accumulation and regulation: bringing life back into economic geography', in Lee and Wills (1997), pp.231–47.

Barnes, T.J. (2000a) 'Inventing Anglo-American economic geography', in Sheppard and Barnes (2000), pp. 11–26.

Barnes, T.J. (2000b) 'Political economy', in Johnston *et al.* (2000), pp.593–4.

Barnes, T.J. (2001) 'Retheorising economic geography: from the quantitative revolution to the "cultural turn"', *Annals of the Association of American Geographers* 91: 546–65.

Barnes, T.J. (2003) 'Introduction: "never mind the economy, here's culture"', in Anderson, K., Domosh, M., Pile, S. and Thrift, N. (eds), *Handbook of Cultural Geography*, London: Sage, pp. 89–97.

Barnes, T.J. (2006) 'Situating economic geographical teaching', *Journal of Geography in Higher Education* 30: 405–9.

Barnes, T.J. and Sheppard, E. (2000) 'The art of economic geography', in Sheppard and Barnes (2000), pp.1–8.

Barnes T.J., Peck, J., Sheppard, E. and Tickell, A. (eds) (2004) *Reading Economic Geography*, Oxford: Blackwell.

Bartlett, C. and Ghoshal, S. (1989) *Managing Across Borders*, Boston: Harvard Business School Press.

Bathelt, H., Malmberg, A. and Maskell, P. (2004) 'Clusters and knowledge: local buzz, global pipelines and the process of knowledge formation', *Progress in Human Geography* 28: 31–57.

BBC (2005) 'Europe agrees cut in sugar subsidies', BBC News at http://news.bbc.co.uk/1/hi/business/4466388.stm.

BBC (2004) 'The Battle over Trade', BBC News at http://news.bbc.co.uk/1/hi/in_depth/business/2004/world_trade#.

BBC (undated) 'Africa's challenges', in 'Africa 2005: time for a change?', BBC News at http://news.bbc.co.uk/2/shared/spl/hi/in_depth/africas_challenges/html/1.stm

Beaverstock, J.V. (2002) 'Transnational elites in global cities: British expatriates in Singapore's financial district', *Geoforum* 33 (4): 525–38.

Beaverstock, J.V., Smith, R.G. and Taylor, P.J. (1999) 'A roster of world cities', *Cities* 16, 6: 445–58.

Bell, D. (1973) *The Coming of Post-Industrial Society*, New York: Basic Books.

Bell, M. (1994) 'Images, myths and alternative geographies of the Third World', in Gregory, D., Martin, R., and Smith, G. (eds) *Human Geography: Society, Space and Social Science*, Basingstoke: Macmillan, pp.174–99.

Benneworth, P., Danson, M., Raines, P., and Whittam, G. (2003) 'Confusing clusters? Making sense of the clusters approach in theory and practice', *European Planning Studies* 11, 511–20.

Benwell Community Project (1978) *The Making of a Ruling Class*, Benwell Community Project, Final Series No. 6.

Berndt, C. (2000) 'The rescaling of labour regulation in Germany: from national and regional corporatism to intrafirm welfare?', *Environment and Planning* A 32 (9): 1569–92

Beynon, H. (1984) *Working for Ford*, 2nd edn, Harmondsworth: Pelican.

Beynon, H. (1997) 'The changing practices of work', in Brown, R. (ed.) *The Changing Shape of Work*, Basingstoke: Macmillan, pp.20–54.

Binns, T. and Nel, E. (2002) 'Tourism as a local development strategy in South Africa', *Geographical Journal* 168: 235–47.

Bishop, P., Gripaios, P. and Bristow, G. (2003) 'Determinants of call centre location: some evidence for UK urban areas', *Urban Studies* 40: 2751–68.

Blanchflower, D. (2006) 'A cross-country study of union membership', Bonn: IZA (Institute for the Study of Labour) Discussion Paper 2016.

Block, F. (2003) 'Karl Polanyi and the writing of *The Great Transformation*', *Theory and Society*, 32: 275–306.

Boyer, R. (1990) *The Regulation School: A Critical Introduction*, New York: Columbia University Press.

Boyle, M. and Hughes, G. (1994) 'The politics of urban entrepreneurialism in Glasgow', *Geoforum* 25: 453–70.

Bradley, H., Erickson, M., Stephenson, C. and Williams, S. (2000) *Myths at Work*, Cambridge: Polity.

Bristow, G. (2005) 'Everyone's a "winner": problematising the discourse of regional competitiveness', *Journal of Economic Geography* 5: 285–304.

Brohmann, J. (1996) 'Postwar development in the Asian NICs: does the neoliberal model fit reality?', *Economic Geography* 72: 107–30.

Brown, J.S. and Duguid, P. (2000) *The Social Life of Information*, Boston: Harvard University Press.

Brown, P. and Hesketh, A. (2004) *The Mismanagement of Talent*, Oxford: Oxford University Press.

Bryson, J. (2006) 'Offshore, onshore, near-shore and blended-shore: understanding the evolving geographies of "new international division of service labour"', paper presented to 2006 annual meeting of the Association of American Geographers, Chicago, IL.

Bryson, J. and Henry, N. (2005) 'The global production system: from Fordism to post-Fordism', in Daniels *et al.* (2005), pp.318–36.

Bryson, J., Henry, N., Keeble, D. and Martin R. (eds) (1999) *The Economic Geography Reader*, Chichester: Wiley.

Bryson, J., Daniels, P. and Warf, B. (2004) *Service Worlds: People, Organisations, Technologies*, London: Routledge.

Bureau of Labor Statistics (2006) NAICS 42–45: wholesale and retail trade, US Department of Labor at http://www.bls.gov/iag/wholeretailtrade.htm

Burnell, P. (2002) 'Foreign aid in a changing world', in Desai, V. and Potter, R.B. (eds) *The Companion to Development Studies*, London: Arnold, pp.473–7.

Burton, I. (1963) 'The quantitative revolution and theoretical geography', *Canadian Geographer*, 7: 151–62.

Cairncross, A. (1997) *The Death of Distance*, Boston: Harvard University Press.

Carlsson, B. (2004) 'The digital economy: what is new and what is not?', *Structural Change and Economic Dynamics* 15: 245–65.

Carney, J., Hudson, R. and Lewis, J. (1977) 'Coal combines and interregional development in the UK', in Massey, D.B. and Batey, P.J. (eds) *London Papers in Regional Science 7: Alternative Frameworks for Analysis*, London: Pion, pp.52–67.

Castells, M. (1989) *The Informational City*, Oxford: Blackwell.

Castells, M. (1996) *The Information Age Volume 1: The Rise of the Network Society*, Oxford: Blackwell.

Castells M. (2000) *The Information Age Volume 3: End of Millennium*, Oxford: Blackwell.

Castells, M. and Hall, P. (1994) *Technopoles of the World*, London: Routledge.

Castree, N., Coe, N., Ward, K. and Samers, M. (2004) *Spaces of Work: Global Capitalism and Geographies of Labour*, London: Sage.

Cater, E. (1995) 'Consuming spaces; global tourism', in Allen, J. and Hamnett, C. (eds) *A Shrinking World? Global Unevenness and Inequality*, Oxford: Oxford University Press, pp.182–231.

Charrie, J. P. (1994) 'The pattern of industry', in Blacksell, M. and Williams, A. (eds) *The European Challenge: Geography and Development in the European Community*, Oxford: Oxford University Press.

Chisholm, G.C. (1889) *Handbook of Commercial Geography*, London: Longmans, Green & Co.

City of London Corporation (undated) 'Key facts about the City of London and the "Square Mile"', at http://www.cityoflondon.gov.uk/Corporation/media_centre/keyfacts.htm. Accessed 8 June 2006.

Clark, G.L. (1994) 'Strategy and structure: corporate restructuring and the scope and characteristics of sunk costs', *Environment and Planning A* 26: 9–32.

Clark, G. L. (1999) *Pension Fund Capitalism*, Oxford: Oxford University Press.

Cloke, P. and Perkins, H. (2005) 'Cetacean performance and tourism in Kaikoura, New Zealand', *Environment and Planning D: Society and Space* 23: 903–24.

Cloke, P., Philo, C. and Sadler, D. (1991) *Approaching Human Geography: An Introduction to Contemporary Theoretical Debates*, London: Paul Chapman.

Cloke, P., Crang, P. and Goodwin, M. (eds) (2005) *Introducing Human Geographies*, 2nd edn, London: Arnold.

Coe, N. and Kelly, P. (2002) 'Languages of labour: representational strategies in Singapore's labour control regime', *Political Geography* 21: 341–71.

Coe, N., Hess, M., Yeung, H.W., Dicken, P. and Henderson, J. (2004) 'Globalising regional development: a global production networks perspective', *Transactions of the Institute of British Geographers* 29: 464–84.

Coffey, W, (1996) 'The newer international division of labour', in Daniels, P.W. and Lever, W.F. (eds) *The Global Economy in Transition*, Harlow: Longman, pp.40–61

Conte-Helm, M. (1999) 'The road from Nissan to Samsung: a historical overview of East Asian investment in a UK region', in Garrahan. P. and Ritchie, J. (eds) *East Asian Direct Investment in Britain*, London: Cass, pp.36–58.

Cook, I. (2002) 'Commodities: the DNA of capitalism', at www.exchange-values.org/. Accessed 14 June 2005.

Cook, I. and Crang, P. (1996) 'The world on a plate: culinary culture, displacement and geographical knowledges', *Journal of Material Culture* 1: 131–53.

Cook, G.A.S., Pandit, N.R., Beaverstock, J.V., Taylor, P.J. and Pain, K. (2003) 'The clustering of financial services in London', *Globalization and World Cities Research Bulletin* 124.

Cooke, P. (1995) 'New wave urban and regional revitalisation strategies in Wales', in Cooke, P. (ed) *The Rise of the Rustbelt*, London: UCL Press, pp.41–51.

Cooke, P. and Morgan, K. (1998) *The Associational Economy: Firms, Regions, and Innovation*, Oxford: Oxford University Press.

Corbridge, S. (2002) 'Third World debt', in Desai, V. and Potter, R.B. (eds) *The Companion to Development Studies*, London: Arnold, pp. 477–80.

Cox, K. and Mair, A. (1988) 'Locality and community in the politics of local economic development', *Annals of the Association of American Geographers* 78: 307–25.

Crang, M. (2002) 'Qualitative methods: the new orthodoxy?', *Progress in Human Geography* 26: 647–55.

Crang, P. (1994) 'It's showtime: on the workplace geographies of display in a restaurant in southeast England', *Environment & Planning D: Society and Space* 12: 675–704.

Crang, P. (1997) 'Introduction: cultural turns and the (re)constitution of economic geography', in Lee and Wills (1997), pp. 3–15.

Crang, P. (2005) 'Consumption and its geographies', in Daniels *et al.* (2005), pp.359–79.

Cresswell, T. (2004) *Place: A Short Introduction*, Oxford: Blackwell.

Crewe, L. and Gregson, N. (1998) 'Tales of the unexpected: exploring car boot sales as marginal spaces of consumption', *Transactions, Institute of British Geographers* NS 23: 39–53.

Cronon, W. (1991) *Nature's Metropolis: Chicago and the Great West*, New York: W.W. Norton.

Cumbers, A. (1996) 'Continuity or change in employment relations: evidence from the UK's old industrial regions', *Capital and Class* 58: 33–57.

Cumbers, A. (1999) 'The transformation of employment relations in the UK's old industrial regions? A regional comparison of the experience of "Japanization"', in Garrahan, P. and Ritchie, J. (eds) *East Asian Direct Investment in Britain*, London: Cass, pp.183–200.

Cumbers, A. (2004) 'Embedded internationalisms: building transnational solidarity in the British and Norwegian trade union movements', *Antipode* 36: 829–50.

Cumbers, A. (2005) 'Genuine renewal or pyrrhic victory? The scale politics of trade union recognition in the UK', *Antipode* 37: 116–38.

Cumbers, A. and Birch, K. (2005) 'Public sector spending and regional economic development: crowding out or adding value?', draft report to Unison, Centre for Public Policy and the Regions, University of Glasgow.

Cumbers, A. and Martin, S. (2001) 'Changing relationships between multinational companies and their host regions? A case study of Aberdeen and the international oil industry', *Scottish Geographical Journal* 117: 31–48.

Currah, J. (2003) 'Digital effects in the spatial economy of film: towards a research agenda', *Area* 35: 64–73.

Danford, A. (1999) *Japanese Management Techniques and British Workers*, London: Mansell.

Daniels, P.W. (1995) 'Services in a shrinking world', *Geography* 80: 87–110.

Daniels, P.W. (2003) ' "Old" economy, "new" economy and services', Working Papers in Services, Space, Society, School of Geography, Earth and Environmental Sciences, University of Birmingham.

Daniels, P.W. and Bryson, J.R. (2002) 'Manufacturing services and servicing manufacturing: knowledge-based cities and changing forms of production', *Urban Studies* 39: 977–91.

Daniels, P., Bradshaw, M., Shaw, D. and Sidaway, J. (eds) (2005) *Human Geography: Issues for the Twenty-first Century*, 2nd edn, Harlow: Pearson.

Dear, M. (2000) 'State', in Johnston *et al.* (2000), pp.788–90.

Department of Culture, Media and Sport (2001) *The Creative Industries Mapping Document*, London: Department of Culture, Media and Sport.

Department of Culture, Media and Sport (2005) *Creative Industries Economic Estimates Statistical Bulletin*, London: Department of Culture, Media and Sport.

Department of Culture, Media and Sport (undated) 'Tourism/Visitor Attractions', at http://www.culture. gov.uk/tourism/tourism_visitor_attractions.htm. Accessed 28 March 2006.

Desforges, L. (2005) 'Travel and tourism', in Cloke *et al.* (2005), pp.517–26.

Dex, S. and McCulloch, J. (1997) 'Unemployment and training histories: findings from the "Family and Working Lives Survey"', *Labour Market Trends* 105: 449–54.

Dicken, P. (1998) *Global Shift: Transforming the World Economy*, 3rd edn, London: Sage.

Dicken, P. (2003a) *Global Shift: Reshaping the Global Economic Map in the Twenty-First Century*, 4th edn, London: Sage.

Dicken, P. (2003b) 'Placing firms: grounding the debate on

the global corporation', in Peck, J. and Yeung, H.W. (eds) *Remaking the Global Economy*, London: Sage, pp.27–44.

Dicken, P., Forsgren, M. and Malmberg, A. (1994) 'The local embeddedness of transnational corporations', in Amin, A. and Thrift, N. (eds) *Globalization, Institutions and Regional Development in Europe*, Oxford: Oxford University Press, pp.23–45.

Dicken, P., Peck, J. and Tickell, A. (1997) 'Unpacking the global', in Lee and Wills (1997), pp.158–66.

Doeringer, P. and Piore, M. (1971) *Internal Labor Markets and Manpower Analysis*, Massachusetts: Heath Lexington Books.

Domosh, M. (1996) 'The feminised retail landscape: gender, ideology and consumer culture in nineteenth-century New York City', in Wrigley, N. and Lowe, M. (eds) *Retailing, Consumption and Capital: Towards the New Retail Geography*, Harlow: Longman, pp.257–70.

Donaghu, M.T. and Barff, R. (1990) 'Nike just did it: international subcontracting and flexibility in athletic footwear production', *Regional Studies* 24: 537–52.

Douglass, M. (1994) 'The "developmental state" and the newly industrialised economies of Asia', *Environment and Planning* A 26: 453–66.

Dove, J. (2003) 'Visitor pressure on Yellowstone', *Geography Review*, 18(4): 18–20.

Dow, S. (1999) 'The stages of banking development and the spatial evolution of financial systems', in Martin, R. (ed) *Money and the Space Economy*, Chichester: Wiley, pp.31–48.

Dowd, D. (2000) *Capitalism and its Economics: A Critical History*, London: Pluto.

Doyle, M. (2005) 'Two countries' contrasting tales', BBC News, Africa 2005: Time for Change?, at http://newsvote.bbc.co.uk/mpapps/pagetools/print/news.bbc.co.uk/2/hi/africa/4337083.

Dunford, M. and Perrons, D. (1994) 'Regional inequality, regimes of accumulation and economic development in contemporary Europe', *Transactions of the Institute of British Geographers* NS 19: 163–82.

Dunning, J.H. (1980) 'Towards an eclectic theory of international production: some empirical tests', *Journal of International Business Studies* 11: 9–31.

Dunning, J.H. (1993) *Multinational Enterprises and the Global Economy*, Wokingham: Addison-Wesley.

The Economist (2001a) 'Internet pioneers: we have lift-off', 3 February, p.69.

The Economist (2001b) 'The WTO under fire', 20 September, pp.29–31.

The Economist (2004) 'Exit strategy: can Silicon Valley's magic last?', 1 May, p.9.

The Economist (2005a) 'Things go from bad to worse in retailing', 14 May, p.34.

The Economist (2005b) 'Consumers have stopped spending: for how long?', 12 November, p.51.

The Economist (2006) 'Podtastic, Apple', 14 January, p.68.

ESRC (2006) 'Poor inner-city areas worst hit by bank branch closures', (The Economic and Social Research Council, Swindon), *The Edge* 21: 12.

Faludi, A. (2004) 'Territorial cohesion: old (French) wine in new bottles?', *Urban Studies* 41: 1349–65.

Financial Times (2005) 'Disruption at the heart of the online revolution', 12 November.

Firn, J. (1975) 'External control and regional development: the case of Scotland', *Environment and Planning A* 7: 393–414.

Florida, R. (1995) 'Toward the learning region', *Futures* 27: 527–36.

Florida, R. and Kenney, M. (1993) *Beyond Mass Production: The Japanese System and its Transfer to the US*, Oxford: Oxford University Press.

Fraser, D. (2003) 'Wake up call for the West', *Sunday Herald*, 30 November.

Freeman, C. (1994) 'The economics of technical change', *Cambridge Journal of Economics* 18: 463–514.

Friedmann, J. (1986) 'The world city hypothesis', *Development and Change* 17: 69–83.

Friedmann, T. (2005) *The World is Flat: A Brief History of the Twenty-first Century*, New York: Farrar, Straus & Giroux.

Froebel, F., Heinrichs, J. and Kreye, O. (1980) *The New International Division of Labour*, Cambridge: Cambridge University Press.

Fukuyama, F. (1989) 'The end of history?', *The National Interest*, Summer.

Galbraith, J.K. (1967) *The New Industrial State*, London: Hamish Hamilton.

Garrahan, P. (1986) 'Nissan in the North East of England', *Capital and Class* 27: 5–13.

Garrahan, P. and Stewart, P. (1992) *The Nissan Enigma: Flexibility at Work in a Local Economy*, London: Cassell.

du Gay, P. (1996) *Consumption and Identity at Work*, London: Sage.

George, S. (1999) *The Lugano Report: On Preserving Capitalism in the Twenty-First Century*, London: Pluto.

Gereffi, G. (1994) 'The organization of buyer-driven commodity chains: how US retailers shape overseas production networks', in Gereffi, G. and Korzeniewicz, M. (eds) *Commodity Chains and Global Capitalism*, Westpost, Conn: Greenwood Press, pp.95–122.

Gertler, M. (1995) ' "Being there": proximity, organisation and culture in the development and adoption of advanced manufacturing technologies', *Economic Geography* 71: 1–26.

Gertler, M. (2003) 'A cultural economic geography of production', in Anderson, K., Domosh, M., Pile, S. and Thrift, N.J. (eds) *The Handbook of Cultural Geography*, London: Sage, pp.131–46.

Gertler, M. (2004) *Manufacturing Culture: The Institutional Geography of Industrial Practice*, Oxford: Oxford University Press.

Gibb, E. (2005) 'International Framework Agreements: Increasing the Effectiveness of Core Labour Standards', Global Labour Institute, Geneva, at http://www.global-labour.org/euan_gibb.htm.

Gibb, R. (2006) 'The European Union's "Everything But Arms" development initiative and sugar: preferential access or continued protectionism?', *Applied Geography* 26: 1–17.

Gibson, C. and Kong, L. (2005) 'Cultural economy: a critical review', *Progress in Human Geography* 29: 451–61.

Gibson-Graham, J.K. (1996) *The End of Capitalism (As We Knew It): A Feminist Critique of Political Economy*, Oxford: Blackwell.

Gibson-Graham, J.K. (2006) *A Postcapitalist Politics*, University of Minnesota Press, Minneapolis.

Gibson-Graham, J.K. (undated) 'A diverse economy: rethinking economy and economic representation', at http://www.communityeconomies/org/papers.php. Accessed 1 May 2006.

Giddens, A. (1985) *The Nation State and Violence*, Cambridge: Polity.

Giddens, A. (1990) *The Consequences of Modernity*, Cambridge: Polity.

Gold, J.R. and Gold, M. (1995) *Imagining Scotland: Tradition, Representation and Promotion in Scottish Tourism since 1750*, Aldershot: Ashgate.

Goodin, R.E. and Le Grand, J. (1987) *Not Only The Poor: The Middle Classes and the Welfare State*, London: Allen & Unwin.

Goodwin, M., Jones, M., Jones, R., Pett, K. and Simpson, G. (2002) 'Devolution and economic governance in the UK: uneven geographies, uneven capacities?', *Local Economy* 17: 200–215.

Gordon, D. (1988) 'The global economy: new edifice or crumbling foundations', *New Left Review* 168: 24–64.

Gordon, D. (1996) *Fat and Mean: The Corporate Squeeze of Working Americans and the Myth of Managerial Downsizing*, New York: Free Press.

Gordon, D., Edwards, R. and Reich, M. (1982) *Segmented Work, Divided Workers*, New York: Basic Books.

Gordon, I. and Goodall, B. (2000) 'Localities and tourism', *Tourism Geographies* 2: 290–311.

Gordon, I., Haslam, C., McCann, P., and Scott-Quinn, B. (2005) *Offshoring and the City of London*, London: The Corporation of London.

Gorz, A. (1982) *Farewell to the Working Class*, London: Pluto.

Goss, J. (1993) 'The "magic of the mall": an analysis of form, function and meaning in the contemporary retail built environment', *Annals of the Association of American Geographers*, 83: 18–47.

Goss, J. (1999) 'Once upon a time in the commodity world: an unofficial guide to the Mall of America', *Annals of the Association of American Geographers* 89: 45–75.

Goss, J. (2005) 'Consumption geographies', in Cloke *et al.* (2005), pp.253–66.

Grabher, G. (2002) 'The project ecology of advertising: tasks, talents and teams', *Regional Studies* 36: 245–62.

Granovetter, M. (1985) 'Economic action and social structure: the problem of embeddedness', *American Journal of Sociology* 91: 481–510.

Gray, J. (1999) *False Dawn: The Delusions of Global Capitalism*, London: Grant.

Green, F. (2001) 'It's been a hard day's night: the concentration and intensification of work in late twentieth-century Britain', *British Journal of Industrial Relations* 39: 53–80.

Green, F. (2004) 'Why has work effort become more intense?', *Industrial Relations* 43: 709–41.

Gregory, D. (1996) 'Areal differentiation and post-modern human geography', in Agnew, J., Livingstone, D.N. and Rogers, A. (eds) *Human Geography: An Essential Anthology*, London: Blackwell, pp.211–32.

Gregory, D. (2000) 'Positivism', in Johnston, *et al.* (2000), pp.606–8.

Gregson, N. (2000) 'Family, work and consumption: mapping the borderlands of economic geography', in Sheppard and Barnes (2000), pp.311–24.

Gregson, N. and Lowe, M. (1994) *Servicing the Middle Classes: Class, Gender and Waged Domestic Labour in Contemporary Britain*, London: Routledge.

Gremple, G (undated) 'The Industrial Revolution', at http://mars.acnet.wnec.edu/~grempel/courses/wc2/lectures/industrialrev.html.

Grimshaw, D., Ward, K.G., Rubery, J. and Beynon, H. (2001) 'Organisations and the transformation of the internal labour market', *Work Employment and Society* 15: 25–54.

Gwynne, R.N (1990) *New Horizons?: Third World Industrialisation in an International Framework*, Harlow: Longman.

Hall, P. (1985) 'The geography of the fifth Kondratieff', in Hall, P. and Markusen, A. (eds) *Silicon Landscapes*, Boston: Allen & Unwin, pp.1–19.

Hall, C.M. and Page, S. (2002) *The Geography of Tourism and Recreation: Environment, Place and Space*, 2nd edn, London: Routledge.

Hamnett, C. (1995) 'Controlling space: global cities', in Allen, J. and Hamnett, C. (eds) *A Shrinking World? Global Unevenness and Inequality*, Oxford: Oxford University Press, pp.103–42.

Hardagon, A. (2005) 'Technology brokering and innovation: linking strategy, practice and people', *Strategy and Leadership* 33: 32–6.

Harding, L. (2001) 'Delhi calling', *The Guardian*, 9 March.

Harrison, B. (1994) *Lean and Mean: The Changing Landscape of Corporate Power in the Age of Flexibility*, New York: Basic Books.

Hartshorne, R. (1939) *The Nature of Geography: A Critical Survey of the Present in the Light of the Past*, Lancaster, Pa: The Association.

Harvey, D. (1973) *Social Justice and the City*, London: Arnold.

Harvey, D. (1982) *The Limits to Capital*, Oxford: Blackwell.

Harvey, D. (1989a) *The Condition of Postmodernity*, Oxford: Blackwell.

Harvey, D. (1989b) 'The transition from managerialism to entrepreneurialism: the transformation of urban governance in late capitalism', *Geografiska Annaler* 71 B: 3–17.

Harvey, D. (1996) *Justice, Nature and the Geography of Difference*, Oxford: Blackwell.

Harvey, D. (2005) *A Brief History of Neoliberalism*, Oxford: Oxford University Press.

Hayter, R. (1997) *The Dynamics of Industrial Location: The Factory, the Firm and the Production System*, London: Wiley.

Healey, M. and Clark, D. (1986) 'Industrial decline and government response in the West Midlands: the case of Coventry', *Regional Studies* 18 (4): 303–18.

Held, D. and McGrew, T. (2002) *Globalization/Anti-Globalization*, Cambridge: Polity.

Held, D., McGrew, A., Goldblatt, D. and Perraton, J. (1999) *Global Transformations: Politics, Economics and Culture*, Cambridge: Polity.

Henderson, J. (1993) 'Against the economic orthodoxy: on the making of the East Asian miracle', *Economy and Society* 22: 200–217.

Henderson, J., Hama, N., Eccleston, B. and Thompson, G. (1998) 'Deciphering the East Asian Crisis', in Garrahan, P. and Ritchie, J. (eds) *East Asian Direct Investment in Britain*, London: Cass, pp.201–19.

Henry, N. and Pinch, S. (2000) 'Spatialising knowledge: placing the knowledge community of Motor Sport Valley', *Geoforum* 31: 191–208.

Henry, N. and Pollard, J. (2000) 'Capitalising on knowledge', *Geoforum* 31: v–vii.

Henwood, D. (1998) 'Talking about work', in Meiskins, E., Wood, P. and Yates, M. (eds) *Rising from the Ashes? Labor in the Age of Global Capitalism*, New York: Monthly Review Press.

Herod, A. (1997) 'From a geography of labor to a labor geography: labor's spatial fix and the geography of capitalism', *Antipode* 29: 1–31.

Herod, A. (2001) *Labor Geographies: Workers and the Landscapes of Capitalism*, New York: Guilford.

Herod, A., Peck, J. and Wills, J. (2003) 'Geography and Industrial Relations', in Ackers, P. and Wilkinson, A. (eds) *Understanding Work and Employment: Industrial Relations in Transition*, Oxford: Oxford University Press, pp.176–94.

Hines, C. (2000) *Localization: A Global Manifesto*, London: Earthscan.

Hirst, P. and Thompson, G. (1999) *Globalization in Question: The International Economy and the Possibilities of Governance*, 2nd edn, Cambridge: Polity.

Hobsbawm, E.J. (1962) *The Age of Revolution, 1789–1848*, London: Weidenfeld & Nicolson.

Hobsbawm, E.J. (1987) *The Age of Empire, 1875–1914*, London: Weidenfeld & Nicolson.

Hobsbawm, E.J. (1999) *Industry and Empire*, new edn, Harmondsworth: Penguin.

Hodgson, G.M. (1993) *Economics and Evolution: Bringing Life Back into Economics*, Cambridge: Polity.

Holmes, J. (2004) 'Rescaling collective bargaining: union responses to restructuring in the North American auto industry', *Geoforum* 35: 9–21.

Howells, J.R.L. (2002) 'Tacit knowledge, innovation and economic geography', *Urban Studies* 39: 871–84.

Hudson, R. (1989) 'Labor-market changes and new forms of work in old industrial regions: maybe flexibility for some but not flexible accumulation', *Environment and Planning D: Society and Space* 7: 5–30.

Hudson, R (1997) 'The end of mass production and of the mass collective worker? Experimenting with production, employment and their geographies', in Lee and Wills (1997), pp.302–10.

Hudson, R. (2001) *Producing Places*, London: Guilford.

Hudson, R. (2003) 'Geographers and the regional problem', in Johnston, R.J. and Williams, M. (eds) *A Century of British Geography*, Oxford: Oxford University Press, pp.583–602.

Hudson, R. (2005) *Economic Geographies: Circuits, Flows and Spaces*, London: Sage.

Hudson, R. (2006) 'On what's right and keeping left: or why geography still needs Marxian political economy', *Antipode* 38: 374–95.

Hudson, R. and Sadler, D. (1986) 'Contesting works closures in Western Europe's old industrial regions: defending place or betraying class?', in Scott, A. and Storper, M. (eds) *Production, Work, Territory: The Geographical Anatomy of Industrial Capitalism*, Boston: Allen & Unwin, pp.172–93.

Huff, W.G. (1995) 'The developmental state, government and Singapore's economic development since 1960', *World Development* 23: 1421–38.

Hunter, C. (2005) 'Theme 1: the nature of tourism: sustainable tourism course notes', School of Geosciences, University of Aberdeen.

Hunter, C. and Green, H. (1995) *Tourism and the Environment: A Sustainable Relationship?*, London: Routledge.

Hutton, W. (1995) *The State We're In*, London: Vintage.

Hutton, W. (2006) 'Today French protest seems more like farce', *The Observer*, 19 March.

Hymer, S. (1972) 'The multinational corporation and the law of uneven development', in Bhagwati. J.H. (ed.) *Economics and World Order*, London: Macmillan, pp.113–40.

IFS (2006) 'Financial services employment 1981–2005', London: International Financial Services, at http://www.ifsl.org.uk/research/index/html.

Illeris, S. (1989) *Services and Regions in Europe*, Aldershot: Avebury.

ILO (2000) *Labour Practices in the Footwear, Leather, Textiles and Clothing Industries*, Geneva: International Labour Organization.

ILO (2005a) *A Global Alliance against Forced Labour*, Geneva: International Labour Organization.

ILO (2005b) *World Employment Report*, Geneva: International Labour Organization.

Imrie, R. and Thomas, H. (eds) (1999) *British Urban Policy and the Urban Development Corporations*, 2nd edn, London: Sage.

International Ecotourism Society (2000) *Ecotourism Statistical Fact Sheet*, Washington, DC, TIES.

International Ecotourism Society (undated) 'What is ecotourism?', at http://www.ecotourism.org/index2.php?what-is-ecotourism.

Ismi, A. (2004) *Impoverishing a Continent: The World Bank and IMF in Africa*, British Columbia, Canadian Centre for Policy Alternatives Report.

Jackson, P. (1989) *Maps of Meaning: An Introduction to Cultural Geography*, London: Unwin Hyman.

Jackson, P., Miller, P., Holbrook, B., Thrift, N.J. and Rowlands, M. (1998) *Consumption, Place and Identity*, London: Routledge.

Jacobs, B. and Dutton, C. (2000) 'Social and community issues', in Roberts, P. and Sykes, H. (eds) *Urban Regeneration: A Handbook*, London: Sage, pp.109–28.

Jacoby, N. (1974) *Multinational Oil: A Study in Industrial Dynamics*, London: Macmillan.

Jain, A. (2005) 'US law firms outsource to India', *Financial Times*, 29 September.

Jeffreys, S. (2001) 'Western European trade unionism at 2000', *Socialist Register* 143–70.

Jenkins, P. (1987) *Mrs Thatcher's Revolution: The End of the Socialist Era*, London: Pan Books.

Jessop, B. (1994) 'Post-Fordism and the State', in Amin, A. (ed.) *Post-Fordism: A Reader*, Oxford: Blackwell, pp.251–79.

Johnston, R.J. (1984) 'The world is our oyster', *Transactions of the Institute of British Geographers* NS 9: 443–59.

Johnston, R.J. (1993) 'The rise and decline of the corporate welfare state: a comparative analysis in global context', in Taylor, P.J. (ed.) *Political Geography of the Twentieth Century*, London: Belhaven, pp.115–70.

Johnston, R.J. (1998) 'The state, the region and the division of labour', in Scott, A.J. and Storper, M. (eds) *Production, Work, Territory: The Geographical Anatomy of Industrial Capitalism*, Boston: Allen & Unwin, pp.265–80.

Johnston, R.J. (2000) 'Central place theory', in Johnston *et al.* (2000) pp.72–3.

Johnston, R.J. and Sidaway, J.D. (2004) *Geography and Geographers: Anglo-American Human Geography Since 1945*, 6th edn, London: Arnold.

Johnston, R.J., Gregory, D., Pratt, G. and Watts, M. (eds) (2000) *The Dictionary of Human Geography*, 4th edn, Oxford: Blackwell.

Jones, M., Jones, R. and Woods, R. (2004) *An Introduction to Political Geography: Space, Place and Politics*, London: Routledge.

Kelly, P. (2001) 'The political economy of local labor control in the Philippines', *Economic Geography* 77: 1–22.

Kempson, E., Whyley, C., Caskey, J. and Collard, S. (2000) *In or Out? Financial Exclusion: A Literature and Research Review*, London: Financial Services Authority, Consumer Research 3.

Khan, M.H. (2002) *When is Economic Growth Pro-Poor? Experiences in Malaysia and Pakistan*, Working Paper WP/02/85, Washington: International Monetary Fund (IMF).

King, R. (1995) 'Migrations, globalization and place', in Massey, D. and Jess, P. (eds) *A Place in the World*, Oxford: Oxford University Press.

Kitson, M. and Michie, J. (2000) *The Political Economy of Competitiveness*, London: Routledge.

Kitson, M., Martin, R. and Tyler, P. (2004) 'Regional competitiveness: an elusive yet critical concept?', *Regional Studies* 38: 991–9.

Klare, M.T. (2002) *Resource Wars: The New Landscape of Global Conflict*, New York: Owl Books.

Knox, P. and Agnew, J. (1989) *The Geography of the World Economy*, London: Arnold.

Knox, P., Agnew, J. and McCarthy, L. (2003) *The Geography of the World Economy*, 4th edn, London: Arnold.

Konadu-Agyemang, K. (2000) 'The best of times and the worst of times: structural adjustment programmes and uneven development in Africa: the case of Ghana', *Professional Geographer* 52: 469–83.

Kong, L. (2000) 'Culture, economy, policy: trends and developments', *Geoforum* 31: 385–90.

Krugman, P. (1991) *Geography and Trade*, Cambridge, Mass.: MIT Press.

Krugman, P. (1996) *Pop Internationalism*, Cambridge, Mass.: MIT Press.

Krugman, P. (1998) 'What's new about the new economic geography?', *Oxford Review of Economic Policy* 14: 7–17.

Krugman, P. (2000) 'Where in the world is the "new economic geography"', in Clark, G., Feldmann, M. and Gertler, M. (eds) *The Oxford Handbook of Economic Geography*, Oxford: Oxford University Press, pp.49–60.

Lash, S. and Urry, J. (1994) *Economies of Signs and Spaces*, London: Sage.

Lazaric, N. and Lorenz, E. (1998) 'The learning dynamics of trust, reputation and confidence', in Lazaric, N. and Lorenz, E. (eds) *Trust and Economic Learning*, Cheltenham: Edward Elgar, pp.1–20.

Leadbetter, C. (1999) *Living on Thin Air: The New Economy*, London: Viking.

Lee, C. (1986) 'Regional structure and change', in Langton, J. and Morris, R.J. (eds) *Atlas of Industrialising Britain 1780–1914*, London and New York: Methuen, pp.30–33.

Lee, R. (2000) 'Economic geography', in Johnston *et al.* (2000), pp.195–99.

Lee, R. (2002) 'Nice maps, shame about the theory? Thinking geographically about the economic', *Progress in Human Geography* 26: 333–55.

Lee, R. (2003) 'The marginalization of everywhere? Emerging geographies of emerging markets', in Peck, J. and Yeung, H.W. (eds) *Remaking the Global Economy*, London: Sage, pp.61–82

Lee, R. and Wills, J. (eds) (1997) *Geographies of Economies*, London: Arnold.

Ley, D. (1994) 'Postmodernism', in Johnston, R.J., Gregory, D and Smith, D.M. (eds) *The Dictionary of Human Geography*, 3rd edn, Oxford: Blackwell, pp.66–8.

Leyshon, A. (1995) 'Annihilating space? The speed-up of communications', in Allen, J. and Hamnett, C. (eds) *A Shrinking World? Global Unevenness and Inequality*, Oxford: Oxford University Press, pp.11–54.

Leyshon, A. (2000) 'Money and finance', in Sheppard and Barnes (2000), pp.432–49.

Leyshon, A. (2001) 'Time–space (and digital) compression: software formats, musical networks and the reorganisation of the music industry', *Environment and Planning A* 33: 49–77.

Leyshon, A. and Thrift, N. (1995) 'Geographies of financial exclusion: financial abandonment in Britain and the

United States', *Transactions of the Institute of British Geographers* NS 20: 312–41.

Leyshon, A., Lee, R. and Williams, C.C. (eds) (2003) *Alternative Economic Spaces*, London: Sage.

Leyshon, A., Webb, P., French, S., Thrift, N. and Crewe, L. (2005) 'On the reproduction of the musical economy after the internet', *Media, Culture and Society* 27: 117–209.

Lipietz, A. (1986) 'New tendencies in the international division of labour: regimes of accumulation and modes of accumulation', in Scott, A. and Storper, M. (eds) *Production, Work, Territory: The Geographical Anatomy of Industrial Capitalism*, Boston: Allen & Unwin, pp.16–40

Livingstone, D. N. (1992) *The Geographical Tradition*, Oxford: Blackwell.

Lloyd, P. and Shutt, J. (1985) 'Recession and restructuring in the North West region 1974–82: the implications of recent events', in Massey, D. and Meegan, R. (eds) *Politics and Method*, London: Methuen, pp.16–60.

London, S. (2004) 'Another step in IBM's march into services', *Financial Times*, 8 December.

Lopez-Levy, M. (2004) *We are Millions: Neoliberalism and New Forms of Political Action in Argentina*, London: Latin American Bureau.

Lovering, J. (1999) 'Theory led by policy: the inadequacies of the "new regionalism" (illustrated from the case of Wales)', *International Journal of Urban and Regional Research* 23: 379–95.

Lowe, M. and Wrigley, N. (1996) 'Towards the new retail geography', in Wrigley, N. and Lowe, M. (eds) *Retailing, Consumption and Capital: Towards the New Retail Geography*, Harlow: Longman, pp.3–20.

Lundvall, B.-A. (1992) *National Systems of Innovation*, London: Pinter.

Lundvall, B.-A. (1994) 'The learning economy: challenges to economic theory and policy', paper presented at the European Association for Evolutionary Political Economy Conference, Copenhagen, October.

Lundvall, B.-A. and Johnson, B. (1994) 'The learning economy', *Journal of Industry Studies* 1: 23–43.

McDowell, L. (1997) *Capital Culture: Gender at Work in the City*, Oxford: Blackwell.

McDowell, L. (2003) *Redundant Masculinities: Employment Change and White Working Class Youth*, Oxford: Blackwell.

McFarlane, C. (2004)' Geographical imaginations and spaces of political engagement: examples from the Indian Alliance', *Antipode* 36: 890–916.

MacLean, K. (1988) 'George Goudie Chisholm 1850–1930', in Freeman, T.W. (ed.) *Geographers Biobibliographic Studies*, vol. 12, London: Mansell Information Publishing, pp.21–33.

MacLeod, G. (1999) 'Place, politics and "scale dependence": exploring the structuration of Euro-regionalism', *European Urban and Regional Studies* 6: 231–53.

Maddison A. (2001) *The World Economy: A Millennial Perspective*, Paris: OECD.

Malcolmson, R.W. (1988) 'Ways of getting a living in eighteenth-century England', in Pahl, R. (ed.) *On Work: Historical, Comparative and Theoretical Approaches*, Oxford: Blackwell, pp.48–60.

Mall of America (2005a) Mall of America facts, at www.mallofamerica.com/about-moa.facts.aspx.

Mall of America (2005b) History of Mall of America, at www.mallofamerica.com/about-moa_history.aspx.

Malmberg, A. and Maskell, P. (1997) 'Towards an explanation of regional specialisation and industry agglomeration', *European Planning Studies* 5: 25–41.

Malmberg, A. and Maskell, P. (2002) 'The elusive concept of localisation economies: towards a knowledge-based theory of spatial clustering', *Environment and Planning A* 34: 429–49.

Mann, M. (1984) 'The autonomous power of the state', *European Journal of Sociology* 25: 185–213.

Mansvelt, J. (2005) *Geographies of Consumption*, London: Sage.

Markusen, A. (1996) 'Sticky places in slippery space: a typology of industrial districts', *Economic Geography* 72: 293–313.

Marshall, J.N, Wood, P., Daniels, P., MacKinnon, A., Bachtler, J., Damesick, P., Thrift, N., Gillespie, A., Green, A. and Leyshon, A. (1988) *Services and Uneven Development*, Oxford: Oxford University Press.

Martin, R. (1988) 'The political economy of Britain's North–South divide', *Transactions of the Institute of British Geographers* NS 13: 389–418.

Martin, R. (1989) 'The new regional economics and the politics of regional restructuring', in Albrechts, L., Moulaert, F., Roberts, P. and Swyngedouw, E. (eds) *Regional Policy at the Crossroads: European Perspectives*, London: Jessica Kingsley, pp.27–51.

Martin, R. (1999a) 'The new "geographical turn" in economics: some critical reflections', *Cambridge Journal of Economics* 23: 65–91.

Martin, R. (1999b) 'The "new economic geography": challenge or irrelevance?', *Transactions of the Institute of British Geographers* NS 24: 387–91.

Martin, R. (2000) 'Institutionalist approaches to economic geography', in Sheppard and Barnes (2000), pp.77–97.

Martin, R. and Sunley, P. (1997) 'The post-Keynesian state and the space economy', in Lee and Wills (1997), pp.278–89.

Martin, R. and Sunley, P. (2001) 'Rethinking the "economic" in economic geography: broadening our vision or losing our focus?', *Antipode* 33: 148–61.

Martin, R. and Sunley, P. (2003) 'Deconstructing clusters: chaotic concept or policy panacea?', *Journal of Economic Geography* 3: 5–35.

Marx, K. [1867] (1976) *Capital*, Vol. 1, New York: International Publishers.

Marx, K. and Engels, F. [1848] (1967) *The Communist Manifesto*, Harmondsworth: Penguin.

Maskell, P. (2001) 'The firm in economic geography', *Economic Geography* 77: 329–43.

Maskell, P. and Malmberg, A. (1999) 'The competitiveness of firms and regions: "ubiquitification" and the importance of localised learning', *European Urban and Regional Studies* 5 (2): 119–37.

Maskell, P., Eskelinen, H., Hannibalsson, I., Malmberg, A. and Varne, E. (1998) *Competitiveness, Localised Learning and Regional Development: Specialisation and Prosperity in Small, Open Economies*, London: Routledge.

Mason, M. (1994) 'Historical perspectives on Japanese Direct Investment in Europe', in Mason, M. and Encarnation, D. (eds) *Does Ownership Matter? Japanese Multinationals in Europe*, Oxford: Clarendon, Chapter 1.

Mason, C.M. and Harrison, R.T. (1999) 'Financing entrepreneurship: venture capital and regional development', in Martin, R. (ed.) *Money and the Space Economy*, Chichester: Wiley, pp.157–84.

Mason, C.M. and Harrison, R.T. (2002) 'The geography of venture capital investments in the UK', *Transactions of the Institute of British Geographers* NS 27: 427–51.

Massey, D. (1984) *Spatial Divisions of Labour: Social Structures and the Geography of Production*, London: Macmillan.

Massey, D. (1988) 'Uneven development: social change and spatial divisions of labour', in Allen, J. and Massey, D. (eds) *Uneven Redevelopment: Cities and Regions in Transition*, London: Hodder & Stoughton, pp.250–76.

Massey, D. (1994) 'A global sense of place', in Massey, D. (ed.) *Place, Space and Gender*, Cambridge: Polity, pp.146–56.

Massey, D. (2001) 'Geography on the agenda', *Progress in Human Geography* 25: 5–17.

Mawdsley, E. and Rigg, J. (2003) 'The World Development Report II: continuity and change in development orthodoxies', *Progress in Development Studies* 3: 271–86.

Maxwell, S. (2005) *The Washington Consensus is Dead: Long Live the Meta-Narrative!*, Working Paper 243, London: Overseas Development Institute.

Meadows, D.H., Meadows, D.L., Randers, J. and Brehens III, W. (1972) *The Limits to Growth*, London: Earth Island.

Meegan, R (1988) 'A crisis of mass production?', in Allen, J. and Massey, D. (eds) *The Economy in Question*, London: Sage, pp.136–83.

Merchant, K. and Johnson, J. (2005) 'India's skills shortage threatens offshore IT', *Financial Times*, 12 December.

Miller, D. (1995) 'Consumption as the vanguard of history: a polemic by way of introduction', in Miller, D. (ed.) *Acknowledging Consumption: A Review of New Studies*, London: Routledge, pp.1–57.

Miller, D. (1998a) 'Coca-Cola: a black sweet drink from Trinidad', in Miller, D. (ed.) *Material Culture: Why Some Things Matter*, Chicago: University of Chicago Press, pp.169–88.

Miller, D. (1998b) *A Theory of Shopping*, Cambridge: Polity.

Mohan, J. (1999) *A United Kingdom? Economic, Social and Political Geographies*, London: Arnold.

Molotch, H. (1976) 'The city as a growth machine: towards a political economy of place', *American Journal of Sociology* 82: 309–31.

Monastiriotis, V. (2005) 'Labour market flexibility in the UK: regional variations and the role of global/local forces', *Economic and Industrial Democracy* 26 (3): 443–77.

Monbiot G. (2003) *The Age of Consent: A Manifesto for a New World Order*, London: Flamingo.

Money News (2006) 'People spending more than they earn', 3 April, at http://www.moneynews.co.uk/558/people-spending.

Moneyweek (2004) 'Who makes money out of Christmas?', *Moneyweek: The Best of the International Financial Media*, 2 January.

Moody, K. (1997) *Workers in a Lean World*, London: Verso.

Moore, G. (2004) 'The fair trade movement: parameters, issues and future research', *Journal of Business Ethics* 53: 73–86.

Morgan, K. (1997) 'The learning region: institutions, innovation and regional renewal', *Regional Studies* 31: 491–504.

Morgan, K. (2004a) 'The exaggerated death of geography; learning, proximity and territorial innovation systems', *Journal of Economic Geography* 4: 3–21

Morgan, K. (2004b) 'Sustainable regions: governance, innovation and scale', *European Planning Studies* 16: 871–89.

Morgan, K. (2006) 'Devolution and development: territorial justice and the North–South divide', *Publius: The Journal of Federalism* 36: 189–206.

Munck, R. (1999) 'Labour dilemmas and labour futures', in Munck, R. and Waterman, P. (eds) *Labour Worldwide in the Era of Globalization*, Basingstoke: Macmillan, pp.3–26.

Munday, M., Morris, J. and Wilkinson, B. (1995) 'Factories or warehouses? A Welsh perspective on Japanese transplant manufacturing', *Regional Studies*, 29: 1–17.

Myrdal, G. (1957) *Economic Theory and the Under-developed Regions*, London: Duckworth.

Nambisan, S (2005) 'How to prepare tomorrow's technologists for global networks of innovation', *Communications of the Association for Computing Machinery* (ACM) 48.5: 29–31.

The National Trust (undated) 'Discover the National Trust', at http://www.nationaltrust.org.uk/main/w-trust/w-thecharity.htm.

Nelson, R.R. and Winter, S.G. (1982) *An Evolutionary Theory of Economic Change*, Cambridge, Mass.: Harvard University Press.

Nonaka, I. and Takeuchi, H. (1995) *The Knowledge-creating Company: How Japanese Companies Create the Dynamics of Innovation*, Oxford: Oxford University Press.

Nonaka, I., Toyama, R. and Konno, N. (2001) 'SECI, *ba* and leadership: a unified model of dynamic knowledge creation', in Nonaka, I. and Teece, D. (eds) *Managing Industrial Knowledge: Creation, Transfer and Utilisation*, London: Sage, pp.13–43.

O'Brien, R. (1992) *Global Financial Integration: The End of Geography*, London: Pinter.

O'Connor, J. (1973) *The Fiscal Crisis of the State*, New York: St Martin's Press.

OECD (2000) *Labour Force Statistics 2000*, Paris: Organization for Economic Cooperation and Development.

OECD (2005a) *Measuring Globalization*, Paris: Organization for Economic Cooperation and Development.

OECD (2005b) *Labour Force Statistics 2005*, Paris: Organization for Economic Cooperation and Development.

OECD (2005c) 'Aid rising sharply, according to the latest OECD figures', news release, Paris: Organization for Economic Cooperation and Development.

OECD (2006a) 'Ghana', in *Africa Economic Outlook 2005–6*, Paris: Organization for Economic Cooperation and Development.

OECD (2006b) *OECD Factbook 2006: Science and Technology Statistics*, Paris: Organization for Economic Cooperation and Development.

Ohmae, K. (1990) *The Borderless World*, London: Collins.

Ohmae, K. (1995) *The End of the Nation State*, London: HarperCollins.

Olesen, A. (2006) 'China's upwards spiral sparks fears of overheating', *The Herald*, 17 April.

Oliver, N. and Wilkinson, B. (1992) *The Japanisation of British Industry: New Developments in the 1990s*, Oxford: Blackwell.

O'Neill, P. (1997) 'Bringing the qualitative state into economic geography', in Lee and Wills (1997), pp.290–301.

ONS (2005a) *Labour Market Statistics*, March, London: Office for National Statistics.

ONS (2005b) *Travel Trends 2004*, London: Office for National Statistics.

Oxfam (2005) *What Happened in Hong Kong? Initial Analysis of the WTO Ministerial, December 2005*, Oxfam Briefing Paper 85, Oxfam International.

Oxfam (2006) *The View From The Summit: Gleneagles G8 One Year On*, Oxfam Briefing Note, Oxfam International, June.

Paddison, R. (1993) 'City marketing, image reconstruction and urban regeneration', *Urban Studies* 30: 339–50.

Pahl, R. (1988) 'Historical aspects of work, employment, unemployment and the sexual division of labour', in Pahl, R. (ed.) *On Work: Historical, Comparative and Theoretical Approaches*, Oxford: Blackwell, pp.1–7.

Painter, J. (1995) *Politics, Geography and Political Geography: A Critical Perspective*, London: Arnold.

Painter, J. (2000) 'State and governance', in Sheppard and Barnes (2000), pp.359–76.

Painter, J. (2002) 'The rise of the workfare state', in Johnston, R.J., Taylor, P. and Watts, M. (eds)

Geographies of Global Change: Remapping The World, 2nd edn, Oxford: Blackwell, pp.158–73.

Painter, J. (2006) 'Prosaic geographies of stateness', *Political Geography* 25: 752–74.

Palmer, M. (2006) 'Internet sales drive growth in retail sector', *Financial Times*, 13 February.

Pavlinek, P. (2004) 'Regional development implications of foreign direct investment in Central Europe', *European Urban and Regional Studies* 11: 47–70.

Peck, J. (1996) *Work-Place: The Social Regulation of Labour Markets*, New York: Guilford.

Peck, J. (1999) 'Grey geography?', *Transactions of the Institute of British Geographers* NS 24: 131–5.

Peck, J. (2000) 'Places of work', in Sheppard and Barnes (2000), pp.133–48.

Peck, J. (2001) 'Neoliberalising states: thin policies/hard outcomes', *Progress in Human Geography* 25: 445–55.

Peck, J. (2005) 'Economic sociologies in space', *Economic Geography* 81: 129–78.

Peck, J. and Theodore, N. (1998) 'The business of contingent work: growth and restructuring in Chicago's temporary employment industry', *Work, Employment and Society* 12: 655–74.

Peck, J. and Theodore, N. (2001) 'Contingent Chicago: restructuring the spaces of temporary labor', *International Journal of Urban and Regional Research* 25 (3): 471–96.

Peck, J. and Tickell, A. (1994) 'Searching for a new institutional fix: the after-Fordist crisis and the global–local disorder', in Amin, A. (ed.) *Post-Fordism: A Reader*, Oxford: Blackwell, pp.280–315.

Peck, J. and Tickell, A. (2000) 'Labour markets', in Gardiner, V. and Matthews, H. (eds) *The Changing Geography of the United Kingdom*, 3rd edn, London: Routledge, pp.150–68.

Peck, J.A. and Tickell, A. (2002) 'Neoliberalising space', *Antipode* 34: 380–404.

Peet, R. and Hardwick, E. (1999) *Theories of Development*, New York: Guilford.

Penrose, E.T. (1959) *The Theory of the Growth of Firms*, New York: Oxford University Press.

Pettifor, A., Cisneros, L. and Olmos Gaona, E. (2001) *It Takes Two to Tango: Creditor Responsibility for Argentina's Crisis and the Need for Independent Resolution*, London: Jubilee Plus Report.

Phelps, N.A., Lovering, J. and Morgan, K. (1998) 'Tying the firm to the region or tying the region to the firm? Early observations on the case of LG in South Wales', *European Urban and Regional Studies* 5: 119–37.

Phelps, N.A., MacKinnon, D., Stone, I. and Braidford, P. (2003) 'Embedding the multinationals? Institutions and the development of overseas manufacturing affiliates in Wales and north-east England', *Regional Studies* 37: 27–40.

Pike, A., Rodríguez-Pose, A. and Tomaney, J. (2006) *Local and Regional Development*, London: Routledge.

Pinch, S. and Henry, N. (1999) 'Paul Krugman's geographical economics, industrial clustering and the British motor sport industry', *Regional Studies* 33: 815–27.

Pinch, S. and Storey, A. (1992) 'Flexibility, gender and part-time work: evidence from a survey of the economically active', *Transactions of the Institute of British Geographers* 17 (2): 198–214.

Piore, M. and Sabel, C. (1984) *The Second Industrial Divide: Prospects for Prosperity*, New York: Basic Books.

Polanyi, K. (1944) *The Great Transformation: The Political and Economic Origins of Our Time*, Boston: Beacon Press.

Polanyi, K. [1959] (1982) 'The economy as an instituted process', in Granovetter, M. and Swelberg, R. (eds) *The Sociology of Economic Life*, Boulder, Col.: Westview Press, pp.29–51.

Pollard, J. (1996) 'Banking at the margins: a geography of financial exclusion in Los Angeles', *Environment and Planning A* 28: 1209–32.

Pollard, J. (2005) 'The global financial system: worlds of monies', in Daniels *et al.* (2005), pp.337–58.

Pollard, S. (1981) *Peaceful Conquest: The Industrialisation of Europe 1760–1870*, Oxford: Oxford University Press.

Pollert, A. (1988) 'The flexible firm: fixation or fact', *Work Employment and Society* 2: 281–316.

Porter, M.E. (1998) 'Clusters and the new economics of competition', *Harvard Business Review*, December, 77–90.

Porter, M.E. (2000) 'Locations, clusters and company strategy', in Clark, G.L., Feldman, M. and Gertler, M. (eds) *The Oxford Handbook of Economic Geography*, Oxford: Oxford University Press, pp.253–74.

Porter, M.E. (2003) 'The economic performance of regions', *Regional Studies* 37: 549–78.

Potter, R.B. (1983) 'Tourism and development: the case of Barbados, West Indies', *Geography* 68: 46–50.

Potter, R.B., Binns, T., Elliott, J.A. and Smith, D. (2004) *Geographies of Development*, 2nd edn, Harlow: Pearson.

Power, D. (2002) ' "Cultural Industries" in Sweden: an assessment of their place in the Swedish economy', *Economic Geography* 78: 103–27.

Power, D. and Scott, A.J. (eds) (2004) *The Cultural Industries and the Production of Culture*, London: Routledge.

Power, M. (2005) 'Worlds apart: global difference and inequality', in Daniels *et al.* (2005), pp.184–97.

Pratt, A. (1997) 'The cultural industries production system: a case study of employment change in Britain', *Environment and Planning A* 29: 1953–74.

Pratt, A. (2000) 'New media, the new economy and new spaces', *Geoforum* 31: 425–36.

Pym, H. (2006) 'More UK jobs set to go to India', BBC News, 7 June, at http://news.bbc.co.uk/go/pr/fr/-/1/hi/business/5057192.stm.

Ramalho, J.R. (1999) 'Restructuring of labour and trade union responses in Brazil', in Munck, R. and Waterman, P. (eds) *Labour Worldwide in the Era of Globalization*, Basingstoke: Macmillan.

Rees, J. (1998) 'The return of Marx?', *International Socialism* 79, at http://pubs.socialistreviewindex.org.uk/isj79.rees.htm.

Reich, R. (1991) *The Work of Nations*, New York: Knopf.

Richards, G. (1996) 'Production and consumption of European cultural tourism', *Annals of Tourism Research* 23: 9–13.

Ritzer, G. (2004) *The Globalization of Nothing*, Thousand Oaks, Calif., and London: Pine Forge Press.

Rock, D. (2002) 'Racking Argentina', *New Left Review* 17: 55–86.

Rodríguez, A., Martinez, A. and Guenga, G. (2001) 'Uneven redevelopment: new urban policies and socio-spatial fragmentation in metropolitan Bilbao', *European Urban and Regional Studies* 8: 161–78.

Rodríguez-Pose, A. and Gill, N. (2004) 'Is there a global link between regional disparities and devolution?', *Environment and Planning A* 36: 2097–117.

Routledge, P. (2002) 'Resisting and reshaping destructive development: social movements and globalising networks', in Johnston, R.J., Taylor, P. and Watts, M. (eds) *Geographies of Global Change: Remapping the World*, Oxford: Blackwell, pp.310–27.

Routledge, P. (2003) 'Voices of the dammed: discursive resistance amidst erasure in the Narmada Valley, India', *Political Geography* 22: 243–70.

Routledge, P. (2005) 'Survival and resistance', in Cloke *et al.* (2005), pp.211–24.

Rutherford, T. and Gertler, M. (2002) 'Labour in "lean" times: geography, scale and national trajectories of workplace change', *Transactions of the Institute of British Geographers* NS 27: 195–212.

Sadler, D. (2004) 'Trade unions, coalitions and communities: Australia's Construction, Forestry, Mining and Energy Union and the international stakeholder campaign against Rio Tinto', *Geoforum* 35: 35–46.

Sassen, S. (1991) *The Global City*, Princeton: Princeton University Press.

Savage, L. (1998) 'Justice for janitors: geographies of organising', in Herod, A. (ed.) *Organising the Landscape: Geographical Perspective on Labour Unionism*, Minneapolis: University of Minnesota Press, pp.225–52.

Sawyer, M.C. and O'Donnell, K. (1999), *A Future for Public Ownership*, London: Lawrence & Wishart.

Saxenian, A.L. (1994) *Regional Advantage: Culture and Competition in Silicon Valley and Route 128*, Cambridge, Mass.: Harvard University Press.

Sayer, A. (1985) 'Industry and space: a sympathetic critique of radical research', *Environment and Planning D: Society and Space* 3: 3–39.

Sayer, A. (1995) *Radical Political Economy: A Critique*, Oxford: Blackwell.

Sayer, A. and Walker, R. (1992) *The New Social Economy: Reworking the Division of Labour*, Oxford: Blackwell.

Schoenberger, E. (1994) 'The firm in the region and the region in the firm', paper presented at Regions, Institutions, and Technology Conference, Toronto, September.

Schoenberger, E. (1997) *The Cultural Crisis of the Firm*, Oxford: Blackwell.

Schumpeter, J.A. (1943) *Capitalism, Socialism and Democracy*, London: Allen & Unwin.

Scott, A.J. (1988) 'Flexible production systems and regional development: the rise of new industrial spaces in Europe and North America', *International Journal of Urban and Regional Research* 12: 171–86.

Scott, A.J. (1999) 'The US recorded music industry: on the relations between organisation, location and creativity in the cultural economy', *Environment and Planning A* 31: 965–84.

Scott, A.J. (2000) 'Economic geography: the great half century', in Clark, G., Feldmann, M. and Gertler, M. (eds) *The Oxford Handbook of Economic Geography*, Oxford: Oxford University Press, pp.18–44.

Scott, A.J. (2001) 'Capitalism, cities and the production of symbolic forms', *Transactions of the Institute of British Geographers* NS 26: 11–24.

Scott, A.J. (2002) 'A new map of Hollywood: the production and distribution of American motion pictures', *Regional Studies* 36: 957–75.

Scottish Enterprise (2003) *Operating Plan 2003–2004*, Glasgow: Scottish Enterprise.

Scottish Executive (2002) 'Five star research gets new investment', Edinburgh: The Scottish Executive, press release, 26 February.

Seddon, M. (2005) 'Kapital gain: Karl Marx is the Home Counties favourite', *The Guardian*, 14 July.

Seldon, A. (ed.) (2001) *The Blair Effect*, London: Little, Brown.

Seldon, A. (2005) *Blair*, London: Free Press.

Sharp, J., Routledge, P., Philo, C. and Paddison, R. (2000) *Entanglements of Power*, London: Routledge.

Shaw, G. and Williams, A.M. (2004) *Tourism and Tourism Spaces*, London: Sage.

Sheppard, E. and Barnes, T. (eds) (2000) *A Companion to Economic Geography*, Oxford: Blackwell.

Sheppard, E., Barnes, T.J., Peck, J. and Tickell, A. (2004) 'Introduction: reading economic geography', in Barnes T.J., Peck, J., Sheppard, E. and Tickell, A. (eds) *Reading Economic Geography*, Oxford: Blackwell, pp.1–11.

Short, J.R. (1999) 'Urban imagineers: boosterism and the representation of cities', in Jonas, A.E.G and Wilson, D. (eds) *The Urban Growth Machines: Critical Perspectives Two Decades Later*, Albany: State University of New York Press, pp.37–54.

Simon, D. (2002) 'Neoliberalism, structural adjustment and poverty reduction strategies', in Desai, V. and Potter, R.B. (eds) *The Companion to Development Studies*, London: Arnold, pp.86–92.

Skapinker, M. (2004) 'Big Blue's PC business is gone: long live IBM', *Financial Times*, 15 December.

Skidelsky, R. (2003) *John Maynard Keynes: Economist, Philosopher, Statesman*, London: Pan Books.

Slater, D. (2003) 'Cultures of consumption', in Anderson, K., Domosh. M., Pile, S. and Thrift, N. (eds) *Handbook of Cultural Geography*, London: Sage, pp.146–63.

Slaven, A. (1986) 'Shipbuilding', in Langton, J. and Morris, R.J. (eds) *Atlas of Industrialising Britain 1780–1914*, London and New York: Methuen, pp. 136–9.

Sloman, J. (1999) *Economics*, 4th edn, New York: FT Prentice Hall.

Smith, A. [1776] (1991) *Wealth of Nations. Inquiry into the Nature and Causes of the Wealth of Nations*, Buffalo, NY: Prometheus Books.

Smith, D.M. (2000) 'Export platform', in Johnston *et al.* (2000), pp.249–50.

Smith, N. (1984) *Uneven Development: Nature, Capital and the Production of Space*, Oxford: Blackwell.

Smith, N. (2001) 'Marxism and geography in the Anglophone world', *Geographische Revue* 3: 5–22, at http://www.geographische-revue.de/gr2-01.htm.

Snow, J. (2005) 'UK call centres: crossroads of an industry', *Journal of Property Investment and Finance* 23: 525–32.

Sorrentino, C. and Moy, J. (2002) 'US labor market performance in international perspective', *Monthly Labor Review*, June, pp.15–35.

Sparke, M., Sidaway, J.D., Bunnell, T. and Grundy-Warr, C.V. (2004) 'Triangulating the borderless world: geographies of power in the Indonesia–Malaysia–Singapore growth triangle', *Transactions of the Institute of British Geographers* NS 29, pp. 485–98.

Standing, G. (1999) *Global Labour Flexibility: Seeking Distributive Justice*, Basingstoke: Macmillan.

Statistics on Tourism and Research (2005) UK Tourist Research Liaison Group, at http://www.staruk.org.uk//default.asp?ID=730&parentid=469.

Stiglitz, J. (2002) *Globalization and its Discontents*, London: Penguin.

Stiglitz, J. (2003) 'Trade imbalances', *The Guardian*, 15 August.

Stoker, G. (ed.) (1999) *The New Politics of British Local Governance*, Basingstoke: Macmillan.

Stone, I. (1998) 'East Asian FDI and the UK periphery working miracles? Regional renewal and East Asian inter-linkages', in Garrahan, P. and Ritchie, J. (eds) *East Asian Direct Investment in Britain*, London: Cass, pp.59–94.

Storey, D., Bullock, H. and Overton, J. (2005) 'The poverty consensus: some limitations of the "popular agenda"', *Progress in Development Studies* 5: 30–44.

Storper, M. (1995) 'The resurgence of regional economies, ten years later: the region as a nexus of untraded interdependencies', *European Urban and Regional Studies* 2: 191–222.

Storper, M. (1997) *The Regional World: Territorial Development in a Global Economy*, London: Guilford.

Storper, M. and Christopherson, S. (1987) 'Flexible specialisation and regional industrial agglomeration: the case of the US motion picture industry', *Annals of the Association of American Geographers* 77: 104–17.

Storper, M. and Venables, A. (2004) 'Buzz: face-to-face

contact and the urban economy', *Journal of Economic Geography* 4: 351–70.

Storper, M. and Walker, R. (1989) *The Capitalist Imperative: Territory, Technology and Industrial Growth*, Oxford: Blackwell.

Swyngedouw, E. (1997) 'Neither global nor local: "globalization" and the politics of scale', in Cox, K. (ed.) *Spaces of Globalization*, New York: Guilford.

Swyngedouw, E. (2000) 'The Marxian alternative: historical–geographical materialism and the political economy of capitalism', in Sheppard and Barnes (2000), pp.40–59.

Taylor, M. and Asheim, B. (2001) 'The concept of the firm in economic geography', *Economic Geography* 77: 315–28.

Taylor, M. and Thrift, N. (eds) (1982) *The Geography of Multinationals*, London: Croom Helm.

Taylor, P.J. and Bain, P. (2004) *Call Centres in Scotland and Outsourced Competition from India*, Report for Scotecon.net, University of Stirling.

Taylor P.J. and Flint, C. (2000) *Political Geography: World Economy, Nation-State and Locality*, Harlow: Pearson.

Thompson, E.P. (1963) *The Making of the English Working Class*, Harmondsworth: Penguin.

Thompson, P. (2003) 'Disconnected capitalism: or why employers can't keep their side of the bargain', *Work Employment and Society* 17 (2): 359–78

Thompson, P. (2004) *Skating on Thin Ice: The Knowledge Economy Myth*, Glasgow: Big Thinking.

Thrift, N. (1994) 'On the social and cultural determinants of international financial centres: the case of the City of London', in Corbridge, S., Thrift, N. and Martin, R. (eds) *Money, Power and Space*, Oxford: Blackwell, pp.327–55.

Thrift, N. (2000) 'Pandora's box?: cultural geographies of economies', in Clark, G., Feldmann, M. and Gertler, M. (eds) *The Oxford Handbook of Economic Geography*, Oxford: Oxford University Press, pp.689–704.

Thrift, N. (2002) 'A hyperactive world', in Johnston, R.J. Taylor, P.J. and Watts, M.J. (eds) *Geographies of Global Change: Remapping the World*, Oxford: Blackwell, pp.29–42.

Thrift, N. and Olds, K. (1996) 'Refiguring the economic in economic geography', *Progress in Human Geography* 20: 311–37.

Tickell, A. (1999) 'Money and finance', in Cloke, P., Crang, P. and Goodwin, M. (eds) *Introducing Human Geographies*, London: Arnold, pp.107–13.

Tickell, A. (2000) 'Dangerous derivatives: controlling and creating risks in international money', *Geoforum* 31: 87–99.

Tickell, A. (2003) 'Cultures of money', in Anderson, K., Domosh, M., Pile, S. and Thrift, N. (eds) *Handbook of Cultural Geography*, London: Sage, pp.116–30.

Tickell, A. (2005) 'Money and finance', in Cloke *et al.* (2005), pp.244–52.

Tormey, S. (2004) *Anti-capitalism: A Beginner's Guide*, Oxford: Oneworld.

Townsend, J.G., Porter, G. and Mawdsley, E. (2004) 'Creating spaces of resistance: development NGOs and their clients in Ghana, India and Mexico', *Antipode* 36: 781–889.

Trends Business Research (2001) *Business clusters in the UK: A First Assessment*, London: Department of Trade and Industry.

Tuppen, J.N. and Thompson, I.B. (1994) 'Industrial restructuring in contemporary France: spatial priorities and policies', *Progress in Planning* 44: 99–172.

Turner, W. (undated) 'The decline of the handloom weaver', at http://www.cottontown.org/page. cfm?pageid=456&language=eng.

Tutor2u (undated) 'Strategy: core competencies', at http://www.tutor2u.net/business/strategy/core_ competencies.htm.

UNCTAD (2002) *World Investment Report 2002: Transnational Corporations and Export Competitiveness*, New York and Geneva: United Nations Conference on Trade and Development.

UNCTAD (2004) *World Investment Report 2004: The Shift Towards Services*, New York and Geneva: United Nations Conference on Trade and Development.

UNCTAD (2005) *World Investment Report 2005: Transnational Corporations and the Internationalisation of R and D*, New York and Geneva: United Nations Conference on Trade and Development.

UNDP (1998) *Consumption for Human Development*, Human Development Report, New York: UNDP, at http://hdr.undp.org/reports.

UNDP (1999) *Globalization With a Human Face*, Human Development Report, New York: UNDP, at http://hdr.undp.org/reports.

UNDP (2001) *Making New Technologies Work For Human Development*, Human Development Report, New York: UNDP, at http://hdr.undp.org/reports.

UNDP (2003b) *Millennium Development Goals: A Compact Among Nations to End Human Poverty*, Human

Development Report, New York: United Nations Development Programme, at http://hdr.undp.org/reports

UNDP (2004) *Cultural Liberty in Today's Diverse World*, Human Development Report, New York: UNDP, at http://hdr.undp.org/reports.

UNDP (2005) *International Cooperation at a Crossroads: Aid, Trade and Security in an Unequal World*, Human Development Report, New York: UNDP, at http://hdr.undp.org/reports.

UNICEF (2006) *Excluded and Invisible: The State of the World's Children*, New York: United Nations Children's Fund.

UNIDO (2004) *The Industrial Development Report 2004: Industrialization, Environment and the Millennium Development Goals in Sub–Saharan Africa*, Vienna: United Nations Industrial Development Organization.

Urry, J. (1995) *Consuming Places*, London: Routledge.

Urry, J. (2000) *Sociology Beyond Societies: Mobilities for the Twenty-First Century*, London: Routledge.

Urry, J. (2002) *The Tourist Gaze*, 2nd edn, London: Sage.

Vernon, R. (1971) *Sovereignty at Bay: The Multinational Spread of US Enterprises*, New York: Basic Books.

Vidal, J. (1999) 'Secret life of a banana', *The Guardian*, 10 November.

Visit Britain (2004) *UK Tourism Facts 2003*.

Visit Britain (undated) *Key Tourism Facts*, at http://www.tourismtrade.org.uk/MarketIntelligenceResearch/KeyTourismFacts.asp.

Visser, G. and Rogerson, C.M. (2004) 'Researching the South African tourism and development nexus', *GeoJournal* 60: 201–15.

Wade, R. (1990) *Governing the Market Economy: Economic Theory and the Role of Government in East Asian Industrialization*, Princeton, NJ: Princeton University Press.

Waitt, G. (2000) 'Consuming heritage: perceived historical authenticity', *Annals of Tourism Research* 27: 835–62.

Walker, R. (2000) 'The geography of production', in Sheppard and Barnes (2000), pp.111–32.

Walsh, J. (2000) 'Organising the scale of labour regulation in the United States: service sector activism in the city', *Environment and Planning A* 32: 1593–610.

Ward, K. (1996) 'Rereading urban regime theory: a sympathetic critique', *Geoforum* 27, 427–38.

Ward, K. (2004) 'Going global? Internationalization and diversification in the temporary staffing industry', *Journal of Economic Geography* 4 (3): 251–73.

Warf, B. (1995) 'Telecommunications and the changing geographies of knowledge transmission in the late twentieth century', *Urban Studies*, 32, pp. 529–55.

Waterman, P. (2000) *Globalization, social movements and the new internationalisms*, New York: Continuum.

Waters, R. (2005) 'From Netscape to the next big thing: how the dotcom decade changed our lives', *Financial Times*, 5 August.

Watts, M. (2005) 'Commodities', in Cloke *et al.* (2005), pp.527–46.

Weiss, L. (1997) 'Globalization and the myth of the powerless state', *New Left Review* 225: 3–27.

Weiss, L. (1998) *The Myth of the Powerless State: Governing the Economy in a Global Era*, Cambridge; Polity.

Weiss, L. (2000) 'Developmental states in transition: adapting, dismantling, innovating, not "normalising"', *The Pacific Review* 13: 21–55.

Whitley, R. (1999) *Divergent Capitalisms: The Social Structuring and Change of Business Systems*, Oxford: Oxford University Press.

Williams, A.M. (1995) *The European Community: The Contradictions of Integration*, 2nd edn, Oxford: Blackwell.

Williams, A.M. (2002) 'Tourism in Portugal: from polarisation to new forms of economic integration?', in Syrett, S. (ed.) *Contemporary Portugal: Dimensions of Economic and Political Change*, Aldershot: Ashgate, pp.83–103.

Williams, K., Haslam, C., Williams, J., Cutler, A. and Johal, S. (1992) 'Against lean production', *Economy and Society* 21: 321–54.

Willis, K. (2005) 'Theories of development', in Cloke *et al.* (2005), pp.187–99.

Wills, J. (2002) 'Bargaining for the space to organize in the global economy: a review of the Accor–IUF trade union rights agreement', *Review of International Political Economy*, 9: 675–700.

Wills, J. (2003) 'Community unionism and trade union renewal in the UK: moving beyond the fragments at last?', *Transactions of the Institute of British Geographers* 26: 465–83.

Wills, J. (2005) 'The geography of union organising in low-paid service industries in the UK: lessons from the T&G's campaign to unionise the Dorchester Hotel, London', *Antipode* 37: 139–59.

Wills, J. and Lee, R. (1997) 'Introduction', in Lee and Wills (1997), pp.302–10.

Wolf, M. (2005) 'On the move: different routes in pursuit

of economic greatness', in *The Financial Times Asia's Emerging Giants. India and China: Prospects for Growth, Cooperation and Competition*, London: *Financial Times*, p.4.

Womack, J.P., Jones, D.T. and Roos, D. (1990) *The Machine That Changed the World*, New York: Macmillan.

Woolfson, C., Foster, J. and Beck, M. (1997) *Paying for the Piper: Capital and Labour in Britain's Offshore Oil Industry*, London: Mansell.

World Bank (2005) *World Development Report 2006: Equity and Development*, Washington, DC: World Bank.

World Bank (2006) *World Development Indicators*, Washington DC: World Bank.

World Tourist Organisation (undated) *Tourism Highlights 2005 Edition*, Madrid: United Nations World Tourist Organisation.

Wright, M. (2001) 'A manifesto against Femicide', *Antipode* 33: 550–66.

Wright, R. (2002) 'Transnational corporations and global divisions of labor', in Johnston, R.J., Taylor, P. and Watts, M. (eds) *Geographies of Global Change: Remapping the World*, Oxford: Blackwell, pp.68–77.

WTO (2004) *World Trade in 2004*, at http://www.wto.org/english/res_e/statis_e/its2005_e/its05_overview_e.htm.

WTO (2005) *International Trade Statistics 2005*, at http://www.wto.org/english/res_e/statis_e/its2005_e/its05_toc_ehtm.

WTO (2006) *Measuring Trade in Services*, Geneva: World Trade Organisation.

WTTC (2002) *South Africa: The Impact of Travel and Tourism on Jobs and the Economy*, London: World Travel and Tourism Council.

WTTC (2003) *The Algarve: The Impact of Travel and Tourism on Jobs and the Economy*, London: World Travel and Tourism Council.

WTTC (2006a) *Portugal: Travel and Tourism: Climbing to New Heights*, London: World Travel and Tourism Council.

WTTC (2006b) *South Africa: Travel and Tourism: Climbing to New Heights*, London: World Travel and Tourism Council.

WTTC (undated) 'Tourism Satellite Accounting' at http://www.wttc.org/2004tsa/frameset2a.htm.

Z/Yen Ltd (2005) *The Competitive Position of London as a Global Financial Centre*, London: The Corporation of London.

Zeller, C. (2000) 'Rescaling power relations between trade unions and corporate management in a globalising pharmaceutical industry: the case of the acquisition of Boehringer Mannheim by Hoffman–La Roche', *Environment and Planning A* 32: 1545–67.

Index